Study Guide for
Introductory Chemistry

Second Edition

Steven S. Zumdahl

Iris Stovall
University of Illinois

D. C. Heath and Company

Lexington, Massachusetts   Toronto

Published simultaneously in Canada.

Printed in the United States of America.

International Standard Book Number: 0-669-32859-6

10 9 8 7 6 5

PREFACE

You might have asked yourself, "What things can I do to help me learn chemistry?" This is a question many students ask when they first begin a chemistry course. The answer to this question is not simple, but there are some things you can do to help make your study of chemistry successful.

First, read the textbook assignment before you go to class so that you will receive maximum benefit from each lecture. Second, keep up with homework assignments. Don't wait until the day before an exam to begin working problems. Solving chemistry problems takes practice--you will need to work steadily to become a good problem solver. Try to work each of the assigned homework problems, even if you don't think you know how to work them all.

This Study Guide can provide additional help. The **Quick Definitions** section provides a list of terms and definitions presented in each chapter. The definitions appear in the same order as they appear in your textbook. The section in the textbook where each term is defined is given following the definition so that you can quickly locate the appropriate section to reread. You can use Quick Definitions to help you review for an exam. Cover the definitions on the right with a sheet of paper and check your knowledge of each.

The **Content Review** summarizes important concepts from each section of the textbook and provides worked-out problems. Use the Content Review when you want a brief review of the subject or you need to refresh your memory.

The **Learning Review** provides end-of-chapter problems. Each one has a worked-out solution in **Answers to Learning Review**. Try to work the problem first, and then check your answer. Looking at the problem solutions in Answers to Learning Review before you work the problem will not benefit you as much as trying to work the problem first, even if your initial answer is wrong. When you use Answers to Learning Review, look at all the steps in the solution, not just the final answer. Make sure you understand the reasoning behind each step.

The **Practice Exam** at the end of each chapter provides ten multiple choice questions similar to those you could encounter on a multiple choice exam. The answers are provided, although the solutions are not worked out. Each answer is followed by a section number in parentheses; this indicates the textbook section where the material corresponding to the question can be found.

If you take advantage of the tools offered to you--your instructor, your textbook, and this Study Guide--you will be able to get the most out of your chemistry course.

Iris Stovall

CONTENTS

CHAPTER 1: CHEMISTRY: AN INTRODUCTION

INTRODUCTION

Your study of chemistry will require work on your part. Use all of the resources available to you so that you can get the most out of the effort you put into learning chemistry. The textbook should be your primary resource, but do not hesitate to turn to this Study Guide for additional help.

This chapter introduces you to how scientists solve problems. Learning how to solve problems is an important part of any chemistry course. Problem solving means more than just calculating a numerical answer. It also includes sifting through the given information, deciding which pieces of information are useful, and finally selecting an approach which will solve the problem. The problem solving skills you develop can be useful to you throughout your life.

AIMS FOR THIS CHAPTER

1. Know how to define the science of chemistry. (Section 1.1)
2. Develop a scientific approach to problem solving. (Section 1.2)
3. Develop an understanding of the scientific method. (Section 1.3)
4. Discover how your textbook can help you learn chemistry by solving many sample problems step by step. (Section 1.4)

QUICK DEFINITIONS

Chemistry — The science which describes matter and attempts to explain changes in matter which the universe undergoes. (Section 1.1)

Scientific method — An organized way to gather information about the universe. (Section 1.3)

Measurement — An observation which tells how much of something is present. (Section 1.3)

Natural law — A statement that describes behavior that is observed over and over in many different situations. (Section 1.3)

Theory — A theory attempts to explain the observations from which natural laws are formulated. (Section 1.3)

CONTENT REVIEW

1.1 WHAT IS CHEMISTRY?

The science of chemistry describes changes which occur in many different kinds of substances many of which are of interest and importance to life as we know it. Understanding why and how these changes occur is crucial to using them to understand our universe and to change the way we live.

1.2 SOLVING PROBLEMS USING A SCIENTIFIC APPROACH

How Do We Solve Problems?

When solving most problems, even problems not related to chemistry, we usually follow a set procedure. We often begin with an observation. This usually means we recognize a problem which we need to do something about. The next step is to propose possible solutions or explanations. From these possibilities we choose the solution most likely to succeed. Sometimes, we do not really know which solution is the best. We need more information before arriving at a solution to the problem. Solving chemistry problems is very similar to the procedure we use to solve problems in life.

1.3 THE SCIENTIFIC METHOD

How Does the Scientific Method Work?

The scientific method usually begins with observations. Someone noticed something happening and was curious enough to ask, "Why?" More detailed observations provide information about what conditions are necessary for the observed phenomenon to occur. After awhile it is possible to devise an initial hypothesis (a guess) which offers a possible explanation for the observations.

The hypothesis attempts to answer the question, "Why?" At this point, no one knows whether or not the hypothesis is correct. To help determine whether the hypothesis should be discarded, or rewritten and modified to form a theory, some testing must be done.

How can we test a hypothesis? We can make additional observations, and decide whether they contradict or support the hypothesis. One way scientists test theories and hypotheses is to make changes to the system they are observing, and then note whether the new observations are consistent with or contradictory to the hypothesis.

If the new observations contradict the explanations given by the hypothesis, then the hypothesis can be completely discarded. If some of the hypothesis fits the new observations, but some does not, then the hypothesis can be restructured to take into consideration the new evidence.

If the new observations support the hypothesis, we now have some evidence in favor of the hypothesis and we can accept the hypothesis. Once several hypotheses have been accepted, we can assemble them into a theory, also called a model. A theory is a set of accepted hypotheses which attempt to provide an overall explanation for some part of nature. A theory is not a finished product, however. It is always undergoing more cycles of testing and revision.

After recording many observations about different systems, sometimes a pattern becomes noticeable. For example, all objects, when dropped, fall toward the earth. This observation holds true for bricks as well as feathers, in China as well as the United States. General patterns of behavior such as these can be designated natural laws. Note that natural laws describe a pattern of behavior, and not the reasons why the behavior is observed.

LEARNING CHEMISTRY

You **can** learn chemistry. It may not always be easy, but it is possible. In order for you to get the most from your textbook, work through the sample worked-out exercises. Make sure you really know the reasons behind each step. Then, try the self-check exercise which follows. Usually, the same principles which are used in the sample exercise are used in the check exercise. Work the assigned problems at the end of each chapter. Some are drill problems which will test your skill with the majority of the assigned material, and some are problems which will require more thought to solve.

How Can This Study Guide Help You?

The **Aims For This Chapter** provides a list of concepts you should have mastered by the end of the chapter. Use this list as a check after you have studied. If there is a topic in the list of aims which you are not comfortable with, spend time reviewing the material in that section.

Quick Definitions gives you a list of words and phrases and their definitions. The wording is not the same as your textbook uses. The alternate wording will provide you with a different perspective and should raise your level of understanding. To make it easier to move between the textbook and the Study Guide, each definition lists the section in the textbook where you can find material related to this definition.

The **Content Review** reviews the major concepts presented in the textbook, and uses examples to illustrate each concept. These additional worked out examples can help you understand topics which are confusing to you.

The **Learning Review** provides additional problems and exercises. The Learning Review is most helpful if you attempt to solve the problems without looking at the answers. Use this section to check your work, or as a last resort if you are unable to approach the problem and the Content Review has not helped. The Learning Review Answers do not just provide answers to the problems, but also explain the logic used to solve the problem.

The **Practice Exam** provides you with the opportunity to test whether or not you thoroughly understand the material presented in the chapter. Try all the questions without referring to the text unless specifically asked. Then check your answers and review any sections or topics that have caused you problems.

LEARNING REVIEW

1. Explain why chemistry is important to you, even if your career is far removed from the sciences.
2. Aside from helping you to get a good grade in chemistry, of what use are the problem solving skills you will learn?
3. Imagine that you are a scientist exploring life on the newly discovered planet, Cryon. Cryon is cold, and is perpetually covered with snow on one side. While exploring the snowy side of Cryon you repeatedly observe that all the birds have white feathers. You hypothesize that all the birds on Cryon have white feathers. Being a good scientist, you:
   a. Declare that all birds on Cryon must be white, since all the ones on the snowy side are white.
   b. Test your hypothesis about all birds on Cryon being white by observing bird color on the non-snow-covered part of the planet as well as the snowy side.
   c. Elevate your hypothesis about white birds to a natural law which states that all life forms on cold planets that are covered with snow on one side are white.

ANSWERS TO LEARNING REVIEW

1. Chemistry will have a different impact on the career of each individual, but even if your career is far removed from the sciences, chemistry plays an important role in each of our everyday lives. We depend on the science of chemistry to provide us with a better standard of living.
2. Problem solving skills can be used throughout your life. Many situations require you to think logically, to propose hypothetical solutions, and to choose the most reasonable one. Chemistry can help develop logical thinking skills.
3. You, as the scientist in this problem, have made some observations. But your information is not complete. You have no information about the color of birds on the other side of Cryon. A good scientist would test the hypothesis about bird color by collecting more data. It would not be appropriate to elevate the hypothesis to a natural law until the hypothesis was more thoroughly tested. Choice c is not correct, so, the correct answer is b.

CHAPTER 2: MEASUREMENTS AND CALCULATIONS

INTRODUCTION

Chemistry is a science which requires observation of the world around us and measurements of the phenomena we observe. In this chapter you will learn how to record your observations and how to perform calculations with measured values. Scientific measurements are usually made using the metric system or the International System. You will need to become familiar with these systems of measurement and know the magnitude of each of the major units.

AIMS FOR THIS CHAPTER

1. Be able to convert numbers written in decimal notation ("normal" numbers) to scientific notation, and numbers written in scientific notation to decimal notation. (Section 2.1)
2. Memorize the commonly used SI system prefixes and their meanings. (Section 2.2)
3. Learn the names of the principal SI units of length, volume and mass and have a general idea how each of them relates to English units you are familiar with. (Section 2.3)
4. Understand that all numbers which arise from measurements have one digit, the last digit, which is not known exactly. Because it is estimated, its exact value is said to be uncertain. (Section 2.4)
5. Learn the rules for determining the correct number of significant figures in a number, and the correct number of significant figures to use in a calculation. (Section 2.5)
6. Be able to solve problems which begin with a quantity with one unit, and have an answer in another unit, using dimensional analysis. The original units cancel, leaving the number expressed in the desired units. (Section 2.6)
7. Be familiar with the three temperature scales. Know the boiling and freezing points of water for each of them. Be able to convert a temperature given in one of the three scales to the other two. (Section 2.7)
8. Know what density means, and be able to calculate density, volume, or mass if given any two items. (Section 2.8)

QUICK DEFINITIONS

Measurement — An observation which always includes a number and a unit. For example, a graduated cylinder contains 38.5 mL of water. (Introduction)

Scientific notation — A way of writing numbers in which a number between 1 and 10 is multiplied by 10 raised to some power. For example, 125 can be written as 1.25×10^2. (Section 2.1)

English system A measurement system used in the U.S. which uses units such as feet, inches and gallons. (Section 2.2)

Metric system A measurement system used by most of the world which has units such as meters, degrees Celsius, and liters. (Section 2.2)

SI system The International System for measurement, based mainly on the metric system. (Section 2.2)

Meter The SI unit of length, a little longer than a yard. One meter equals 39.37 inches. (Section 2.3)

Volume A measure of the amount of space occupied by a substance. (Section 2.3)

Liter A metric unit of volume which is equal to 1.06 qt. (Section 2.3)

Milliliter A metric unit of volume. There are one thousand milliliters in one liter. (Section 2.3)

Mass Quantity of material present. The mass of an object is the same no matter where it is measured. (Section 2.3)

Weight The response of mass to gravity. Weight depends upon how strong the gravitational field is, that is, where it is measured. (Section 2.3)

Kilogram The SI unit of mass. One kilogram equals 2.205 pounds. (Section 2.3)

Significant figures The proper number of figures to record when making a measurement. Record all measured values, plus one uncertain value. Also refers to the correct number of digits to use in a calculation. (Section 2.4)

Rounding off When your calculator provides more digits than you have significant figures, you must get rid of the extra digits by rounding off. Rules for rounding off are given in the Content Review. (Section 2.5)

Dimensional analysis A way of solving a problem by multiplying a quantity which includes unwanted units by a factor which cancels the unwanted units and leaves a new quantity and the desired units. (Section 2.6)

Equivalence statement A statement that shows how two different units are related to each other. For example, 2.54 cm = 1 in. (Section 2.6)

Fahrenheit Temperature scale expressed as °F. Water boils at 212 °F and freezes at 32 °F. This system is used primarily in the U.S. and Britain. (Section 2.7)

Celsius The metric system temperature scale expressed as °C. Water boils at 100 °C and freezes at 0 °C. (Section 2.7)

Kelvin Temperature scale, also called the absolute scale, expressed as K. Water freezes at 273 K and boils at 373 K. (Section 2.7)

Density How much mass there is in a substance, relative to the space (or volume) it takes up. (Section 2.8)

CONTENT REVIEW

2.1 SCIENTIFIC NOTATION

You can tell whether a number written in scientific notation will be greater than one or less than one by looking at the sign associated with the power of ten. If there is no sign, the number will be greater than one. If there is a minus sign, the number will be less than one.

How Can You Convert Large Decimal Numbers to Scientific Notation?

Let's convert 42,515 to **scientific notation**. There are five digits in the number. We need to convert this number to a small number between one and ten and to ten raised to some power. To find the small number, begin moving the decimal point, which is understood to be just to the right of the last number on the right but not written, to the left until you are left with one digit to the left of the decimal point. In this case the small number would be 4.2515. To determine how many times ten is multiplied by itself, count the number of places the decimal point was moved to produce 4.2515. In this case, we moved the decimal point four times. The power of 10 is 10^4. The exponent is a positive four because we moved the decimal point to the left. The entire number expressed in scientific notation is 4.2515×10^4.

How Can You Convert Small Decimal Numbers to Scientific Notation?

A number such as 0.00125 is less than one and can also be expressed in scientific notation. To convert a small number to scientific notation, begin moving the decimal place to the right until you have one number to the left of the decimal point. The new number, 1.25, will be the small part of the total expression. To get the power of ten count the number of times you moved the decimal place. Put a minus sign in front of the three to indicate that we have moved the decimal place to the right. The number is expressed in scientific notation as 1.25×10^{-3}.

How Can You Convert Scientific Notation to Decimal Notation?

It is also possible to convert numbers in scientific notation to numbers in decimal format. 9.43×10^5 would be equivalent to 943,000. Begin by noting the power ten is raised to, in this case, five. This means we will move the decimal place five times. Which direction do we move it? There is no minus sign associated with the five so the number is greater than one. Therefore we move the decimal place in the direction which will produce a number greater than one, to the right. If we moved the decimal to the left five times we would have as an answer 0.0000943, which is not correct. When we move the decimal to the right, we produce the number 943,000. Notice that as we move the decimal past the 3 we add some zeros to show where the decimal is actually located.

Now let's convert 1.6443×10^{-2} to decimal notation. The minus 2 tells us that the decimal number will be smaller than 1, and the 2 tells us how many places to move the decimal point. Since we want to produce a number smaller than one, we move the decimal point to the left, two times. We need to add a zero in front to keep track of the decimal point, so our decimal number is 0.016443.

2.2 UNITS

Each number in chemical problems should have a unit associated with it. The unit tells you what kind of measurement has been made, that is, whether the number represents mass, or length, or volume, or temperature. A number without a unit is not very useful. Try to include a unit with each number you use. The amount of extra effort it takes will pay off later when you are unsure how to work a problem. The units can often give you a clue.

Learn the prefixes for the metric and SI systems now, or you will be eternally confused about the relative sizes of the various units. If you are not sure whether a decimeter is smaller or larger than a centimeter, deciding whether or not your answer is reasonable will be very difficult.

What Systems of Measurement Are in Common Use?

There are three systems of measurement in common use, the **English system** we are all familiar with, the **metric system**, and the **SI** or **International System**. The metric system and the SI system are similar, since the SI is based on metric units. You will need to memorize the fundamental SI units, and each of the prefixes which modifies the size of the fundamental unit. If you do not memorize these units and prefixes now and have a general idea of their relative sizes, working problems later in the course will be difficult. The four fundamental units in the SI system are the kilogram (mass), meter (length), second (time) and Kelvin (temperature). Note that there is one fundamental unit for each physical quantity. Smaller and larger quantities are indicated by using the prefixes. A useful feature of the prefixes is that they always mean the same thing. For example, a centimeter means one hundredth of a meter, and a centigram means one hundredth of a gram. Prefixes are used with all of the fundamental units, except temperature.

2.3 MEASUREMENTS OF LENGTH, VOLUME, AND MASS

In science we often want to know the quantity of matter present, the **mass**. We call the process of obtaining the mass "weighing". A more accurate term for this would be "massing", but we do not use it. Weighing is used when we determine the mass of a substance.

What Units of Length Are Used?

The fundamental SI unit for **length** is the meter. A meter is 39.37 inches, a little longer than a yard. When smaller (or larger) units of measure are required, prefixes can be combined with meter. For example, a centimeter is one hundredth (0.01) of a meter. Many of the smaller units are used in science, because the quantities we have available to measure are often very small.

What Units of Volume Are Used?

The fundamental SI unit of **volume** is the cubic meter, which is represented by a cube whose width, height, and breadth are each one meter. A cubic meter, or m^3, takes up a lot of space. We do not commonly measure substances in the chemistry laboratory with units of cubic meters, because the volume is so large. The cube you can draw to symbolize a cubic meter can be broken down into 1000 smaller cubes, with ten cubes along each edge of the big cube. These smaller cubes are $decimeters^3$ or dm^3. This volume is usually called the **liter**. A liter is a useful amount of volume, a little larger than a quart. A dm^3 can be broken down into still smaller cubes. A decimeter is equal to 10 centimeters, so a dm^3 is equal to 10 x 10 x 10 centimeters or 1000 cm^3. This is equivalent to breaking up the liter into 1000 smaller cubes, called **milliliters**.

What Units of Mass Are Used?

The fundamental SI unit for mass is the **kilogram**. Remember that "weighing" in the scientific sense means determining the mass, or quantity of matter in a substance. Although the kilogram is the fundamental unit, the smaller unit, the gram, is often more convenient to use in the laboratory. The kilogram is usually too large, 2.2 pounds.

2.4 UNCERTAINTY IN MEASUREMENT

When using a piece of equipment to make a measurement, first look at the graduations. What unit is represented by each number? Usually, the units associated with the numbers are given somewhere on the measuring device. Once you know what the major numbers represent, you can determine what the smaller divisions represent by reference to the whole numbers.

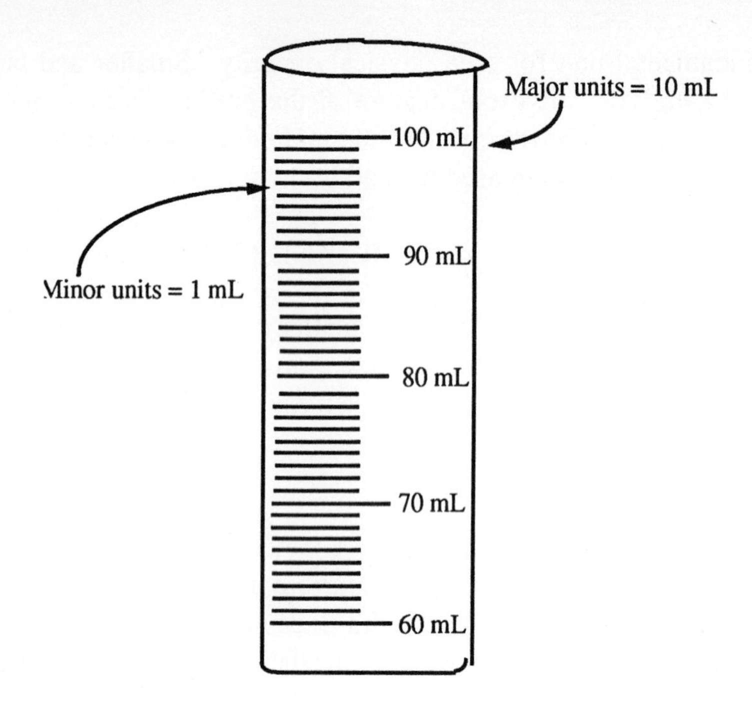

What Is the Correct Way to Measure?

Uncertainty in measurement comes about because the quantity we are measuring often falls between the marks on the measuring device. For example, on the ruler below the arrow is longer than 15 cm because it passes the 15 cm mark, but it stops short of the 16 cm mark.

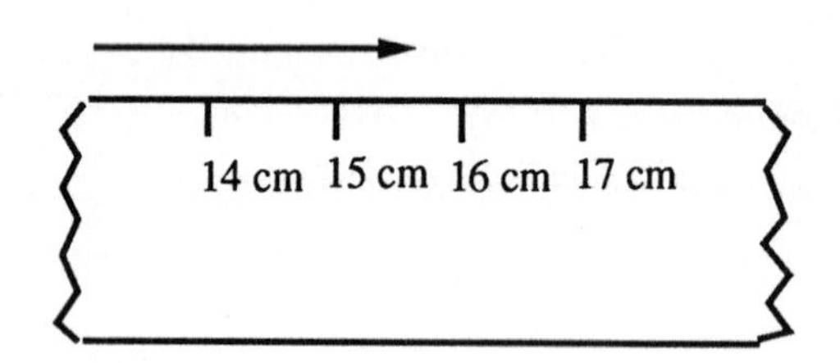

How can we tell how much greater than 15 cm the length really is if there are no graduations between 15 and 16 cm? We have to estimate, and estimates are uncertain. They may depend on who does the measuring, or how you hold the ruler. Your best guess might be that the length is between 15.6 and 15.7 cm. When you are making a measurement and presenting the results, always give the numbers you know for sure, and one estimated number. For example, you know that the length is 15 cm and some fraction, so 15 cm represents a number you know for sure. The estimated number is the one between the graduations, in this case 0.6 cm. The numbers you finally decide to record are called the **significant figures**.

2.5 SIGNIFICANT FIGURES

Zeros cause the most trouble when counting the number of significant figures.

Some measuring devices are more sensitive than others. For example, a graduated cylinder may measure a volume to 21.5 mL. The estimated digit is five. Another piece of laboratory glassware, the buret, can measure the same quantity to 21.52 mL, to an extra decimal place. The estimated digit in this measurement is the two. The total number of measured digits is the number of significant figures. The graduated cylinder measurement has three, while the buret measurement has four significant figures. Before we can do arithmetic with the results of measurements which each contain an uncertain number, we need to be able to correctly determine the number of significant figures in any number. Usually this is easy, as with the graduated cylinder and buret measurements above, but some situations require the application of a set of rules. You will need to learn these rules, or the results of your calculations will be inaccurate.

How Can You Determine the Correct Number of Significant Figures?

Here is a brief review of the rules. Non-zero digits are always significant. They always count. There are three classes of zeroes, however. Some count and some do not. Zeroes in the middle always count. 101.2 kg has four significant figures, and 100.2 m also has four significant figures. Zeroes at the beginning and end of numbers cause the most trouble. Zeroes which come before non-zero digits are **not** significant. 0.00204 has three significant figures, the 204 part. All zeroes which come before the 2 do not add to the accuracy of the measurement. They only fix the decimal place. Zeroes to the right of non-zero digits are not significant unless the number contains a decimal place. For example, 12,400 contains three significant figures, and so does 1.00×10^2. But if there is a decimal point as in 12,400. then all the numbers are significant, So 12,400. has five significant figures.

What happens when you try to perform calculations with measurements which contain uncertain numbers? The resulting number contains some significant figures, and some which are not significant because the arithmetic was performed with uncertain numbers. There are some rules to help you decide which of the calculated digits are significant and which are not.

Do All Numbers Contain a Limited Number Of Significant Figures?

When using equivalent statements such as 1 in = 2.54 cm, the use of 1 in would appear to limit the results of all calculations to one significant figure, because the 1 looks as though it has one significant figure. However, 1 is an exact number because it is part of a definition. Exactly one inch equals 2.54 centimeters. Definitions are considered to be exact numbers and do not affect the number of significant figures in a calculation.

How Should You Use Significant Figures in Multiplication and Division?

When you multiply or divide numbers, the number of significant figures in the answer should be equal to the number of significant figures in the number which has the smallest number of significant figures. 1.06 x 8.8 = 9.328 on your calculator, but the answer as calculated has too many significant figures. Because 8.8 has only two significant figures, the answer can only have two significant figures. The correct value is 9.3.

How Should You Use Significant Figures in Addition and Subtraction?

When you add or subtract numbers, the strategy is a little different. Keep the same number of decimal places as you have in the number with the least number of decimal places. For example, 456.0914 - 35.21 = 420.8814 on your calculator, but because 35.21 has two decimal places, the correct answer should be 420.88.

How Can You Round Off Numbers?

Very often when you use a calculator, the number on the display contains many unneeded and insignificant digits. You should get rid of the extras and produce an answer with the correct number of significant figures. The process of getting rid of the extra numbers is called **rounding off**. There are some rules which determine which digits to keep, and which to discard. First, determine how many significant figures your answer should have. You will drop all the insignificant figures. Look at the digit to the right of the last digit you will keep. If the first digit you are discarding is less than five, then drop it and all others to the right of it, and the last digit you are keeping remains the same. If the digit you are discarding is greater than or equal to five, then drop it, and the last digit you are keeping is increased by one. When a calculation involves several steps, use all your calculator numbers until you get to the final step, then correct for significant figures and round off. 5.32 / 6.23 equals 0.853932584 on your calculator, but the correct number of significant figures is 3. We want to keep three significant figures and discard the rest. The first digit we discard is a 9. Because this digit is greater than five, we drop it (and all the others to the right of it) and increase the three by 1. The correct result is 0.854.

2.6 PROBLEM SOLVING AND DIMENSIONAL ANALYSIS

Solving dimensional analysis problems requires the use of one or more **unit factors**. The unit factors can be given to you in the problem, or they can be ones you should have learned. A unit factor such as 1 in/2.54 cm means 1 in equals 2.54 cm, or another way to view it is 1 in per 2.54 cm. Note that you can use a unit factor as 1 in/2.54 cm or 2.54 cm/1 in, depending upon which unit you need to cancel.

Example:

How can you convert 186.2 g to lbs? The technique you use can be used to solve virtually all unit conversion problems. You can solve this problem by multiplying the number of grams (given in the problem) by a unit factor. The unit factor in this problem is $\frac{1 \text{ lb}}{453.6 \text{ g}}$ and is read 1 lb equals 453.6 g, or 1 lb per 453.6 g. Because 1 lb and 453.6 g are equivalent, multiplying 186.2 g by $\frac{1 \text{ lb}}{453.6 \text{ g}}$ is just like multiplying 186.2 g by 1. Notice that gram appears in both the numerator and the denominator of the expression, and cancels out, leaving pounds, which is the unit we want. Unit factors are sometimes provided in the problem, but some common equivalencies you will be expected to memorize. We are changing the units, but we are not actually changing the quantity of material in the equation.

$$186.2 \not{g} \times \frac{1 \text{ lb}}{453.6 \not{g}} = 0.4105 \text{ lb}$$

We can see that the quantity of matter has not changed by converting 0.4105 lb back to grams. In this conversion, we will use the conversion $\frac{453.6 \text{ g}}{1 \text{ lb}}$ so that lb cancels.

$$0.4105 \not{\text{lb}} \times \frac{453.6 \text{ g}}{1 \not{\text{lb}}} = 186.2 \text{ g}$$

Example:

How many milliliters are in 2.31 qt? You can solve this problem just as we did the example above, by multiplying the given units, 2.31 qt, by a unit factor. Two possible unit factors are $\frac{1 \text{ L}}{1000 \text{ ml}}$ and $\frac{1 \text{ L}}{1.057 \text{ qt}}$. Since there is no unit factor given which converts directly from milliliters to quarts, you can use two unit factors to arrive at an answer. Multiply 2.31 qt by the unit factor which has qts, $\frac{1 \text{ L}}{1.057 \text{ qt}}$. Note that the unit factor can be written $\frac{1 \text{ L}}{1.057 \text{ qt}}$ or $\frac{1.057 \text{ qt}}{1 \text{ L}}$. Both are correct, and the one you use depends on which unit you want to cancel out. In this problem, use $\frac{1 \text{ L}}{1.057 \text{ qt}}$ to cancel quarts.

$$2.31 \not{\text{qt}} \times \frac{1\text{L}}{1.057 \not{\text{qt}}} = 2.185 \text{ L}$$

Now you have a number expressed in liters. We have a unit factor which can convert between liters and milliliters.

$$2.185\ \cancel{L} \times \frac{1000\ ml}{1\ \cancel{L}} = 2185\ ml$$

Is this number expressed to the correct number of significant figures? There are three significant figures in 2.31. The two conversion factors, 1000 mL and 1.057 qt, do not have significant figures because they are definitions. So, the answer should be expressed to three significant figures. In 2185, the number we want to drop is 5, but if we drop it and round up, we are left with 219, which is not large enough. We need to add a zero on the right, since zeros on the right in numbers without a decimal place are not significant. The correct answer is 2190 mL.

2.7 TEMPERATURE CONVERSIONS: AN APPROACH TO PROBLEM SOLVING

Converting between temperature scales, and checking your answers will be a little easier if you can remember the boiling and freezing points of water in degrees Fahrenheit, in kelvins, and in degrees Celsius.

How Can You Convert Between Celsius and Kelvin?

Example:

One common temperature setting for hot water heaters is 60 °C. What is this on the Kelvin scale? The Celsius scale and the Kelvin scale are related because each of them has 100 degrees between the boiling and freezing points of water. The size of the degree is the same, but the zero point is different. The formula $^{t}K = {}^{t\circ}C + 273$ can be used to convert from Celsius to Kelvin. We add 273 degrees to the Celsius temperature because the only difference between the scales is that the Kelvin scale has the freezing point of water 273 degrees above that in the Celsius scale.

$$^{t}K = 60 + 273$$

The correct answer is 330 K.

How Can You Convert Between Kelvin and Celsius?

Example:

To convert 15 K to degrees Celsius, we can rearrange the equation $^{t}K = {}^{t\circ}C + 273$ so that the quantity we wish to solve for, degrees Celsius, is isolated on the left side of the equation.

$$^{t}K - 273 = {}^{t\circ}C + 273 - 273$$

Subtract 273 from both sides of the equation and we are left with

$$t°C = tK - 273$$

Now, substitute into the new equation.

$$t°C = 15 - 273$$

$$t°C = -258$$

You do not need to memorize two equations. If you remember one, you can derive the other by rearranging the terms.

How Can You Convert Between Fahrenheit and Celsius?

Example:

A common temperature for cooking beef roasts is 350 °F. What is this on the Celsius scale? We can use the formula $t°C = \frac{(t°F - 32)}{1.80}$ to calculate °C. In this example, beef would roast at 177 °C. Why does this formula work? On the Celsius scale, 0 °C is equivalent to the freezing point of water, while water freezes at 32 °F on the Fahrenheit scale. There are 32 units difference between the two scales. Because there are fewer Celsius degrees between the freezing and boiling points of water, the degrees are larger than Fahrenheit degrees. Subtract 32 from the Fahrenheit temperature, which adjusts the Fahrenheit temperature for the difference in the zero points between the two temperature scales.

$$350\ °F - 32 = 318\ °F$$

There are 100 degrees Celsius between the freezing and boiling points of water, and 180 degrees Fahrenheit between the freezing and boiling points of water. The ratio of Fahrenheit to Celsius is 1.80. So, to adjust for the different degree sizes, we divide degrees Fahrenheit by 1.80.

$$\frac{318\ °F}{1.80} = 176.7$$

There are two significant figures in our measurement, 350 °F, so our answer should also have two significant figures. The correct answer is 180 °F.

2.8 DENSITY

A common "trick" question that one child will ask another is "Which is heavier, a pound of lead or pound of feathers?" A pound of feathers takes up a lot of volume (ever had a feather pillow fight?) per unit mass, so the density is low. It takes a lot of feathers to make a pound. Lead takes up very little volume for the same amount of mass, so its density is high compared with an equal mass of feathers.

How Can You Calculate Density Given Mass and Volume?

Example:

What is the density of a sample of cocoa butter which weighs 10.32 g and occupies a volume of 11.72 mL? To solve this problem, you must know that density = $\frac{\text{mass}}{\text{volume}}$, or $d = \frac{g}{ml}$. In this problem,

$$d = \frac{10.32\text{ g}}{11.72\text{ mL}} = 0.8805\text{ g/mL}$$

The answer expressed to the correct number of significant figures is 0.881 g/mL.

How Can You Calculate Volume Given Density and Mass?

Example:

How much volume would a 480 lb piece of sculptor's marble occupy? The density of marble is 2.84 g/mL. In this problem, we are given the mass and density of a substance, and asked for the volume. We can find the volume because we know both the mass and the density. Mass is given in pounds, and we need grams. First we need to convert 480 lbs to grams by using the unit conversion factor which converts from lbs to grams.

$$480\text{ }\cancel{\text{lb}} \times \frac{453.6\text{ g}}{1\text{ }\cancel{\text{lb}}} = 217{,}728\text{ g}$$

Now we know both density and mass, expressed in the correct units. We can rearrange $d = \frac{\text{mass}}{\text{volume}}$ to solve for volume. Divide both sides of the equation by mass so that mass cancels on the right side.

$$\frac{d}{\text{mass}} = \frac{\text{mass}}{\text{volume} \times \text{mass}}$$

We are left with

$$\frac{d}{\text{mass}} = \frac{1}{\text{volume}}$$

Taking the inverse of both sides leaves the equation expressed in the desired form.

$$\text{volume} = \frac{\text{mass}}{\text{d}}$$

The volume of the sample is

$$\text{volume} = \frac{217{,}728\ \text{g}}{2.84\frac{\text{g}}{\text{mL}}} = 76{,}664.79\ \text{mL}$$

The answer expressed to the correct number of significant figures is 77,000 mL. If you were trying to explain to a sculptor how much volume his marble would occupy, you might want to convert your answer to liters, which is a unit more in scale with a large piece of stone.

$$77{,}000\ \text{mL} \times \frac{1\ \text{liter}}{1000\ \text{mL}} = 77\ \text{L}$$

LEARNING REVIEW

1. To express each of the following numbers in scientific notation, would you move the decimal point to the right or to the left? Would the power of 10 be positive or would it be negative (have a minus sign)?

   a. 0.001362
   b. 146,218
   c. 342.016
   d. 0.986
   e. 18.8

2. Complete the table below and convert the numbers to scientific notation.

		coefficient		exponent
a.	0.00602	6.02	x	
b.	48,190,001		x	10^7
c.	60,000	6	x	
d.	49		x	10^1
e.	1.002	1.002	x	

3. Convert the numbers below to scientific notation.

a.	1,999,945	e.	0.0068
b.	650,700	f.	0.042001
c.	0.1109	g.	1.2
d.	545	h.	13.921

4. To express the following numbers in decimal notation, would you move the decimal point to the right or to the left? How many places?

a. 1.02×10^3
b. 4.1×10^{-6}
c. 5×10^5
d. 4.31×10^2
e. 9.31×10^{-2}

5. Convert the numbers below to decimal notation.

a.	4.91×10^{10}	e.	1.009×10^{-4}
b.	1.1×10^{-3}	f.	9.2×10^1
c.	5.42×10^{-6}	g.	4.395×10^5
d.	2.07×10^3	h.	7.03×10^{-2}

6. How can you convert -1235.1 to scientific notation?

7. Which quantity in each pair is larger?

a. 1 meter or 1 millimeter
b. 10 seconds or 1 microsecond
c. 1 mg or 1 kg
d. 1 cm or 1 Mm
e. 1 kilogram or 1 decigram

8. Which quantity in each pair is larger?

a. 1 mile or 1 kilometer
b. 1 liter or 1 cubic meter
c. 1 kilogram or 1 pound
d. 1 quart or 1 milliliter
e. 1 micrometer or 12 inches

9. What metric or SI unit would you be likely to use in place of the English units given below?

a. Bathroom scales commonly provide weight in pounds.
b. A convenient way to purchase small quantities of milk is by the quart.
c. A cheesecake recipe calls for 1 teaspoon of vanilla extract.
d. Carpeting is usually priced by the square yard.
e. "An ounce of prevention is worth a pound of cure."

10. What number would you record for each of the following measurements?

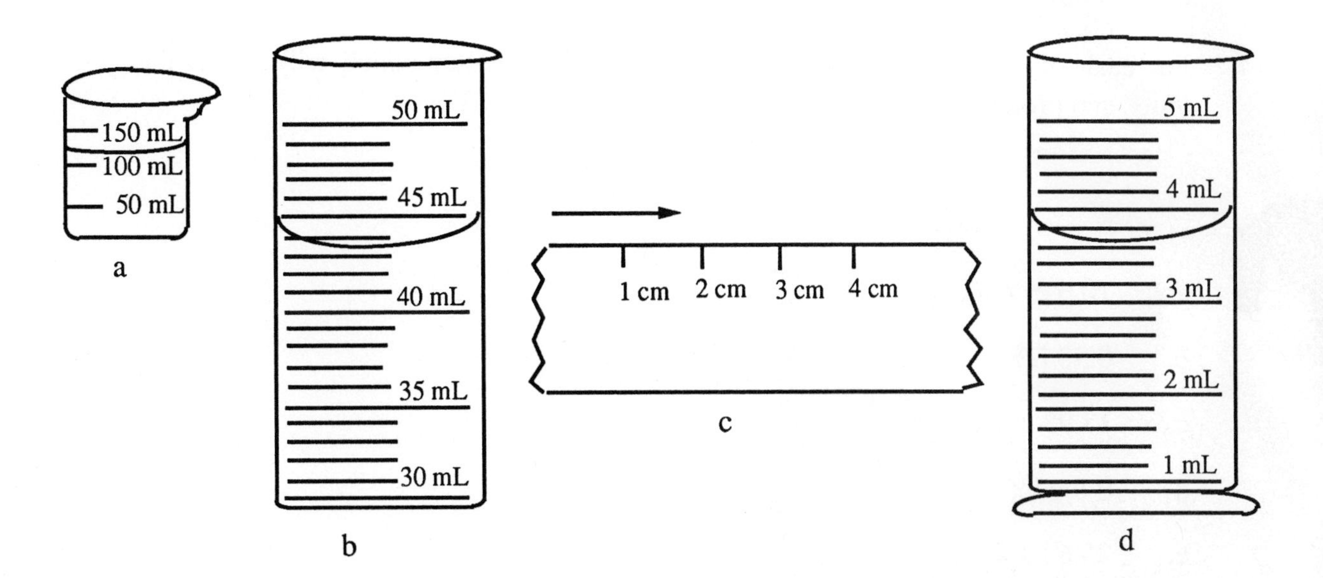

11. How many significant figures are in each of the following numbers?

a. 100
b. 1180.3
c. 0.00198
d. 1.001
e. 45.00
f. 67,342
g. 0.0103
h. 4.10×10^4

12. Express the results of each calculation to the correct number of significant figures.

a. 1.8 x 2.93
b. 0.002/0.041
c. 483.21 x 5.00
d. 0.00031 x 4.030
e. 495.0/390
f. 5024 x 19.2
g. 91.3 x 2.10 x 7.7
h. 8.003 x 4.93/61.05

13. Round off the following numbers to the number of significant figures indicated.

		number of significant figures
a.	0.58333333	4
b.	451.0324	3
c.	942.359	4
d.	0.0090060	2
e.	6.8	1
f.	1346	3
g.	490,000.423	6
h.	0.06295	3

14. For each of the quantities below, give a conversion factor which will cancel the given units, and produce a number which has the desired units. For example:

$$8.6\ \not{g} \times \frac{1\ kg}{1000\ \not{g}}$$

a. 10.6 m x $\frac{cm}{m}$

b. 0.98 L x $\frac{qt}{L}$

c. 18.98 cm x $\frac{in}{cm}$

d. 0.5 yd x $\frac{m}{yd}$

e. 25.6 kg x $\frac{lb}{kg}$

15. Perform the following conversions.

a. 5.43 kg to g
b. 65.5 in to cm
c. 0.62 L to ft^3
d. 111.3 g to lbs
e. 40.0 qts to L
f. 850 yds to m
g. 2.83 g to lbs
h. 0.21 cm to in

16. Fill in the important reference temperature on each of the temperature scales.

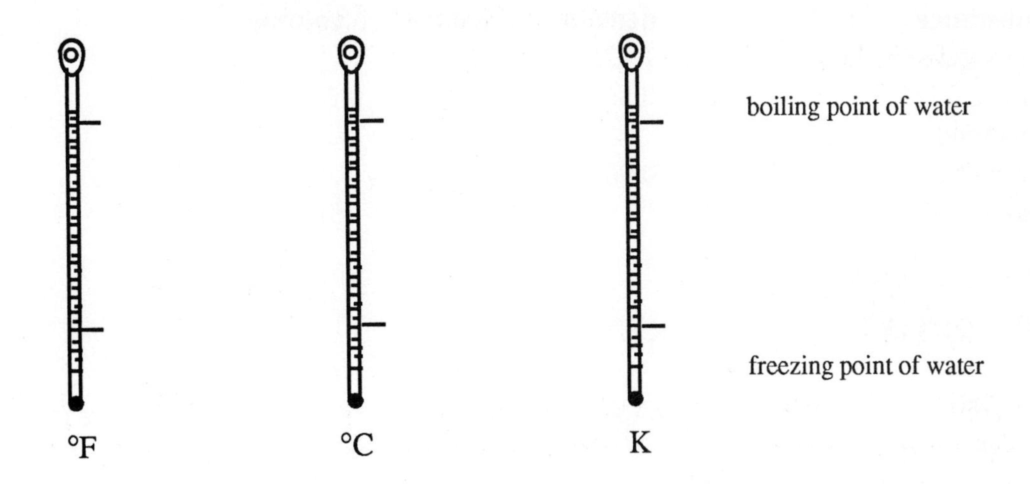

17. How many degrees are there between the freezing point and the boiling point of water on the Fahrenheit and on the Celsius scales?

    a. Calculate the ratio of the number of degrees Fahrenheit to the number of degrees Celsius between the freezing and boiling points of water.
    b. Calculate the ratio of the number of degrees Celsius to the number of degrees Kelvin between the freezing and boiling points of water.
    c. Calculate the ratio of the number of degrees Fahrenheit to the number of degrees Kelvin between the freezing and boiling points of water.

18. Comfortable room temperature for houses is 70 °F. What is this on the Celsius scale?

19. Ethyl alcohol boils at 78 °C. What is this on the Fahrenheit scale?

20. In some parts of the Midwest, temperatures may drop as low as -22 °F in winter. What is this on the Kelvin scale?

21. Perform the temperature conversions below.

    a. 180 °F to °C
    b. -10.8 °C to K
    c. 244 K to °C
    d. 0.5 °C to °F
    e. 25.1 °F to °C

22. Fill in the missing quantities in the table below.

substance	density	mass	volume
seawater	1.025	52.6 g	
butter	0.86		527.4 mL
diamond		2.13 g	0.65 mL
beeswax	0.96	125.5 g	
oak wood		4.63 kg	6173.3 mL

ANSWERS TO LEARNING REVIEW

1. To convert to scientific notation for numbers which are greater than 0 but less than 1, move the decimal point to the right. For numbers which are greater than 1, move the decimal point to the left. Make sure that your final answer has only 1 number to the left of the decimal point.

   a. right 0.001362

   b. left 146218

   c. left 342.016

   d. right 0.986

   e. left 18.8

2. Remember that numbers written in scientific notation are divided into two parts. The coefficient on the left is a small number between one and nine, and the exponent on the right is ten raised to some power.

		coefficient		exponent
a.	0.00602	6.02	x	10^{-3}
b.	48,190,001	4.8190001	x	10^{7}
c.	60,000	6	x	10^{4}
d.	49	4.9	x	10^{1}
e.	1.002	1.002	x	10^{0}

3. The answer for g, 1.2×10^0, means that we do not need to move the decimal point of the coefficient. 1.2×10^0 is the same as writing 1.2.

a	1.999945×10^6	e.	6.8×10^{-3}
b.	6.507×10^5	f.	4.2001×10^{-2}
c.	1.109×10^{-1}	g.	1.2×10^0
d.	5.45×10^2	h.	1.3921×10^1

4. When converting from scientific notation to decimal, look first at the exponent. If the exponent is positive (has no negative sign) move the decimal point to the right. If the exponent is negative, move the decimal point to the left.

a.	right	1020
b.	left	0.0000041
c.	right	500,000
d.	right	431
e.	left	0.0931

5. A large number such as 49,100,000,000 has only three significant figures. The trailing zeros are not significant because there is no decimal point at the end.

a	49,100,000,000	e.	0.0001009
b.	0.0011	f.	92
c.	0.00000542	g.	439,500
d.	2070	h.	0.0703

6. This number is different from others we have seèn. It is smaller than one, and also smaller than zero. You can convert these numbers to scientific notation in much the same way as you convert numbers which are greater than one. First, move the decimal point to the left as you normally would.

$$-1235.1$$

Then, count the number of times the decimal point was moved and add the correct exponent.

$$1.2351 \times 10^3$$

Just keep the minus sign in front of the entire number.

$$-1.2351 \times 10^3$$

The minus sign goes in front of 1.235 because this number is less than zero. The exponent is negative only for numbers which are between 0 and 1.

7. To work this problem, you need to have learned the SI prefixes and how they modify the size of the base unit.

   a. meter is larger than millimeter
   b. 10 seconds are larger than 1 microsecond
   c. 1 kg is larger than 1 mg
   d. 1 Mm is larger than 1 cm
   e. 1 kilogram is larger than 1 decigram

8. This problem asks about the relationship between English units and SI units. You need to know the relative sizes of English and SI units.

   a. 1 mile is larger than 1 kilometer
   b. 1 cubic meter is larger than 1 liter
   c. 1 kilogram is larger than 1 pound
   d. 1 quart is larger than 1 milliliter
   e. 12 inches are larger than 1 micrometer

9. a. kilograms
   b. liter
   c. cubic meter (m^3)
   d. square meter (m^2)
   e. "A gram of prevention is worth a kilogram of cure."

10. a. This measuring device is a beaker. Each division represents 50 mL. The volume of liquid in the beaker is somewhere between 100 mL and 150 mL. We estimate that the volume is 120 mL.

    b. This measuring device is a graduated cylinder. The numbers tell us that each major graduation is 5 mL, so each of the smaller lines must be 1 mL. We can accurately measure the volume to the nearest 1 mL. The volume in this cylinder is between 43 and 44 mL. We estimate the volume to be 43.5 mL.

    c. The length of the arrow lies between 1 cm and 2 cm. We estimate that the arrow lies 0.8 of the way between the two marks. So the reported measurement would be 1.8 cm.

    d. This graduated cylinder has major divisions of 1 mL. The smaller marks represent 0.2 mL. The liquid lies between 3.6 and 3.8 mL. We estimate that the volume is 0.25 of the way between the two marks, so the volume would be reported as 3.65 mL.

11. Remember that all nonzero numbers count as significant figures, and zeros in the middle of a number are always significant. Zeros to the right of some nonzero numbers are only significant if they are followed by a decimal point.

a. 1
b. 5
c. 3
d. 4
e. 4
f. 5
g. 3
h. 3

12. For problems involving multiplication and division, your answer should have the same number of decimal points as the measurement with the least number of significant figures. For problems involving addition and subtraction, your answer should have the same number of significant figures as the measurement with the least number of digits to the right of the decimal point.

a. 5.3
b. 0.05
c. 2420
d. 0.0012
e. 1.3
f. 96,500
g. 1500
h. 0.646

13. You can answer problems such as 13.g by putting the decimal point at the end to show that all 6 digits are significant, or use scientific notation with a coefficient which contains 6 digits. Either form is acceptable.

a. 0.5833
b. 451
c. 942.4
d. 0.0090
e. 7
f. 1350
g. 490,000. or 4.90000×10^5
h. 0.0630

14. To answer this question, you need to know the common equivalencies and how to write them as a unit factor. For problem a, we have a measurement in meters. The conversion factor relates cm to m. You need to know that 100 cm equals 1 m.

a. $10.6\ \cancel{\text{m}} \times \frac{100\ \text{cm}}{1\ \cancel{\text{m}}}$

b. $0.98\ \cancel{\text{L}} \times \frac{1.06\ \text{qt}}{1\ \cancel{\text{L}}}$

c. $18.98\ \cancel{\text{cm}} \times \frac{1\ \text{in}}{2.54\ \cancel{\text{cm}}}$

d. $0.5\ \cancel{\text{yd}} \times \frac{1\ \text{m}}{1.094\ \cancel{\text{yd}}}$

e. $25.6\ \cancel{\text{kg}} \times \frac{2.205\ \text{lb}}{1\ \cancel{\text{kg}}}$

15. a. $5.43\ \cancel{\text{kg}} \times \dfrac{1000\ \text{g}}{1\ \cancel{\text{kg}}} = 5430\ \text{g}$

b. $65.5\ \cancel{\text{in}} \times \dfrac{2.54\ \text{cm}}{1\ \cancel{\text{in}}} = 166\ \text{cm}$

c. $0.62\ \cancel{\text{L}} \times \dfrac{1\ \text{ft}^3}{28.32\ \cancel{\text{L}}} = 0.022\ \text{ft}^3$

d. $111.3\ \cancel{\text{g}} \times \dfrac{1\ \text{lb}}{453.6\ \cancel{\text{g}}} = 0.2454\ \text{lb}$

e. $40.0\ \cancel{\text{qt}} \times \dfrac{1\ \text{L}}{1.06\ \cancel{\text{qt}}} = 38\ \text{L}$

f. $850\ \cancel{\text{yd}} \times \dfrac{1\ \text{m}}{1.094\ \cancel{\text{yd}}} = 780\ \text{m}$

g. $2.83\ \cancel{\text{g}} \times \dfrac{1\ \text{lb}}{453.6\ \cancel{\text{g}}} = 6.24 \times 10^{-3}\ \text{lb}$

h. $0.21\ \cancel{\text{cm}} \times \dfrac{1\ \text{in}}{2.54\ \cancel{\text{cm}}} = 0.083\ \text{in}$

16.

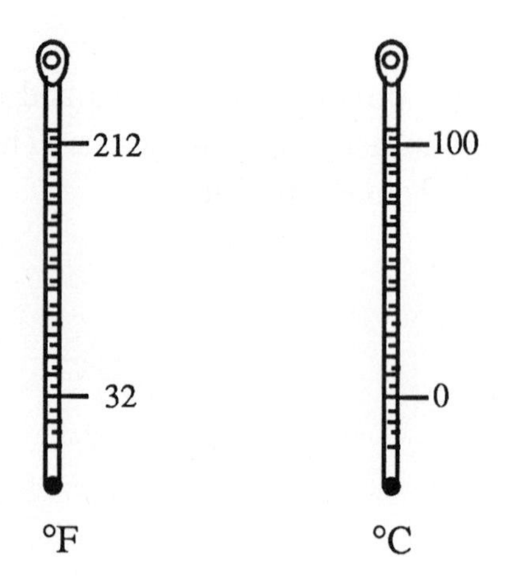

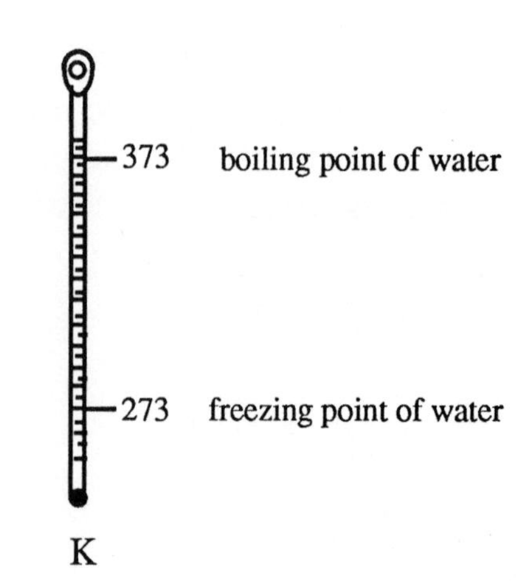

17. There are 180 degrees between the freezing and boiling points of water on the Fahrenheit scale, and 100 degrees on the Celsius scale.

a. $$\frac{°F}{°C} = \frac{180}{100} = 1.80$$

b. $$\frac{°C}{K} = \frac{100}{100} = 1$$

c. $$\frac{°F}{K} = \frac{180}{100} = 1.80$$

18. We want to convert from degrees Fahrenheit to degrees Celsius.

$t\,°F = 70$

We can use the formula below to calculate degrees Celsius.

$$t\,°C = \frac{(t\,°F - 32)}{1.80}$$

$$t\,°C = \frac{(70. - 32)}{1.80}$$

$t\,°C = 21$

70 degrees Fahrenheit is equivalent to 21 degrees Celsius.

19. We want to convert from degrees Celsius to degrees Fahrenheit.

$t\,°C = 78$

We can use the formula below to calculate degrees Fahrenheit.

$t\,°F = 1.80(t\,°C) + 32$

$t\,°F = 1.80(78) + 32$

$t\,°F = 170$

78 degrees Celsius is equivalent to 170 degrees Fahrenheit.

20. We want to convert from degrees Fahrenheit to Kelvin.

$t\,°F = -22$

We do not have a formula to directly convert degrees Fahrenheit to Kelvins, but we can convert from degrees Fahrenheit to degrees Celsius, then from degrees Celsius to Kelvins.

Convert $t\,^{\circ}F$ to $t\,^{\circ}C$ first.

$$t\,^{\circ}C = \frac{(t\,^{\circ}F - 32)}{1.80}$$

$$t\,^{\circ}C = \frac{(-22 - 32)}{1.80}$$

$$t\,^{\circ}C = -30.$$

Now, calculate kelvins.

$$t_K = t\,^{\circ}C + 273$$

$$t_K = -30. + 273$$

$$t_K = 243$$

21. a. $t\,^{\circ}F = 180$

$$t\,^{\circ}C = \frac{t\,^{\circ}F - 32}{1.80}$$

$$t\,^{\circ}C = \frac{180 - 32}{1.80}$$

$$t\,^{\circ}C = 82$$

b. $t\,^{\circ}C = -10.8$

$$t_K = t\,^{\circ}C + 273$$

$$t_K = -10.8 + 273$$

$$t_K = 262$$

c. $t_K = 244$

$$t_K = t\,^{\circ}C + 273$$

Rearrange this equation to isolate $t\,^{\circ}C$.

$$t\,^{\circ}C = t_K - 273$$

$$t\,^{\circ}C = 244 - 273$$

$$t\,^{\circ}C = -29$$

d. $t\,^{\circ}C = 0.5$

$$t\,^{\circ}F = 1.80(t\,^{\circ}C) + 32$$

$$t\,^{\circ}F = 1.80(0.5) + 32$$

$$t\,^{\circ}F = 32.9$$

e. $t\,^{\circ}F = 25.1$

$$t\,^{\circ}C = \frac{(t\,^{\circ}F - 32)}{1.80}$$

$$t\,^{\circ}C = \frac{(25.1 - 32)}{1.80}$$

$$t\,^{\circ}C = -3.8$$

22.

substance	density	mass	volume
seawater	1.025 g/mL	52.6 g	51.3 mL
butter	0.86 g/mL	450 g	527.4 mL
diamond	3.3 g/mL	2.13 g	0.65 mL
beeswax	0.96 g/mL	125.5 g	130 mL
oak wood	0.750 g/mL	4.63 kg	6173.3 mL

PRACTICE EXAM

1. Which choice is correct scientific notation for 1200.5?

   a. 1.2×10^3
   b. 1.2×10^4
   c. 1.2005×10^4
   d. 1.2005×10^3
   e. 1.2×10^{-3}

2. Which choice is correct decimal notation for 6.02×10^{-4}?

   a. 60,200
   b. 0.602
   c. 0.000602
   d. 0.0006
   e. 0.0060

3. Which of the numbers below contains 4 significant figures?

   a. 420.1
   b. 5300
   c. 0.302
   d. 0.0002
   e. 6.20×10^4

4. Which of the choices below shows 489,620 rounded to three significant figures, and 0.00043239 rounded to four significant figures?

   a. 490, 0.0004
   b. 489,000; 4324
   c. 4.89×10^5, 0.0004323
   d. 489, 4.323×10^{-4}
   e. 4.90×10^5, 0.0004324

5. Correctly evaluate the expression below.

$$\frac{3.20 \times 0.02019}{4.901}$$

a. 0.01318
b. 0.0132
c. 0.013
d. 0.0131826
e. 0.0131

6. Which is the largest unit of mass?

a. 0.5 kg
b. 100 mg
c. 10 μg
d. 1 g
e. 10 dg

7. Ethyl ether melts at -116.3 °C. What is this on the Fahrenheit scale?

a. -209.3 °F
b. -151.7 °F
c. -177 °F
d. -177.3 °F
e. 151.7 °F

8. A glass pitcher holds 732 mL of milk. How many quarts is this?

a. 0.776 qt
b. 0.732 qt
c. 0.73 qt
d. 0.69 qt
e. 0.691 qt

9. Which unit of length is the smallest?

a 4.10×10^{3} ft
b. 560 m
c. 5.60×10^{-2} km
d. 0.8 mi
e. 1400 yd

10. If 16.56 g of liquid bromine has a volume of 5.34 mL, what is the density of bromine?

   a. 0.322 g/mL
   b. 0.3225 g/mL
   c. 3.22 g/mL
   d. 3.10 g/mL
   e. 3.101 g/mL

PRACTICE EXAM ANSWERS

1. d (2.1)
2. c (2.1)
3. a (2.5)
4. e (2.5)
5. b (2.5)
6. a (2.3)
7. d (2.7)
8. a (2.6)
9. c (2.6)
10. d (2.8)

CHAPTER 3: MATTER AND ENERGY

INTRODUCTION

This chapter provides you with a basic foundation of facts and concepts about matter you will need throughout your chemistry course. There are fewer mathematical calculations in this chapter than in other chapters you will study.

Pay careful attention to Sections 3.2 (Physical and Chemical Properties and Changes) and 3.4 (Mixtures and Pure Substances) in your textbook. The concepts in these sections often seem confusing when you are first introduced to them. Look carefully at the examples in your text which will help you distinguish between physical and chemical changes and between mixtures and pure substances.

Section 3.6 introduces you to heat energy, and you will calculate the amount of energy needed to heat water and other substances. Practice setting up good dimensional analysis equations with the energy problems. Make sure all numbers have a unit and that all the units cancel. Now is a good time to gain some confidence with your problem solving skills.

AIMS FOR THIS CHAPTER

1. To define matter and be able to state the distinguishing characteristics of the three states of matter. (Section 3.1)
2. Know the difference between chemical and physical properties, and chemical and physical changes. (Section 3.2)
3. Know what elements and compounds are, and be able to identify which substances are elements and which are compounds. (Section 3.3)
4. Be able to tell whether a substance is a pure substance or a mixture, and if a mixture, whether it is homogeneous or heterogeneous. (Section 3.4)
5. Be able to describe how distillation and filtration work, and how you would use them to separate a mixture. (Section 3.5)
6. Know the relative amounts of energy associated with the three states of matter, and which units are commonly used when discussing heat energy. (Section 3.6)
7. Be able to use the formula $Q = s \times m \times \Delta T$ to calculate Q. Rearrange the equation to calculate any one of the quantities, if you are given any of the other three. (Section 3.6)

QUICK DEFINITIONS

Matter Everything that has mass and takes up space. (Section 3.1)

Physical property A characteristic such as odor, color, or physical state. You can observe a physical property of a substance without changing the composition of the substance. (Section 3.2)

Chemical property A characteristic such as the kind of elements a substance will react with to form new substances. When you observe a chemical property of a substance, the composition and properties of the substance changes. (Section 3.2)

Physical change A change that does not affect the composition of the material undergoing the change, although it can affect the physical properties or physical states. Some examples are changes in temperature, and change from the solid state to the liquid state. (Section 3.2)

Chemical change A change which causes the substance to become a new substance with a different composition. When a chemical change occurs we say that a chemical reaction has taken place. (Section 3.2)

Element A fundamental unit which cannot be broken into smaller units by chemical means. (Section 3.3)

Compound A unit composed of two or more elements which can be broken into elements by chemical means. A compound always has the same relative numbers and kinds of elements. (Section 3.3)

Mixture Anything whose composition varies if you sample different parts of it. (Section 3.4)

Pure substance Anything whose composition stays the same, even if you examine samples taken from different locations within the substance. (Section 3.4)

Homogeneous mixture A mixture whose composition is constant throughout. Also called a solution. (Section 3.4)

Heterogeneous mixture A mixture whose composition is different depending upon where you sample it. (Section 3.4)

Distillation A method for separating mixtures. Heat is applied to a mixture until one of the substances begins to boil and becomes a vapor. The second substance stays behind because its boiling point is higher. The vapor of the first substance moves through the apparatus until it becomes cooled and recondenses to a liquid. Distillation is a good method for separating homogeneous mixtures. (Section 3.5)

Filtration A method for separating mixtures which depends upon differences in the sizes of particles to work. Pour a heterogeneous mixture through a mesh which catches the particles, but allows the liquid to pass through. Often the mesh is made of paper, but it can be made of other materials as well. (Section 3.5)

Energy A measure of the ability to do work. (Section 3.6)

Calorie A unit of energy which is equal to the amount of heat energy it takes to raise the temperature of one gram water one degree Celsius. (Section 3.6)

Joule The SI unit of energy. 1 calorie = 4.184 joules. (Section 3.6)

Specific heat capacity The amount of heat energy it takes to raise the temperature of one gram of some substance by one degree Celsius. Each substance has a different specific heat capacity. (Section 3.6)

CONTENT REVIEW

3.1 MATTER

What Is Matter?

Everything in the universe that has mass and occupies space is made of **matter**. We can see large masses of matter everywhere we look, from automobiles to tree trunks. Some pieces of matter are so small we can't see them. When many of the small pieces of matter are together in one spot, they form a mass large enough to see; for example, a gold nugget is made of many small pieces of matter, each one too small to see. Much of the science of chemistry concerns what happens to matter.

3.2 PHYSICAL AND CHEMICAL PROPERTIES AND CHANGES

How Can You Tell Physical Properties From Chemical Properties?

A **physical property** is anything you can observe without destroying or changing the composition of a substance. Odor, color, and the temperature at which a substance melts are examples of physical properties. A **chemical property** describes how a substance(s) can change chemically to form a new substance. A chemical property of the element sodium is that it will combine with the element chlorine to form a new substance, sodium chloride.

How Can You Tell Physical Changes From Chemical Changes?

Sometimes it can be difficult to tell the difference between a physical change and a chemical change. Section 6.1 of Chapter 6 will provide more information that can help you tell whether or not a chemical change has occurred.

A **physical change** often involves a change in the state of matter. Ice is made of water molecules in the solid state. When ice melts, the resulting liquid is also made of water molecules, but now in the liquid state. The difference between ice and liquid water is the difference in physical state. Other physical changes involve changes in appearance. For example, dissolving salt crystals in water doesn't change the composition of the salt. Although the salt has dissolved into particles so small you can no longer see them, they are still salt particles. A taste of the water will tell you the salt is still there. If you evaporate the water in a hot oven, the salt remains on the bottom of the container. You can recover it in its original state.

A **chemical change** involves a change in composition. Burning a piece of oak firewood produces substances such as water vapor, carbon dioxide and ash as well as heat and light. The new products are chemically different from the original oak firewood. When fresh milk is stored for a long time in the refrigerator, it spoils. Spoiled milk has a sour taste and smell. A chemical change has occurred which caused the characteristics and the composition of the milk to change.

3.3 ELEMENTS AND COMPOUNDS

What Are Elements and Compounds?

Matter is composed of over 100 fundamental units called **elements**. Elements cannot be decomposed into smaller units by chemical means. Some examples of elements are oxygen, iron, and gold. Any sample of the element oxygen contains only oxygen atoms, and any sample of the element gold contains only gold atoms.

Some elements can combine with each other to form **compounds**. Any particular compound is always made from the same relative numbers and kinds of elements, that is, its composition is constant. For example, the compound water is made from the elements oxygen and hydrogen. Regardless of the source of the water, there are always two hydrogen atoms and one oxygen atom in each molecule of water. You will learn more about atoms and molecules in Chapter 4.

3.4 MIXTURES AND PURE SUBSTANCES

How Can You Tell Mixtures From Pure Substances?

Because they have constant composition, compounds are pure substances even though they are composed of a combination of elements. A **pure substance** contains only 1 compound, or 1 kind of element or molecule. Water is a pure substance because it is always composed of one oxygen atom and two hydrogen atoms.

Sodium bicarbonate, called baking soda, is another example of a pure substance. It contains one sodium atom, one hydrogen atom, one carbon atom and three oxygen atoms. Pure baking soda is always the same, no matter where you buy it, or what brand you buy, or where in the box you take a sample.

Most of the substances in our environment are **mixtures**. The quantity and types of ingredients in a mixture change from time to time, and a mixture can have several different areas with different ingredients. Milk is a mixture, although we can't see each of the individual ingredients. Milk contains sugars, proteins, fats, and other components. Whole milk has approximately four percent fat, while other milk has two percent fat. The ingredients in milk will also depend upon the diet of the cow. When you order a glass of milk at a restaurant, you are not guaranteed to get exactly the same product each time, because milk is a mixture.

How Can You Distinguish Homogeneous and Heterogeneous Mixtures?

Some mixtures contain pure substances evenly distributed throughout the mixture, so that if you took different samples from the same mixture, you would find the same quantities of pure substances in each sample. These are called **homogeneous mixtures**. Table sugar dissolved in water is a homogeneous mixture since the sugar is evenly distributed throughout the water. You can't see areas in the sugar and water mixture which look different from other areas.

Some other mixtures have an uneven distribution of pure substances. These mixtures are called **heterogeneous mixtures**. For example, a mixture of oil and water is a heterogeneous mixture. The top part of the mixture contains more oil, and the bottom part of the mixture has more water. The two components do not dissolve in each other. If you take samples of the mixture from both the top and the bottom, the two samples will contain completely different amounts of oil and water.

3.5 SEPARATION OF MIXTURES

Remember that even the best separation methods are not 100 percent complete. There is always a little bit of one substance left with the other substance. The better your laboratory skill, and the better the equipment, the better the final separation will be.

How Can You Separate a Mixture by Distillation?

Distillation is useful as a separation method only when the two materials to be separated boil at different temperatures. Table salt boils at 1413 °C while water boils at 100 °C. When you separate a mixture of salt and water by distillation, you apply heat to the mixture and raise the temperature until the water starts to boil. The salt has a much higher boiling point than water does, and does not boil at the temperature which boils the water. As the water boils, it is converted to steam, which rises in the distillation apparatus. You can collect the steam in a separate container, and cool it until water vapor condenses to liquid water. The salt remains behind in the original container because the temperature is never high enough for the salt to boil. Both have retained their original chemical properties. You haven't changed either of them by the separation process. Therefore you can say that distillation causes a physical change. Distillation is only useful as a separation method when the two materials to be separated have different boiling points.

How To Separate a Mixture by Filtration.

Filtration is another common way to separate mixtures containing liquids and solids. Filtration is non-destructive because after filtration, the two substances you have separated are the same chemically as they were before the separation. Filtration is a good way to separate heterogeneous mixtures. If we pour a mixture of water and aquarium sand through a mesh of paper, the water molecules pass through the paper, but the grains of sand are too large to pass through, and are trapped on the paper.

3.6 ENERGY AND ENERGY CHANGES

Energy is the capacity to do work. We often use energy to change the temperature of various substances. When substances change temperature, heat energy is either absorbed or lost. One unit used to measure the amount of heat gained or lost is the **calorie**, abbreviated cal. A calorie is the amount of heat it takes to raise the temperature of one gram of water by one degree Celsius. Another common unit of heat energy is the joule, abbreviated J.

How Can You Convert Energy Given in Joules to Calories?

Many energy problems require conversion from a given energy unit to another energy unit. The conversion factor for calories to joules is 1 cal / 4.184 J. Using this conversion factor you can convert from one unit to the other.

Example:

How can you convert 2.31 J to calories?

$$2.31\ \not{J} \times \frac{1\ \text{cal}}{4.184\ \not{J}} = 0.552\ \text{cal}$$

Set up the problem so that the joules cancel and you are left with calories, the desired unit. The answer should be expressed to three significant figures because the given 2.31 J has three significant figures. Remember that 1 cal and 4.184 J are definitions and do not affect the number of significant figures in the answer.

How Can You Calculate the Amount of Energy Needed to Change the Temperature of Liquid Water a Given Number of Degrees?

Example:

How many calories does it take to convert 1.0 kg water from 12.9 °C to 14.2 °C? From the definition of a calorie you know that it takes one calorie to raise the temperature of one gram of water by one degree Celsius. You can write the definition in the form of a conversion factor: 1 cal/g °C which reads "one calorie per gram of water per degree Celsius increase in temperature". The initial temperature in the example is 12.9 °C and the final temperature is 14.2 °C, a temperature increase of 14.2 minus 12.9 which is equal to 1.3 °C.

$$1.3\ \text{°}\cancel{\text{C}} \times \frac{1\ \text{cal}}{\text{g} \times \text{°}\cancel{\text{C}}} = 1.3\ \text{cal/g}$$

But in this example we have 1.0 kg of water, not 1 g, so we must convert kg to g. The appropriate conversion factor is 1000 g/1 kg. The correct answer is 1.3 cal/g x 1000 g = 1.3 x 10^3. If we solve the problem in one equation,

$$1.0\ \cancel{\text{kg}} \times \frac{1000\ \cancel{\text{g}}}{1\ \cancel{\text{kg}}} \times 1.3\ \text{°}\cancel{\text{C}} \times \frac{1\ \text{cal}}{\cancel{\text{g}} \times \text{°}\cancel{\text{C}}} = 1.3 \times 10^3\ \text{cal}$$

Remember that using scientific notation leaves no doubt that we mean to indicate two significant figures. 1.0 kg has two significant figures so our answer should also have two significant figures.

How Can You Calculate the Amount of Energy Needed to Change the Temperature of Any Substance?

You have seen that it takes one calorie to raise the temperature of 1 gram of water by 1 degree Celsius. How much energy does it take to raise the temperature of 1 gram of other substances by 1 degree Celsius? Each substance takes a different amount of energy. This amount of energy is known as the **specific heat capacity**, or sometimes as **specific heat**. Specific heats are given in units of joules.

Example:

Solid lead has a specific heat capacity of 0.16 J/g x °C. How much energy, in joules, is required to raise the temperature of 32.8 g solid lead from -24.4 °C to 56.4 °C? The specific heat capacity of solid lead is 0.16 J for **each** gram lead and for **each** °C. In this problem you are presented with 32.8 g lead, and a temperature increase of 80.8 °C. You can set the problem up as

$$\frac{0.16\ \text{J}}{\cancel{\text{g}} \times \text{°}\cancel{\text{C}}} \times 32.8\ \cancel{\text{g}} \times 80.8\ \text{°}\cancel{\text{C}} = 420\ \text{J}$$

A generalized way of solving problems like the one above would be to state the following. The heat energy needed (Q) equals the specific heat capacity of the substance (s) times the mass of the substance in g (m) times the change in temperature in degrees Celsius (ΔT). To write the equation with symbols only:

$$Q = s \times m \times \Delta T.$$

If you know any three of the quantities, you can find the fourth by rearranging the equation to isolate the quantity you want to find on the left side of the equation.

LEARNING REVIEW

1. Which of the properties below is/are physical properties, and which are chemical properties?

   a. Oxygen atoms can combine with hydrogen atoms to form water molecules.
   b. Ethyl alcohol boils at 78 °C.
   c. Liquid oxygen is pale blue in color.

2. Which of the changes below are physical changes, and which are chemical changes?

   a. A copper strip is hammered flat to make a bracelet.
   b. Copper and sulfur react to form a new substance, copper sulfide.
   c. Liquid water freezes at 273 K.
   d. Oxygen gas condenses to a liquid at -183 °C.
   e. You prepare a 3 minute egg for breakfast.

3. Which of the symbols below represent elements, and which represent compounds?

   a. S
   b. H_2O
   c C
   d. N_2O_5
   e. NaOH

4. Is each of the properties below a physical or chemical property?

   a. temperature at which a solid is converted to a liquid
   b. odor
   c. temperature at which a compound breaks down into its elements
   d. oxygen reacts with a substance to produce energy

5. Which of the substances below are mixtures, and which are pure substances?

   a. gasoline
   b. table sugar (sucrose)
   c. garden soil
   d. sterling silver necklace

6. Which of the mixtures is homogeneous and which is heterogeneous?

   a. sweetened hot tea
   b. plastic bag filled with leaves and grass clippings
   c. a weak solution of rubbing alcohol in water
   d. devil's food cake mix

7. Describe how you can separate a mixture by filtration.

8. What type of mixture is best separated by filtration, a homogeneous mixture or a heterogeneous mixture?

9. Describe how you would separate the following mixtures.

   a. sand from gravel
   b. salt from sand
   c. sugar from water

10. Label each part of the distillation apparatus below.

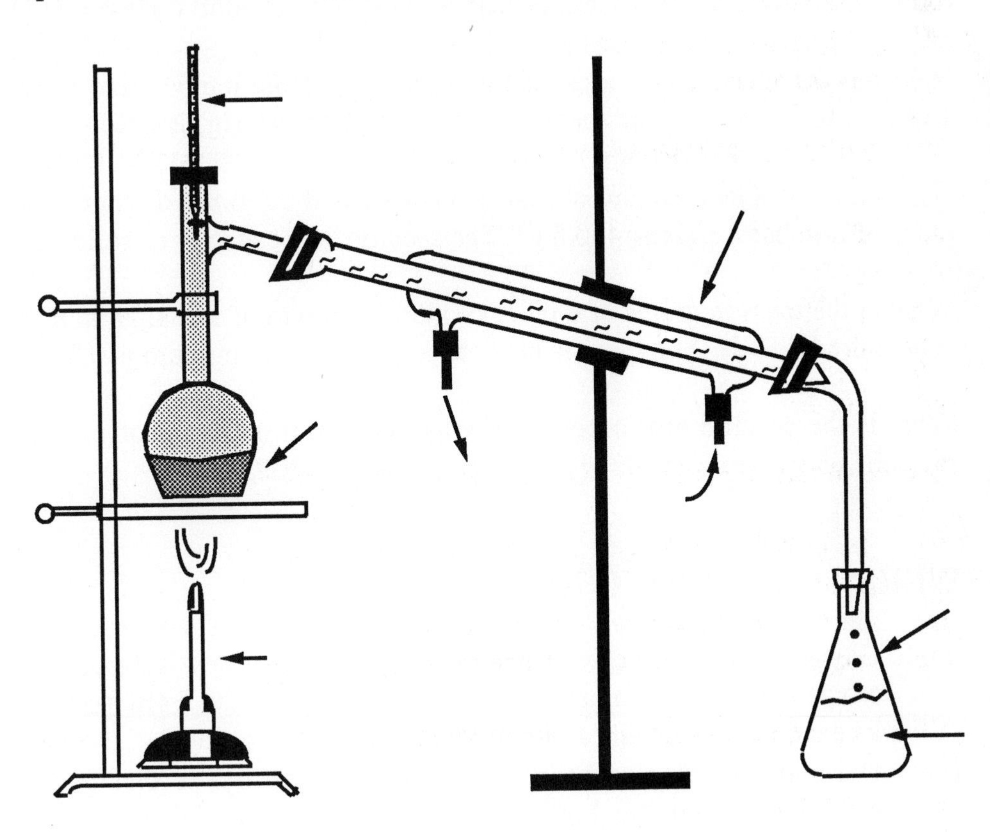

11. Convert the energy values below to the desired units.

   a. 45.8 cal to J
   b. 0.561 cal to J
   c. 5.96 J to cal
   d. 76 J to cal

12. Calculate the number of calories required to change the temperature of each of the quantities of water below.

   a. 100.1 g of water from 6 °C to 25 °C
   b. 2.32 g of water from 36 °F to 42 °F
   c. 40. g of water by 12 °C
   d. 16.9 g of water from 75.0 °C to 80.0 °C

13. How much energy (in joules) is required to raise the temperature of 25.2 g of solid carbon rod from 25 °C to 50. °C? The specific heat capacity of solid carbon is 0.71 J/g °C.

14. How much energy (in calories) is required to raise the temperature of 10. g steam from 122.2 °C to 130.4 °C? The specific heat capacity of water(g) is 2.0 J/g °C.

15. How much of a temperature change would occur if 2736.8 J of energy were applied to a piece of iron bar weighing 450.5 g? The specific heat capacity of solid iron is 0.45 J/g °C.

16. What is the mass in grams of a piece of aluminum wire if a change in temperature of 5.67 °C required 8.53 J? The specific heat capacity of solid aluminum is 0.89 J/g °C.

17. What is the specific heat capacity of ethyl alcohol if 1972.4 J of energy is necessary to raise the temperature of 53.4 g ethyl alcohol by 15.2 °C?

ANSWERS TO LEARNING REVIEW

1. a. Oxygen can combine with hydrogen to produce a new substance, water. Because oxygen and hydrogen have the potential to combine to form a new substance, this is an example of a chemical property.
   b. Observing ethyl alcohol boiling does not destroy or change the ethyl alcohol molecules. Therefore, boiling point is a physical property.
   c. Observing the color of a substance does not change its composition. The pale blue color of liquid oxygen is a physical property.

2. a. When a copper strip is hammered into a bracelet, the shape of the copper is changed, but not the composition. This is a physical change.
   b. The new substance, copper sulfide, is a black solid which does not have any of the characteristics of copper metal or yellow elemental sulfur. This is a chemical change.
   c. When liquid water freezes to become solid ice, the molecules are still those of water. No change in composition has occurred. This is a physical change.
   d. Liquid oxygen molecules become solid oxygen molecules at -183 °C. This is a physical change.
   e. Cooking an egg causes changes to the egg's composition. Heating changes the structure of large egg proteins, causing them to form solids. This is a chemical change.

3. a. S is the symbol for the element sulfur.
   b. H_2O has two kinds of atoms, O and H. It is a compound.
   c. C is the symbol for the element carbon.
   d. N_2O_5 has two kinds of atoms, N and O. It is a compound.
   e. NaOH has three kinds of atoms, Na, O and H. It is a compound.

4. a. Physical property, because a melted substance has the same compostion as a solid substance.
   b. Physical property, because observing the odor of a substance does not change the composition of the substance.
   c. Chemical property, because the temperature which causes a substance to break down into elements causes destruction of the substance.
   d. Chemical property, because oxygen reacting with a substance describes a chemical change occurring.

5. a. Gasoline is a mixture of complex organic chemicals, detergents, and additives.
   b. Table sugar is a pure substance. It contains only sucrose molecules.
   c. Garden soil is a mixture. It contains sand, water, clay, dead plant leaves, and other components.
   d. Sterling silver is made from 93% silver and 7% copper by mass. It is a mixture.

6. a. Sweetened tea is a homogeneous mixture. When sugar is added to hot tea, it dissolves. All of the tea is equally sweetened.
   b. A bag full of leaves and grass clippings is a heterogeneous mixture. Some parts of the bag will contain more leaves than grass, and other parts will contain more grass than leaves.
   c. Rubbing alcohol and water mix freely with each other. The molecules of one completely disperse in molecules of the other. This produces a homogeneous solution.
   d. A devil's food cake mix is a homogeneous mixture. There are no lumps (usually) of sugar, or clumps of flour. All of the ingredients are distributed equally throughout the mix.

7. Separating mixtures by filtration depends upon a difference in physical properties of the mixture. Pour the mixture onto a mesh. One common mesh is filter paper, which is made from a mesh of cellulose fibers. The liquid in the mixture can pass through the cellulose fibers into a container below. The particles remain behind, trapped in the fibers of the mesh.

8. Filtration can separate a heterogeneous mixture which contains a liquid and a solid component.

9. a. Sand and gravel can be separated by filtration. Use a mesh which allows the sand particles to pass through, but retains the gravel, for example, a piece of wire screen.
   b. A salt and sand mixture can be separated by adding water and filtering. The salt will dissolve in the water, while the sand will not. By filtering the sand and salt water mixture, the pure sand remains behind, while the salt passes through the mesh with the water.
   c. Sugar and water cannot be separated by filtration because the sugar molecules dissolve in the water and will pass through a mesh. They can, however, be separated by distillation. Heat the mixture until the water boils (at 100 °C) and is converted to steam. The sugar has a higher boiling point than water does and is not vaporized. The steam rises in the distillation apparatus and can be captured and recondensed to liquid water. The sugar remains behind in the distillation flask.

10.

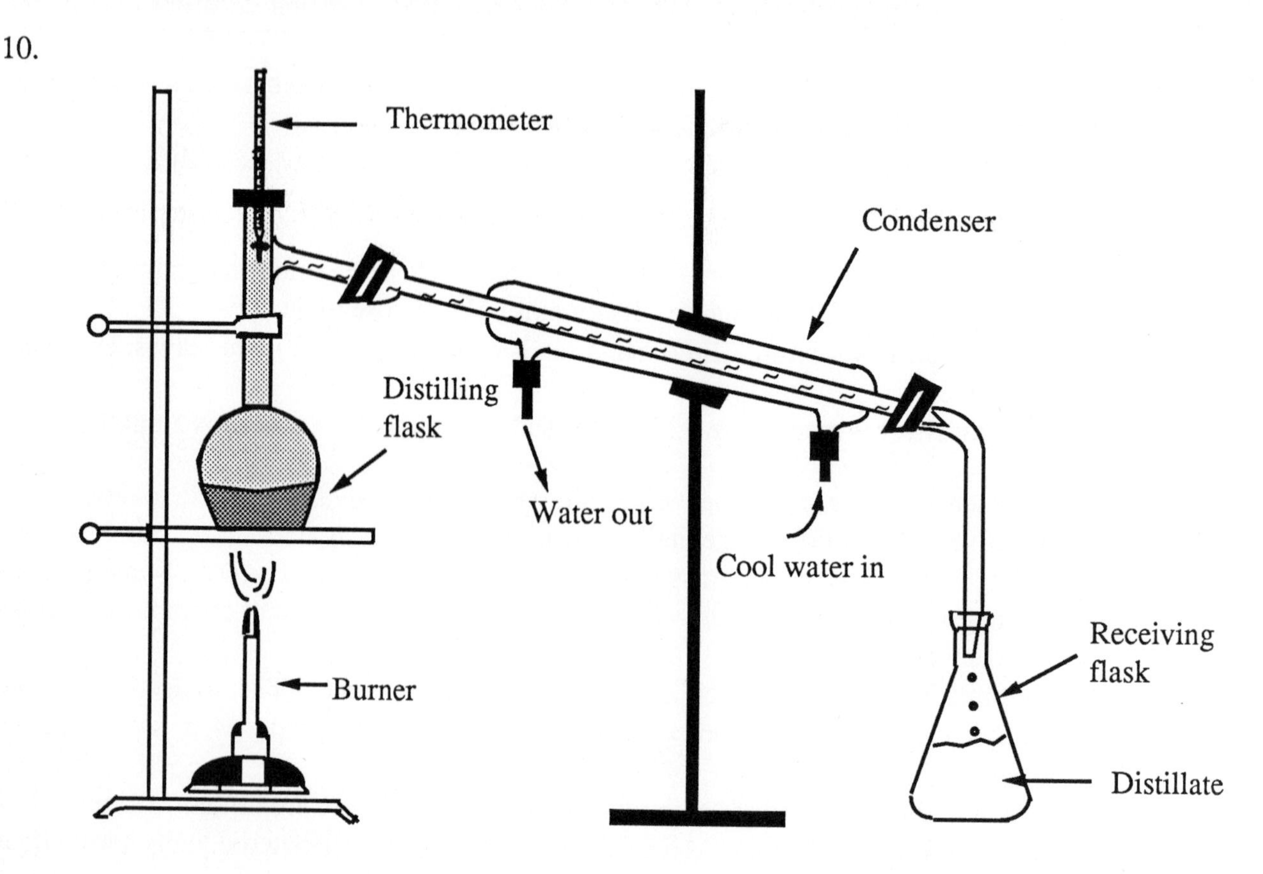

11. Converting calories to joules or joules to calories requires knowing that 1 cal = 4.184 J. If calories are the given unit, then use the conversion factor $\frac{4.184\ \text{J}}{1\ \text{cal}}$. If joules are given, then use $\frac{1\ \text{cal}}{4.184\ \text{J}}$.

a. $45.8\ \text{cal} \times \frac{4.184\ \text{J}}{1\ \text{cal}} = 192\ \text{J}$

b. $0.561\ \text{cal} \times \frac{4.184\ \text{J}}{1\ \text{cal}} = 2.35\ \text{J}$

c. $5.96\ \text{J} \times \frac{1\ \text{cal}}{4.184\ \text{J}} = 1.42\ \text{cal}$

d. $76\ \text{J} \times \frac{1\ \text{cal}}{4.184\ \text{J}} = 18\ \text{cal}$

12. These problems ask you to calculate the calories required to heat a quantity of water. One calorie is defined as the amount of heat required to raise the temperature of 1 gram of water by 1 degree Celsius. To solve these problems, you need to multiply the number of grams of water to be heated by the number of degrees Celsius change in the temperature of the water.

a. $100.1\ \text{g water} \times 19\ ^\circ\text{C} \times \frac{1\ \text{cal}}{\text{g water} \times ^\circ\text{C}} = 1900\ \text{cal}$

b. The initial and final temperatures of water are given in °F. We must convert to °C before solving the problem.

Initial temperature:

$$t\,^\circ\text{C} = \frac{(t\,^\circ\text{F} - 32)}{1.80}$$

$$t\,^\circ\text{C} = \frac{(36 - 32)}{1.80}$$

$$t\,^\circ\text{C} = 2.2$$

Final temperature:

$$t\,^\circ\text{C} = \frac{(t\,^\circ\text{F} - 32)}{1.80}$$

$$t\,°C = \frac{(42 - 32)}{1.80}$$

$$t\,°C = 5.6$$

Temperature change:

$$5.6\ °C - 2.2\ °C = 3.4\ °C$$

Solution:

$$2.32\ \cancel{g}\ \text{water} \times 3.4\ °\cancel{C} \times \frac{1\ \text{cal}}{\cancel{g}\ \text{water} \times °\cancel{C}} = 7.9\ \text{cal}$$

c. $$40.\ \cancel{g}\ \text{water} \times 12\ °\cancel{C} \times \frac{1\ \text{cal}}{\cancel{g}\ \text{water} \times °\cancel{C}} = 480\ \text{cal}$$

d. $$16.9\ \cancel{g}\ \text{water} \times 5.0\ °\cancel{C} \times \frac{1\ \text{cal}}{\cancel{g}\ \text{water} \times °\cancel{C}} = 85\ \text{cal}$$

13. In this problem we want to calculate the heat energy needed to raise the temperature of a substance other than water. To do this, we need to know the specific heat capacity of the substance. The specific heat capacity tells us the amount of heat energy required to change the temperature of 1 gram of a substance by 1 degree Celsius. Every substance has its own specific heat capacity. That of solid carbon is 0.71 J/g °C.

    If it takes 0.71 J to raise the temperature of 1 gram of carbon 1 degree Celsius, then it will take 0.71 J x 25.2 to raise 25.2 g carbon by 1 degree Celsius. We wish to raise the temperature of the carbon rod by 25 °C, not 1 °C. We will need twenty-five times the heat energy needed to raise the temperature of 25.2 g carbon by 1 degree Celsius.

$$\text{Joules} = \frac{0.71\ \text{J}}{\cancel{g}\ \text{carbon} \times °\cancel{C}} \times 25.2\ \cancel{g}\ \text{carbon} \times 25\ °\cancel{C}$$

    The joules required to raise the temperature of 25.2 g carbon by 25 °C = 450 J.

14. When you are given the number of grams of a substance, a change in temperature in degrees Celsius, and the specific heat capacity for that substance, and are asked to calculate the heat energy required, you can use the formula $Q = s \times m \times \Delta T$. The specific heat capacity is s, m equals the mass in grams, ΔT is the change in temperature in degrees Celsius, and Q is the heat energy required. We can solve this problem with the formula, although we will need to convert Q from joules to calories, since calories are asked for.

$$Q = s \times m \times \Delta T$$

$$Q = \frac{2.0\ \text{J}}{\text{g water(g)} \times {}^\circ\text{C}} \times 10.\ \text{g water(g)} \times 8.2\ {}^\circ\text{C}$$

$$Q = 160\ \text{J}$$

The answer should be expressed in calories:

$$160\ \text{J} \times \frac{1\ \text{cal}}{4.184\ \text{J}} = 38\ \text{cal}$$

15. In this problem, we are given Q, the heat energy in joules; m, the mass in grams of a piece of iron; and s, the specific heat capacity of iron. We are asked for ΔT, the change in temperature. If we rearrange the equation $Q = s \times m \times \Delta T$, we can solve for ΔT.

Divide both sides by ΔT.

$$\frac{Q}{\Delta T} = s \times m$$

Now, divide both sides by Q (same as multiplying by $\frac{1}{Q}$).

$$\frac{Q}{\Delta T} \times \frac{1}{Q} = \frac{s \times m}{Q}$$

We now have $1/\Delta T$ (the inverse of ΔT) isolated on one side of the equation.

$$\frac{1}{\Delta T} = \frac{s \times m}{Q}$$

Invert both sides of the equation.

$$\frac{\Delta T}{1} = \frac{Q}{s \times m}$$

$$\frac{\Delta T}{1} = \Delta T = \frac{Q}{s \times m}$$

Now, find ΔT.

$$\Delta T = \frac{2736.8\ \text{J}}{\frac{0.45\ \text{J}}{\text{g}\ {}^\circ\text{C}} \times 450.5\ \text{g}}$$

$$\Delta T = 14\ {}^\circ\text{C}$$

16. In this problem we are asked to solve for mass, m. We are given ΔT, Q, and s. We can rearrange the equation as illustrated below.

Divide both sides of the equation by m.

$$\frac{Q}{m} = \frac{s \times \Delta T \times \cancel{m}}{\cancel{m}}$$

Divide both sides of the equation by Q.

$$\frac{\cancel{Q}}{m} \times \frac{1}{\cancel{Q}} = s \times \Delta T \times \frac{1}{Q}$$

$$\frac{1}{m} = \frac{s \times \Delta T}{Q}$$

Invert both sides of the equation.

$$\frac{m}{1} = m = \frac{Q}{s \times \Delta T}$$

Now, find m.

$$m = \frac{8.53\ \cancel{J}}{\frac{0.89\ \cancel{J}}{g \times {}^{\circ}\cancel{C}} \times 5.67\ {}^{\circ}\cancel{C}}$$

$$m = 1.7\ g$$

17. We are asked to find the specific heat capacity, s, when given ΔT, Q, and m. Rearrange the equation to isolate s on one side of the equation.

$$s = \frac{Q}{m \times \Delta T}$$

Now, substitute values into the equation.

$$s = \frac{1972.4\ J}{53.4\ g \times 15.2\ {}^{\circ}C}$$

$$s = \frac{2.43\ J}{g \times {}^{\circ}C}$$

The specific heat capacity of ethyl alcohol is 2.43 J/g °C.

PRACTICE EXAM

1. Which of the following is not a physical property of table salt, sodium chloride?

   a. It dissolves in water.
   b. It is white in color.
   c. It decomposes when heated to produce sodium metal and chlorine gas.
   d. It is a solid at room temperature.
   e. It will melt to produce liquid sodium chloride at a very high temperature.

2. Cesium metal melts at 28.5 °C and boils at 705 °C. If the temperature of a piece of cesium is 152 °C, what is the physical state of the metal?

   a. Solid
   b. Liquid
   c. Gas
   d. Not enough information
   e. Mixture of all three states

3. Which of the following is a physical change?

   a. Nitrogen gas is cooled until it becomes a liquid.
   b. H_2 and O_2 combine to produce H_2O_2.
   c. Grape juice is fermented to produce wine.
   d. Natural gas is burned to produce energy.
   e. Mercuric oxide is heated until liquid mercury and oxygen gas are formed.

4. Which of the choices below is an element?

   a. N_2O_5
   b. A unit which can be broken down into smaller units.
   c. Na
   d. Contains at least two substances mixed together.
   e. Upon heating, produces two new substances.

5. Which of the following is a pure substance?

   a. Lemon flavored jello
   b. River water clouded with silt particles
   c. Piece of oak firewood
   d. Skim milk
   e. Copper metal

6. When sodium bicarbonate (baking soda) is dissolved in water, a mixture results. Which statement below is true?

   a. The mixture is homogeneous.
   b. The mixture is heterogeneous.
   c. The mixture can be separated by filtration.
   d. a and c
   e. b and c

7. Which statement about filtration is true?

   a. The particles pass through the mesh.
   b. Filtration is best for separating homogeneous mixtures.
   c. Filtration cannot be used when both components of the mixture are solids.
   d. Filtration is based on a difference in size of the particles of the mixture.
   e. Filtration causes destruction of the filtered materials.

8. How many calories are required to raise the temperature of 0.78 g water by 14 °C?

   a. 11 cal
   b. 0.056 cal
   c. 18 cal
   d. 46 cal
   e. 2.6 cal

9. How many joules are required to raise the temperature of 105 g water from 85.0 °C to 93.0 °C?

   a. 3500 J
   b. 840 J
   c. 200 J
   d. 2100 J
   e. 2300 J

10. What is the mass of a piece of silver wire if the temperature changes from 20 °C to 65 °C, the energy required is 107.1 J, and the specific heat capacity of silver is 0.238 J/g °C?

   a. 0.10 g
   b. 0.57 g
   c. 20,250 g
   d. 1.8 g
   e. 10. g

PRACTICE EXAM ANSWERS

1. c (3.2)
2. b (3.1)
3. a (3.2)
4. c (3.3)
5. e (3.4)
6. a (3.4)
7. d (3.5)
8. a (3.6)
9. a (3.6)
10. e (3.6)

CHAPTER 4: CHEMICAL FOUNDATIONS: ELEMENTS AND ATOMS

INTRODUCTION

In Chapter 4 you are introduced to the names and symbols of the common elements. Make sure you learn the names and symbols now. Most of the chemistry covered in subsequent chapters depends upon a knowledge of these elements and symbols. If you are having difficulty remembering the names, use Table 4.3 in your textbook to make up some flash cards with the element name on the front side, and the symbol on the back. Find several minutes each day to drill yourself with the cards until you have committed the symbols and names to memory.

The remainder of the chapter covers the individual parts of an atom and the categories of elements which are organized into the periodic table.

AIMS FOR THIS CHAPTER

1. Know which elements are common on earth, and their relative abundances. (Section 4.1)
2. Learn the names and symbols for the common elements given in Table 4.3 of your text. (Section 4.2)
3. Learn Dalton's atomic theory and the law of constant composition. (Section 4.3)
4. Know what a compound is, and how to write and interpret chemical formulas. (Section 4.4)
5. Know the names and locations within the atom of the major subatomic particles. (Section 4.5)
6. Understand how Rutherford's gold foil experiment was performed and what information it provided about subatomic structure. (Section 4.5)
7. Know the sizes, relative masses and relative charges of the major subatomic particles. (Section 4.6)
8. Understand why elements composed of the same subatomic particles can have different properties. (Section 4.6)
9. Know how the isotopes of an element differ, and how to write and interpret $^{A}_{Z}X$ symbols for isotopes. (Section 4.7)
10. Know how to determine the mass number, number of protons, number of electrons or number of neutrons for any isotope. (Section 4.7)
11. Know what the symbols and numbers on a periodic table mean, what the major families of elements are, and where to find them on the periodic table. (Section 4.8)

QUICK DEFINITIONS

Element symbols	One or two letter abbreviations based on modern or ancient element names. (Section 4.2)
Law of constant composition	States that a compound always contains the same proportions by mass of different atoms. This means that the composition of a compound is always the same. (Section 4.3)
Dalton's atomic theory	All elements are made of atoms. For any one element, all the atoms are the same. Different elements are made from different kinds of atoms. Atoms from different elements can combine to make compounds. Each compound always has the same relative numbers and kinds of atoms. Chemical reactions do not cause atoms to break apart into subatomic particles or cause new elements to form. (Section 4.3)
Compound	A pure substance which always contains the same relative numbers and kinds of atoms. (Section 4.4)
Chemical formula	A representation of a compound that shows the symbol for each element and how many atoms of each element are present by using subscript numbers. (Section 4.4)
Subatomic particles	The individual units of which atoms are made. (Section 4.5)
Electron	A small subatomic particle with small mass and a negative charge. Located outside the nucleus. (Section 4.5)
α-particle	An alpha particle consists of two protons and two neutrons. It has a 2^+ charge and is relatively heavy, compared with a single proton or neutron. (Section 4.5)
Nuclear atom	A model of the atom in which the positive charge is concentrated in one location, the nucleus, instead of being spread over the entire atom. (Section 4.5)
Nucleus	The area where the positive charge of an atom is concentrated, and where the protons and neutrons are located. (Section 4.5)
Neutron	A subatomic particle that is only slightly more massive than a proton, and that has no charge. Found inside the atomic nucleus. (Section 4.6)

Proton	A subatomic particle which has a positive charge equal in size to the negative charge on an electron, but is much more massive than an electron. Found inside the nucleus. (Section 4.6)
Isotopes	Different kinds of atoms of the same element. Each isotope of an element has the same number of protons and electrons, but a different number of neutrons. (Section 4.7)
Atomic number	The number of protons in the nucleus of an atom. (Section 4.7)
Mass number	The total number of protons plus neutrons. (Section 4.7)
Periodic table	A way to organize the elements so that families of elements with similar properties are found grouped together in columns of the table. (Section 4.8)
Groups	Families of elements with similar properties. Certain columns in the periodic table are referred to as groups. (Section 4.8)
Alkali metals	Elements in the first column (group 1) of the periodic table. (Section 4.8)
Alkaline earth metals	Elements in the second column (group 2) of the periodic table. (Section 4.8)
Halogens	Elements in group 7 of the periodic table. (Section 4.8)
Noble gases	Elements in group 8 of the periodic table. (Section 4.8)
Transition metals	The transition metals are the elements in the middle columns of the periodic table, to the left of the heavy jagged line, and to the right of the alkaline earth metals are transition metals. (Section 4.8)
Metals	All elements to the left of the heavy jagged line in the periodic table. Most metals appear shiny, can conduct electricity, can be beaten into thin sheets, and can be drawn into fine wires. (Section 4.8)
Nonmetals	Any of the elements to the right of the heavy jagged line of the periodic table. The characteristics of the nonmetals vary from element to element. (Section 4.8)

Metalloids — Elements which border the heavy jagged line on the right side of the periodic table. They have some characteristics of metals, but not all. They are also called semi-metals. (Section 4.8)

CONTENT REVIEW

4.1 THE ELEMENTS

How Many Elements Are There?

There are 108 elements at the present time, but that number will increase when scientists synthesize new ones in the laboratory. Eighty-eight elements occur naturally. The remainder are made in the laboratory.

Which Elements Are Most Abundant On Earth?

Oxygen (49.2 %), silicon (25.7 %), aluminum (7.50 %) and iron (4.71 %) are the four most abundant elements on earth. The table below lists mass percents of many of the earth's elements.

Table 4.1 Common Elements found on earth

Distribution (Mass Percent) of the 18 Most Abundant Elements in the Earth's Crust, Oceans, and Atmosphere			
Element	Mass percent	Element	Mass Percent
Oxygen	49.2	Titanium	0.58
Silicon	25.7	Chlorine	0.19
Aluminum	7.50	Phosphorus	0.11
Iron	4.71	Manganese	0.09
Calcium	3.39	Carbon	0.08
Sodium	2.63	Sulfur	0.06
Potassium	2.40	Barium	0.04
Magnesium	1.93	Nitrogen	0.03
Hydrogen	0.87	Fluorine	0.03
		All others	0.49

Which Elements Are Most Abundant In the Human Body?

Oxygen (65.0 %), carbon (18.0 %), hydrogen (10.0 %), and nitrogen (3.0 %) are the most abundant elements in the human body. There are many other elements present in smaller amounts, and some trace elements are present in very small, but detectable amounts. The table below lists elements which are found in the human body.

Table 4.2 Common Elements in the Human Body

Abundance of Elements in the Human Body		
Major Elements	Percent by Mass	Trace Elements (in alphabetical order)
Oxygen	65.0	Arsenic
Carbon	18.0	Chromium
Hydrogen	10.0	Cobalt
Nitrogen	3.0	Fluorine
Calcium	1.4	Iodine
Phosphorus	1.0	Manganese
Magnesium	0.50	Molybdenum
Potassium	0.34	Nickel
Sulfur	0.26	Selenium
Sodium	0.14	Silicon
Chlorine	0.14	Vanadium
Iron	0.004	
Zinc	0.003	

4.2 SYMBOLS FOR THE ELEMENTS

You must learn the names and symbols for the common elements now, or you will have trouble understanding material later in the course. It is easier to learn the names and symbols if you first break them up into smaller groups. Three categories of symbols are presented in the tables below and in Table 4.3 in your textbook. Try learning the symbols and names one category at a time.

Symbols with one letter

Element	Symbol	Element	Symbol
Boron	B	Phosphorus	P
Carbon	C	Potassium	K
Fluorine	F	Sulfur	S
Iodine	I	Tungsten	W
Nitrogen	N	Uranium	U
Oxygen	O		

Symbols with the first two letters of the element name			
Aluminum	Al	Helium	He
Argon	Ar	Lithium	Li
Barium	Ba	Neon	Ne
Bismuth	Bi	Nickel	Ni
Bromine	Br	Radium	Ra
Calcium	Ca	Silicon	Si
Cobalt	Co	Titanium	Ti

Symbols with two letters based on the original name			
Antimony	Sb	Manganese	Mn
Arsenic	As	Mercury	Hg
Cadmium	Cd	Platinum	Pt
Chlorine	Cl	Silver	Ag
Copper	Cu	Sodium	Na
Gold	Au	Strontium	Sr
Iron	Fe	Tin	Sn
Lead	Pb	Zinc	Zn
Magnesium	Mg		

4.3 DALTON'S ATOMIC THEORY

What Is Dalton's Atomic Theory?

All elements are made of atoms. For any one element, all the atoms are the same. Different elements are made of different kinds of atoms. Atoms from different elements can combine to make compounds. Each compound always has the same number and kind of atoms. Chemical reactions don't cause atoms to break apart into subatomic particles.

4.4 FORMULAS OF COMPOUNDS

How Can You Write and Interpret Chemical Formulas?

Compounds are substances composed of more than one kind of element. A particular compound always contains the same relative numbers and kinds of elements. **Chemical formulas** contain the one or two letter symbols for the elements to indicate what kinds of elements are present in the compound. Small numbers written below the level of the letters (subscript numbers) indicate how many atoms of each element the compound contains. For example, the compound dinitrogen tetroxide contains two nitrogen atoms and four oxygen atoms. You can write its formula by using the symbol for nitrogen followed by a subscript 2 and the symbol for oxygen, followed by a subscript 4. N_2O_4. Don't use a subscript number when a compound contains only one atom of a particular kind. Nitrogen monoxide contains one atom of nitrogen and one atom of oxygen. Its formula is written NO. We do not need subscripts for nitrogen and oxygen because when compounds contain only one of a kind of atom, the subscript 1 is understood. When you see the symbol for an atom which is not followed by a subscript, it means that the compound contains only one of that particular atom.

4.5 THE STRUCTURE OF THE ATOM

How Did J.J. Thomson Demonstrate the Existence of Subatomic Particles?

For a long time scientists did not know what the atom was like. They did not have much information to work with. In 1895, J.J. Thomson was able to make atoms emit particles with a negative charge. He could get all the elements he tried to emit these small particles, so he concluded that atoms of all elements contain them. Thomson realized that overall, atoms are not positively or negatively charged; they are electrically neutral. This meant that the small, negatively charged particles must be counterbalanced in the atom by something with a positive charge. Thomson's experiments showed the existence of negatively charged particles (electrons), and showed that positively charged particles must exist as well.

What Is the Plum Pudding Model Of the Atom?

Scientists knew that there were negatively and positively charged particles in the atom, but until 1911 did not have any firm evidence about how they were arranged. A model by Lord Kelvin stated that the electrons (negatively charged particles) could be evenly dispersed in a positively charged matrix, like raisins in a pudding. This is often called the **plum pudding model** of the atom.

How Did Rutherford Perform the Gold Foil Experiment?

In 1911 Ernest Rutherford performed some experiments which gave a much clearer understanding of what an atom is really like and how the subatomic particles are arranged. Rutherford used a heavy, positively charged particle called an α-particle to bombard a very thin (only several atoms thick) piece of gold foil. α-particles are tiny and cannot be seen with the naked eye, but Rutherford could detect their presence by using a detector which produced a flash of light when struck by the α-particle. Each flash of light showed where an α-particle hit the detector and allowed Rutherford to follow the path of the α-particles.

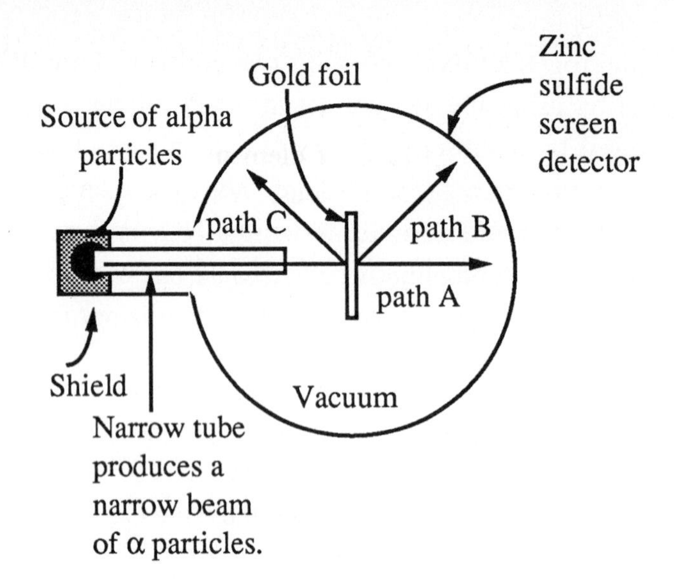

Before the experiment, Rutherford predicted what he thought would happen when the α-particle hit the gold foil. His predictions were correct, but something unexpected also happened. He predicted that most of the α-particles would go right through the gold foil and hit the detector on the other side. Most of the α-particles penetrated the foil and were not deflected from their straight paths. Rutherford thought that some α-particles could come close to the part of an atom containing the positive charge and be repelled slightly, causing the positively charged α-particle to be somewhat deflected from its straight path. Rutherford was correct in this prediction too. However, he did not predict that some of the α-particles would hit an area in the gold foil so dense and massive that the α-particle would bounce almost straight back, as though it had hit a brick wall.

What Does Rutherford's Experiment Show?

Rutherford's experiment shows that atoms are mostly empty space, since most of the α-particles passed through the gold foil without hitting anything. Some of the α-particles were deflected slightly from their straight path. The particles came close to but did not hit an area in the atom which contained a positive charge. The α-particles were deflected because two areas of positive charge tend to repel each other.

A few α-particles bounced almost straight back. This was surprising because the α-particle is more massive than an individual proton or neutron. How could a heavy particle hit a relatively light proton and bounce back? It meant that the α-particle hit a small area within a gold atom which was very massive because it contained all the protons in one place. This evidence led Rutherford to postulate that the positive charge is not spread all over the atom, as the plum pudding model stated, but is concentrated in one area, the **nucleus**.

4.6 INTRODUCTION TO THE MODERN CONCEPT OF ATOMIC STRUCTURE

One part of Dalton's model states that all atoms of a single kind of element are exactly the same. We now know that some atoms of each kind of element are different from the others. The difference is in the number of neutrons in the nucleus. Atoms of a single kind of element which contain different numbers of neutrons in the nucleus are called **isotopes**.

What Is the Size Of An Atom?

Recent information has provided us with data about the size of an atom and an individual nucleus. Atoms themselves are very small. The nucleus is on average 10^{-13} cm in diameter, and the electrons are 10^{-8} cm away from the nucleus. Every atom has lots of empty space, and in fact, if the nucleus of an atom were the size of a grape, the electrons would be found a mile away from it.

What Are the Charges and Masses Of Subatomic Particles?

The relative masses and charges of the subatomic particles are important. The electron has been assigned a relative mass of one and a charge of 1^-. The proton has a relative mass of 1836, much more than an electron. The relative charge on a proton is 1^+, equal in magnitude to an electron, but opposite in sign. The neutron has a relative mass the same as a proton, but no charge.

How Do Different Atoms, All With the Same Subatomic Particles,
Have Such Different Properties?

Each kind of atom has a different number of protons and electrons. The chemistry of an atom is determined by the number and arrangements of the electrons, so each kind of atom can have different chemical properties. The way elements react with other elements is determined by the number and arrangements of the electrons.

4.7 ISOTOPES

What Are Isotopes?

Atoms of all elements contain one or more electrons. Each different kind of element contains the same number of electrons as it does protons, because atoms are electrically neutral. All atoms of each element always have the same number of electrons and the same number of protons. The number of protons and electrons never changes. For example, all boron atoms have five electrons and five protons. Most atoms also have neutrons in their nucleus. Unlike protons and electrons, the number of neutrons in different atoms of the same kind of element can vary. Atoms of the same kind of element with different numbers of neutrons in the nucleus are called **isotopes**. One isotope of boron has five protons and five electrons and five neutrons. Another isotope of boron has five protons and five electrons and six neutrons. Only the number of neutrons is different between the two isotopes of boron.

How Can We Show Different Isotopes of An Element?

Chemists use the symbols $^{A}_{Z}X$ to help show which isotope is under discussion. X is the one or two letter symbol for the element. Z represents the number of protons in the nucleus, also called the **atomic number**. A is the **mass number**, equal to the number of protons plus the number of neutrons. For any element, Z will remain the same, regardless of which isotope we are discussing. A, which is the number of protons plus neutrons, will vary. For example, $^{11}_{5}B$ represents the isotope of boron (called B-eleven) which has a mass number of eleven and an atomic number of five. $^{10}_{5}B$ represents the isotope of boron with a mass number of ten and an atomic number of five.

How Can We Use $^{A}_{Z}X$ *Information About Isotopes?*

Z tells us how many protons there are, and because there are always the same number of protons as electrons, we know the number of electrons too. For example, $^{11}_{5}B$ has an atomic number of five so it also contains five electrons in addition to its five protons. How can we tell how many neutrons are in an isotope? The mass number, A, for boron eleven is eleven. This means that the number of protons and neutrons adds up to eleven. From the atomic number we know that five of the eleven are protons. Therefore the rest of the mass number must be neutrons. A minus Z equals number of neutrons. Eleven minus five equals six.

4.8 INTRODUCTION TO THE PERIODIC TABLE

The periodic table has lots of information useful to a chemist. Let's see what kind of basic information you can get from the periodic table now, and in later chapters, you will see that the periodic table contains even more useful information.

What Basic Information Is In the Periodic Table?

The table is divided up into boxes, each of which contains the one or two letter symbol for an element. Every element has a place in the periodic table. There are usually numbers written above and below each of the symbols. The periodic table below has been simplified to show only one set of numbers, written above the element symbols. The numbers represent the atomic number, the number of protons (and the number of electrons) for each element. As you move from left to right along the periodic table, the atomic number of subsequent elements increases by one. When you reach the end of a row, if you move to the beginning of the next row, the elements again begin increasing in atomic number by one. When the elements are arranged in this way, it's possible to group elements with similar properties together in columns, which are often called families. For example, the elements He, Ne, Ar, Kr, Xe, and Rn, called the noble gases, all have similar properties. They are all gases which do not tend to combine with other elements to produce compounds. Many of the columns, or **groups**, have numbers. These numbers are always written along the top of the periodic table. The first group (group 1) on the left side of the periodic table is called the **alkali metals**, and the second (group 2), the **alkaline earth metals**. The group with the unreactive elements He, Ne, Ar, Kr, Xe and Rn is numbered group 8.

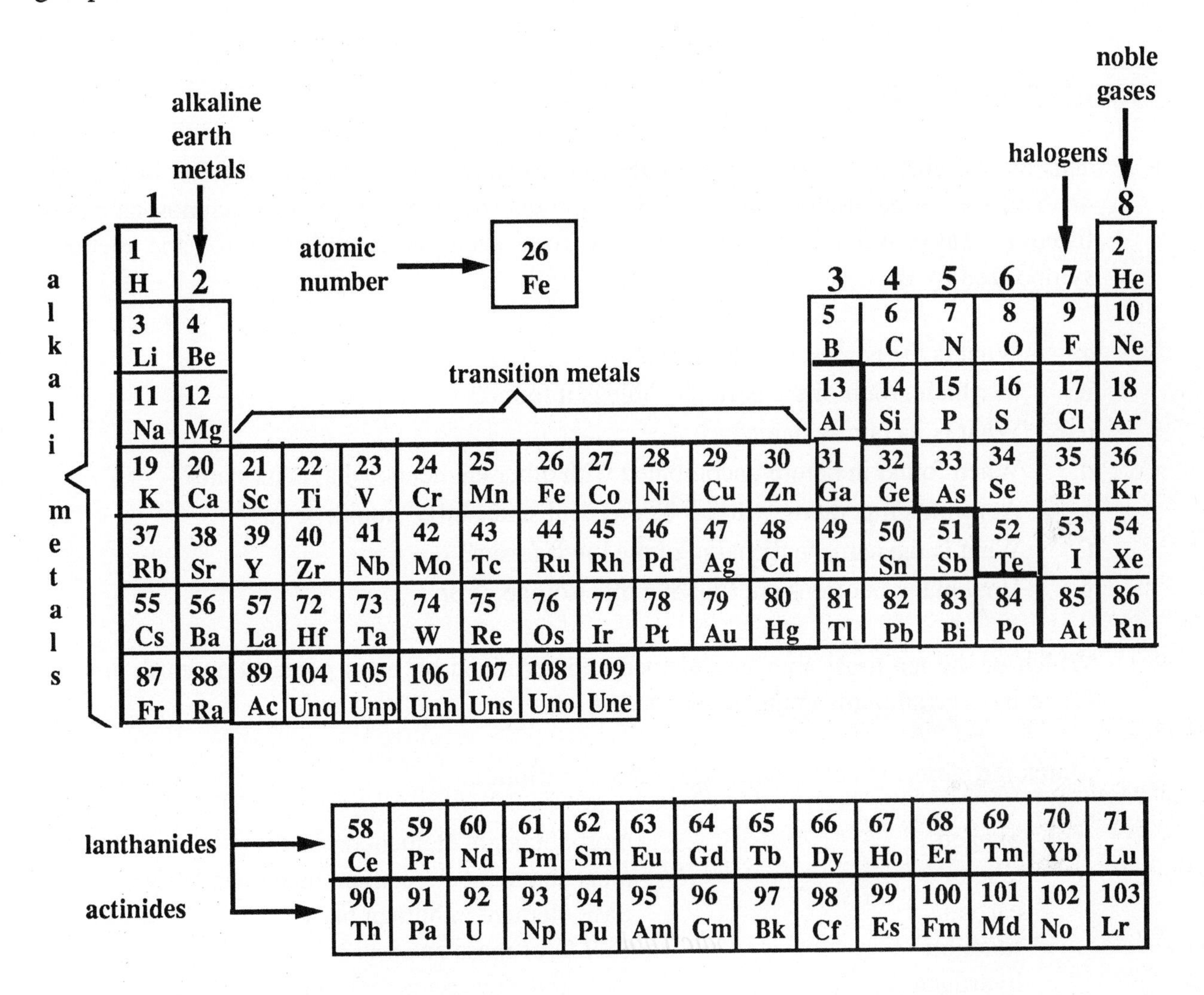

On the right side of the periodic table is a heavy jagged line running from the top to the bottom of the table. This line separates elements which are nonmetals from those which are metals. The **metals** lie on the left side of the jagged line, and the **nonmetals** on the right. There are many more metals than nonmetals. Many of the metals in the short columns in the middle of the table are called **transition metals**. Metals can be distinguished from nonmetals by the following characteristics. Metals appear shiny. They conduct both heat and electricity. They can be pulled into wire without crumbling or breaking, and can be beaten into thin sheets.

The elements to the right of the jagged line are the nonmetals. The nonmetals do not have a fixed set of characteristics. Their properties are more variable than the metals.

There are some elements which lie right on the jagged line which separates metals from nonmetals. These elements have some of the properties of both metals and nonmetals. They are boron, B; silicon, Si; germanium, Ge; arsenic, As; antimony, Sb; and tellurium, Te and are often called **metalloids**, or **semi-metals**.

LEARNING REVIEW

1. This review question can help you to determine your progress with the material in Chapter 4. You should be able to answer each of the questions below for the common elements listed in Table 4.3 of the textbook. Answer each question below for the element symbolized by Br.

   a. What is the name of the element?
   b. In which group of the periodic table is it found?
   c. What is its family name?
   d. When found in nature uncombined with other elements, what is its state?
   e. At room temperature, what is its physical state, solid, liquid, or gas?
   f. What is the name and charge of the ion it forms?
   g. How many neutrons are found in this isotope, ${}^{80}_{35}\text{Br}$?

2. Which of the ten most abundant elements (determined by mass percent) on earth are not found in large amounts in the human body?

3. Match the elements below with the correct description.

oxygen	most abundant element on earth
silicon	most abundant element in the human body
carbon	trace element in human body
titanium	25.7 % of mass on earth
hydrogen	these three elements make up 93% of mass in the human body
molybdenum	less than 1% of the mass on earth

4. Write symbols for the following elements.

a.	arsenic	f.	lead
b.	fluorine	g.	potassium
c.	magnesium	h.	chromium
d.	iron	i.	nitrogen
e.	neon	j.	calcium

5. Which of the common elements in Table 4.3 of your textbook have a one-letter symbol?

6. Some of the element symbols are not related to the modern name of the element. What are the elements represented by the following element symbols?

a.	W	e.	Fe
b.	Hg	f.	Pb
c.	Cu	g.	Sb
d.	K	h.	Na

7. Match the element name with the correct element symbol.

cadmium	Cl
carbon	Cr
calcium	C
chlorine	Co
cobalt	Cu
copper	Cd
chromium	Ca

8. Match the element symbol with the correct element name.

Na	silver
Sr	sulfur
S	sodium
Ag	silicon
Si	strontium

9. Describe the main parts of Dalton's atomic theory.

10. How does Dalton's atomic theory relate to the law of constant composition?

11. Dalton's model became more widely accepted when the existence of NO, NO_2, and N_2O became known. What aspect of Dalton's model allowed Dalton to predict the existence of these compounds?

12. Write chemical formulas for the following compounds.

a. ethyl alcohol, which contains 2 carbon atoms, 6 hydrogen atoms, and 1 oxygen atom
b. a compound which contains 1 atom of magnesium and 2 atoms of bromine
c. a compound which contains 4 atoms of phosphorus and 10 atoms of oxygen
d. a compound which contains 1 atom of arsenic and 3 atoms of hydrogen

13. What is the total number of atoms found in each of the following compounds? What is the total number of elements found in each?

a. KOH
b. N_2O_3
c. CCl_4
d. H_2O_2
e. Na_3PO_4

14. A physicist named J. J. Thomson showed that all atoms can be made to emit tiny particles which are repelled by the negative pole of an electric field. Which subatomic particle was this evidence for?

a. proton
b. neutron
c. electron
d. nucleus
e. isotope

15. Match the scientist with the discovery.

Ernest Rutherford	demonstrated the existence of electrons
J. J. Thomson	demonstrated the existence of neutrons
Lord Kelvin	developed the plum pudding model of the atom
Rutherford & Chadwick	developed the nuclear atom model from gold foil experiments

16. Label the parts of the experimental apparatus used to develop the model of the nuclear atom.

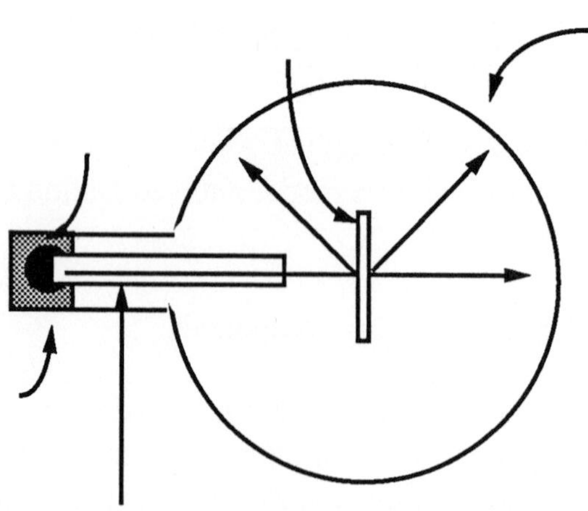

17. In the gold foil experiment, how did Rutherford interpret each of the following observations?

a. Most of the α-particles traveled unimpeded through the foil.
b. Some of the α-particles were deflected slightly from the straight path when they entered the foil.
c. A few of the α-particles bounced back when they entered the foil.

18. Fill in the missing relative masses and relative charges for each of the subatomic particles.

		Relative mass	Relative charge
a.	Electron		1-
b.	Proton		
c.	Neutron	1839	

19. Is the following statement true or false? An isotope of sodium could contain 12 protons, 12 neutrons and 11 electrons.

20. Label the parts of the symbol below.

$$^{A}_{Z}X$$

21. Write the symbols for the isotopes below in $^{A}_{Z}X$ notation.

a. An isotope of hydrogen has an atomic number of 1, and a mass number of 3.
b. An isotope of chlorine has an atomic number of 17 and a mass number of 37.
c. An isotope of oxygen has 8 protons and 10 neutrons.
d. An isotope of uranium has 92 electrons and 143 neutrons.
e. An isotope of sulfur has an atomic number of 16 and 16 neutrons.

22. An isotope of titanium contains 24 neutrons and has a mass number of 46.

a. How many protons does it contain?
b. How many electrons does it contain?

23. Aluminum-29 has an atomic number of 13.

a. What is its mass number?
b. How many neutrons does it have?

24. Match the group name on the left with an element found in that group.

halogen	Ca
transition metal	Ne
alkali metal	Fe
alkaline earth metal	K
noble gas	F

25. Fill in the boxes of the periodic table with element symbols for each of the families below. The number at the top of each box represents atomic number.

a. halogens

b. alkaline earth metals

c. noble gases

d. alkali metals

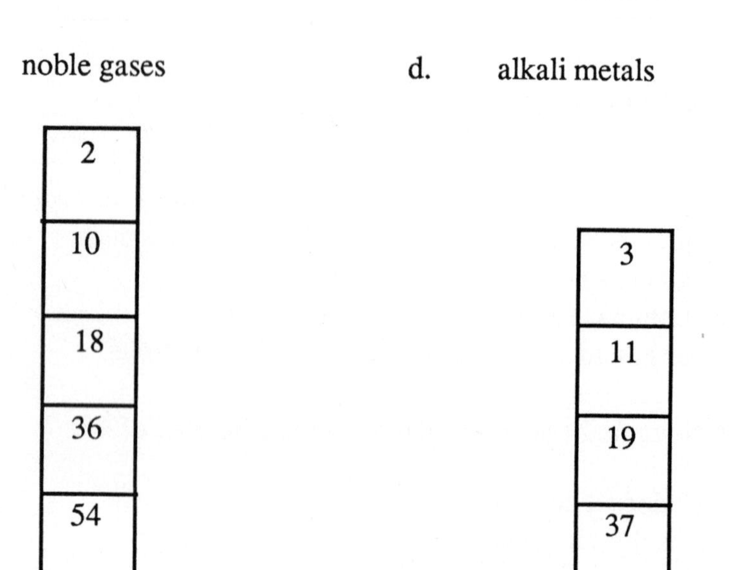

26. Which of the following elements are nonmetals?

a. Al
b. C
c. Cr
d. P
e. Br
f. I

27. Some of the elements along the jagged line on the right side of the periodic table have properties of both metals and nonmetals. Fill in the elemental symbols for these metalloids.

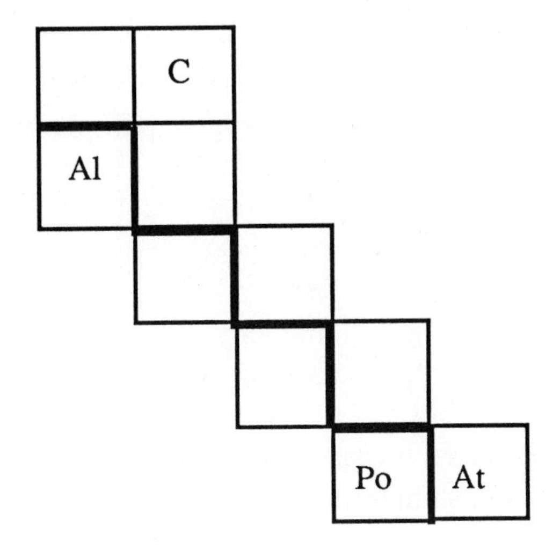

ANSWERS TO LEARNING REVIEW

1. a. bromine
   b. group 7
   c. halogens
   d. Br_2
   e. liquid
   f. bromide ion, Br^-
   g. 45

2. Silicon, aluminum and iron are found in large amounts on earth, but in small amounts in the human body.

3. Note that some of the elements are found in more than one category.

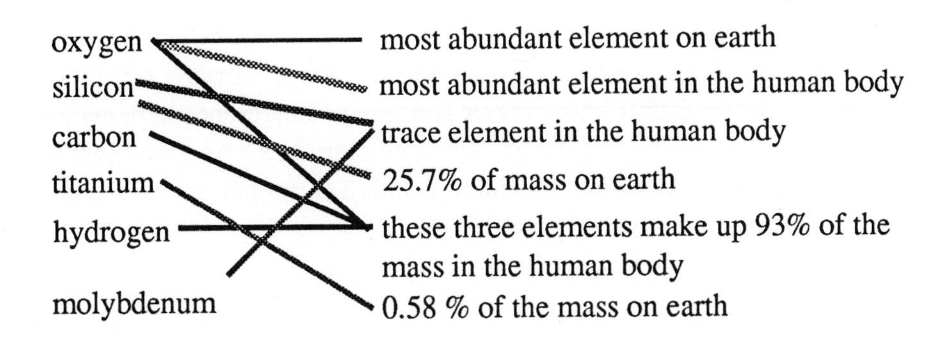

4.
   a. As f. Pb
   b. F g. K
   c. Mg h. Cr
   d. Fe i. N
   e. Ne j. Ca

5. Boron, carbon, fluorine, iodine, nitrogen, oxygen, phosphorus, potassium, sulfur, tungsten and uranium all have one letter symbols.

6.
   a. tungsten e. iron
   b. mercury f. lead
   c. copper g. antimony
   d. potassium h. sodium

7. There are quite a few elements whose symbols begin with the letter "c". The symbols for these elements are therefore similar to each other.

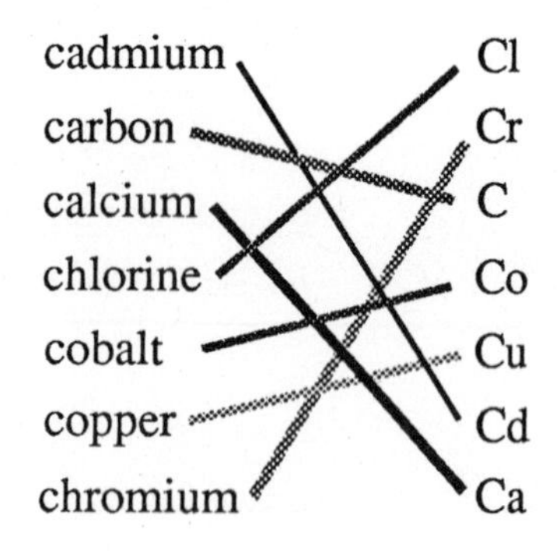

8.

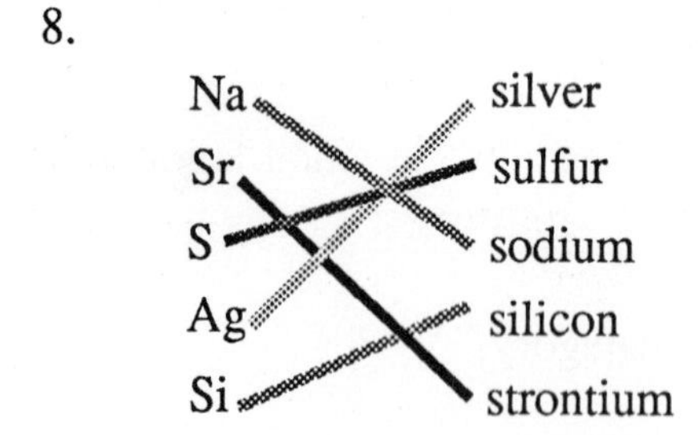

9. Dalton's atomic theory states that all elements are made of atoms. For any one element, all the atoms are the same (Dalton didn't know about isotopes). Different elements are made from different kinds of atoms. Atoms from different elements can combine to make compounds. Each compound always has the same relative numbers and kinds of atoms. Chemical reactions do not cause new elements to form.

10. The law of constant composition states that compounds always have the same proportions of each element by mass. Dalton's model states that compounds always have the same relative numbers and kinds of atoms. If compounds always have the same relative numbers of atoms, they will also have a constant proportion by mass. For example, the compound carbon dioxide always has 1 carbon atom for 2 oxygen atoms. The ratio of the mass of a carbon atom to the mass of two oxygen atoms also stays constant for molecules of carbon dioxide. This relationship was predicted by Dalton's model.

11. Dalton's model states that atoms from different elements can combine to produce compounds and that each compound always has the same relative numbers and kinds of atoms. Dalton predicted that different compounds would be found which were made of the same kinds of atoms, but combined in different numbers. The discovery of NO, NO_2, and N_2O confirmed Dalton's prediction and supported his model.

12. a. C_2H_6O
    b. $MgBr_2$
    c. P_4O_{10}
    d. AsH_3

13. Remember that when an element symbol has no subscript, only 1 atom of that element is present. Subscript numbers always refer to the element to the left of the subscript number.

    a. There are 3 atoms total and 3 different elements.
    b. There are 5 atoms total and 2 different elements.
    c. There are 5 atoms total and 2 different elements.
    d. There are 4 atoms total and 2 different elements.
    e. There are 8 atoms total and 3 different elements.

14. Because the particles were repelled by the negative pole, it was believed that the particles were negatively charged because like charges repel each other. The electron is a subatomic particle with a negative charge, so the correct answer is c.

15.

Ernest Rutherford
J. J. Thomson
Lord Kelvin
Rutherford & Chadwick

demonstrated the existence of electrons
demonstrated the existence of neutrons
developed the plum pudding model of the atom
developed the nuclear atom model from gold foil experiments

16.

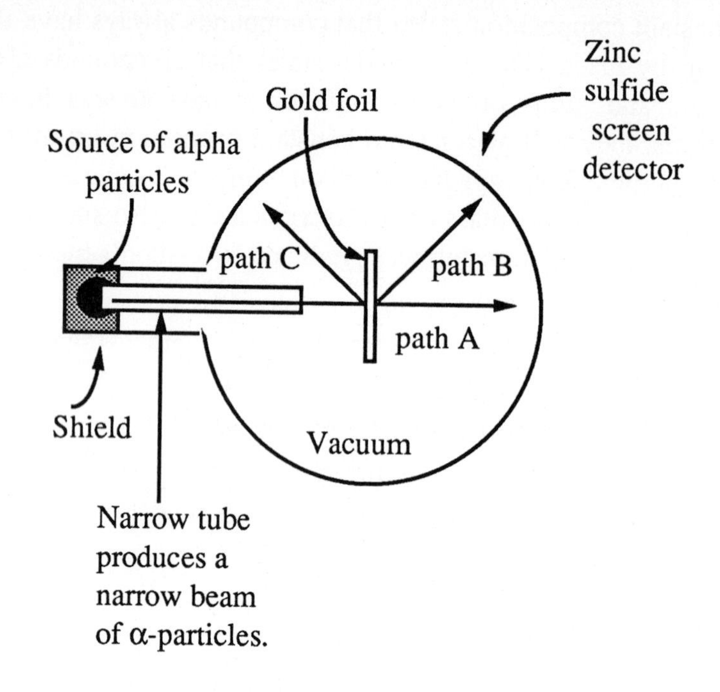

17. a. An atom consists mostly of empty space.

    b. α-particles have a positive charge, because they contain 2 protons. When a moving α-particle travels close to the nucleus of an atom which itself contains protons, the α-particle is deflected from its path because two areas of positive charge repel each other.

    c. Some of the α-particles scored a direct hit and bounced straight back. The particle which the α-particle hit must be an area within the atom which is very massive for the heavy α-particle to bounce straight back.

18.

		Relative mass	Relative charge
a.	Electron	1	1-
b.	Proton	1836	1+
c.	Neutron	1839	0

19. Isotopes of all atoms have the same number of protons as they do electrons. All sodium isotopes have 11 protons and 11 electrons. So the answer is false.

20.

A ← mass number

X ← element symbol

Z ← atomic number

21. a. $^{3}_{1}H$

b. $^{37}_{17}Cl$

c. $^{18}_{8}O$

d. $^{235}_{92}U$

e. $^{32}_{16}S$

22. a. The mass number provides the number of protons plus the number of neutrons. If the number of neutrons is 24, then the number of protons is 46 minus 24, or 22 protons.

b. The number of protons always equals the number of electrons, so there are 22 electrons in this isotope.

23. a. When isotopes are designated with the element name followed by a number, as in aluminum-29, the number is equal to the mass number. 29 is the mass number for aluminum-29.

b. The mass number of any isotope minus the atomic number is equal to the number of neutrons. So 29 minus 13 equals 16 neutrons.

24.

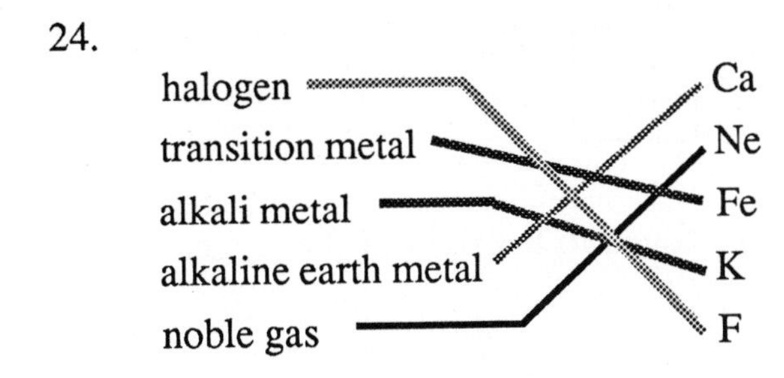

25. a. halogens

9 F
17 Cl
35 Br
53 I

b. alkaline earth metals

12 Mg
20 Ca
38 Sr
56 Ba

c. noble gases

2 He
10 Ne
18 Ar
36 Kr
54 Xe

d. alkali metals

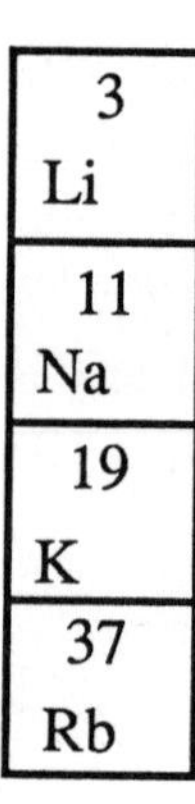

26. C, P, Br, and I are nonmetals. The nonmetals are found to the right of the jagged line on the right side of the periodic table.

27.

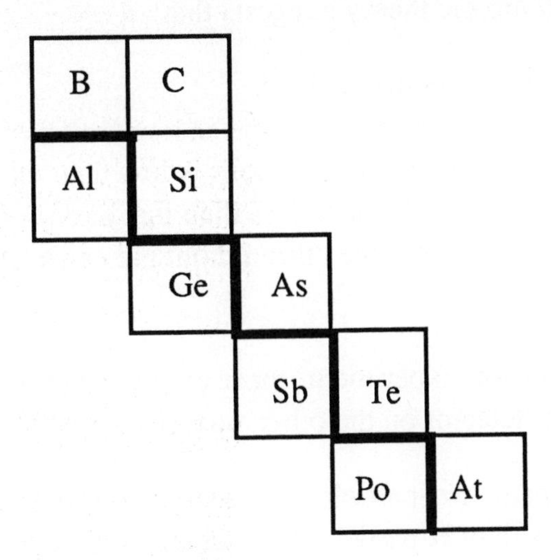

PRACTICE EXAM

1. Oxygen is the most abundant element on earth. What element is the second most abundant element on earth?

   a. carbon
   b. iron
   c. silicon
   d. hydrogen
   e. aluminum

2. Which element name and symbol pair is **not** correct?

   a. silver, Ag
   b. chlorine, Co
   c. potassium, K
   d. manganese, Mn
   e. iron, Fe

3. Which of the following statements is not a part of Dalton's atomic theory?

   a. Different elements are composed of different kinds of atoms.
   b. The mass of a proton is greater than the mass of an electron.
   c. All elements are made of atoms.
   d. In a chemical reaction, no new elements are formed.
   e. Atoms of different elements combine to produce compounds.

4. The current view of atomic theory suggests that:

   a. Electrons are found in the nucleus.
   b. Not all atoms of an element have the same number of neutrons.
   c. In some atoms, the number of protons is less than the number of electrons.
   d. The mass of a neutron is much less than that of a proton.
   e. Atoms have protons scattered throughout the entire atom.

5. In Rutherford's gold foil experiment, most of the α-particles traveled straight through the gold foil and hit the detector on the other side. How was this observation interpreted?

   a. The straight path of α-particles provided evidence for the location of protons.
   b. The straight path of α-particles provided evidence for the nuclear atom.
   c. The straight path of α-particles showed that an atom is mostly empty space.
   d. The straight path of α-particles demonstrated that the charge on an electron was 1^-.
   e. The straight path of α-particles showed that the atom was electrically neutral.

6. An atom of silicon is $^{30}_{14}Si$. The nucleus of the atom contains:

   a. 14 protons, 14 electrons, and 16 neutrons
   b. 14 protons, 14 electrons, and 30 neutrons
   c. 16 protons, 16 electrons, and 14 neutrons
   d. 30 protons, 30 electrons, and 14 neutrons
   e. 14 protons and 14 neutrons

7. The formula for potassium permanganate contains 1 potassium atom, 1 manganese atom, and 4 oxygen atoms. Which of the following formulas for potassium permanganate is correctly written?

   a. $PMnO_4$
   b. $K_1Mn_1O_4$
   c. $KMgO_4$
   d. $KMnO^4$
   e. $KMnO_4$

For questions 8-10, choose one of the following:

a. antimony
b. gold
c. argon
d. cobalt
e. copper

8. Which choice would exhibit none of the characteristics of the metals?

9. Which choice has the highest atomic number?

10. Which choice would show properties of both the metals and the nonmetals?

PRACTICE EXAM ANSWERS

1. c (4.1)
2. b (4.2)
3. b (4.3)
4. b (4.6)
5. c (4.4)
6. a (4.7)
7. e (4.4)
8. c (4.8)
9. b (4.8)
10. a (4.8)

CHAPTER 5: ELEMENTS, IONS, AND NOMENCLATURE

INTRODUCTION

This chapter will prepare you for learning how to determine the product of a chemical reaction in Chapter 6. In this chapter, you will learn how elements can form ions and combine to form compounds. You will be asked to learn rules for naming ions and compounds and you will need to memorize some names. Many of the problems at the end of this chapter have as their goal the application of the rules you will learn. Work on the naming problems until you have developed your skill at naming chemical compounds.

GOALS FOR THIS CHAPTER

1. Know how some elements exist when uncombined with other elements, and some properties of common elements. (Section 5.1)
2. Know how atoms form anions, and how they form cations. Be able to predict what type of ion is formed, and its charge based on the location of elements within the periodic table. (Section 5.2)
3. Realize that compounds can be composed of ions and that in a molecule, the number of negative charges on the anions must equal the positive charges on cations. (Section 5.3)
4. Be able to write the chemical formula for compounds made from ions. (Section 5.3)
5. Know how to name compounds where the cation can have only one charge (Type I), and how to name compounds where the cation can have more than one charge (Type II). (Section 5.5)
6. Be able to name compounds composed of two nonmetals (Type III). (Section 5.6)
7. Learn the names, formulas, and charges for the common polyatomic ions and how to name compounds which contain polyatomic ions. (Section 5.8)
8. Know the names of the common acids, and how to name less common ones. (Section 5.9)
9. Be able to write the chemical formula of a compound given the name. (Section 5.10)

QUICK DEFINITIONS

Noble gases — Elements which are found in group 8 of the periodic table. They do not readily react with other elements to form compounds. (Section 5.1)

Diatomic molecules — Molecules which are composed of two atoms of the same elements. (Section 5.1)

Ion — An atom or group of atoms which has either lost one or more electrons (has a positive charge) or gained one or more electrons (has a negative charge). (Section 5.2)

Cation	An ion with a positive charge. It has lost one or more electrons. (Section 5.2)
Anion	An ion with a negative charge. It has gained one or more electrons. (Section 5.2)
Transition metals	A large group of metals in the middle of the periodic table which can lose different numbers of electrons, and can form cations with different positive charges. (Section 5.2)
Ionic compounds	Compounds made by reacting a metal with a nonmetal. The compound contains cations of the metal which are attracted to the anions of the nonmetal. (Section 5.3)
Binary compound	Compound made from two different elements. (Section 5.4)
Binary ionic compound	Compound made from two different elements, one of which is a metal, and the other is a nonmetal. (Section 5.5)
Type I binary ionic compound	Contains a metal which can only form one kind of cation. (Section 5.5)
Type II binary ionic compound	Contains a metal which can form more than one kind of cation. (Section 5.5)
Type III binary compound	Contains only elements which are nonmetals. (Section 5.6)
Polyatomic ions	Groups of atoms which bear positive or negative charges. (Section 5.8)
Oxyanions	A series of polyatomic ions which differ from each other by the number of oxygen atoms they contain. (Section 5.8)
Acids	Acids can be defined as anything which produces hydrogen ions, H^+, when they are dissolved in water. (Section 5.9)

CONTENT REVIEW

5.1 NATURAL STATES OF THE ELEMENTS

How Are the Common Elements Normally Found in Nature?

Only a few elements found in nature are not combined with other atoms. Those few elements include the noble gases, which as the name suggests, are gases at room temperature. Some of the metals, such as gold, silver and platinum, can be found as free metals, too.

Some elements are commonly found in nature as **diatomic molecules**, that is, two atoms of the same element combined to form a molecule. The nitrogen found in the atmosphere is present as N_2 molecules, and oxygen is found as O_2. Hydrogen gas, although not usually present in the atmosphere, exists as molecules of H_2.

The halogens, including fluorine, chlorine, bromine and iodine, when present as free elements also exist as diatomic molecules.

Metals are found neither as single atoms, nor as diatomic molecules. Rather, they exist as large clusters of atoms.

At room temperature there are two liquid elements, the metal mercury, and the halogen bromine. Several elements are gases at room temperature- H_2, O_2, N_2, F_2, and Cl_2, and the noble gases. All the other elements are solids at room temperature.

The structures of the nonmetal solids do not fall into any easily described categories.

5.2 IONS

What Are Ions and How Do They Form?

An **ion** is formed when an atom with equal numbers of protons and electrons either loses or gains electrons. The number of protons remains the same. The ion which is formed has either an excess or a deficiency of electrons, which means it has one or more positive or negative charges.

A **cation** is an ion with one or more positive charges. Cations form when atoms lose electrons. For example, a potassium atom has nineteen protons and nineteen electrons. Potassium forms a cation when the potassium atom loses an electron. The cation has nineteen protons and eighteen electrons, one more proton than electron. So the potassium cation has a 1^+ charge.

An **anion** is an ion with one or more negative charges. Anions form when atoms gain electrons. For example, a sulfur atom has sixteen protons and sixteen electrons. An anion of sulfur forms when sulfur gains two electrons. The anion which forms has sixteen protons and eighteen electrons. So the sulfur anion has a 2^- charge.

How Can the Periodic Table Tell You the Charge On An Ion?

The periodic table is organized so that the group number also helps you know the charge on the ion. This relationship works for elements in groups 1, 2, and 3. Elements in group 1 form cations with a 1^+ charge, those in group 2 form cations with 2^+ charge, group 3 elements form cations with 3^+ charge. Elements in group 6 form anions with a 2^- charge, and group 7 elements form anions with 1^- charge. Elements which are not in groups 1, 2, 3, 6, or 7 either do not form ions, or the charge on the ion varies.

5.3 COMPOUNDS THAT CONTAIN IONS

How Can We Show That Compounds Contain Ions?

Not only can atoms form ions, compounds exist which are made of ions. Chemists believe that compounds such as sodium chloride are made of sodium ions with a 1^+ charge, and chloride ions with a 1^- charge. The evidence which suggests the presence of ions is that solid sodium chloride does not conduct an electric current, but molten sodium chloride and sodium chloride dissolved in water do. It is believed that movement of the ions is necessary for an electric current. The ions in solid sodium chloride are not able to move around because solids have a very rigid structure, and thus there is no electric current. With molten sodium chloride, or sodium chloride dissolved in water, ions are free to move around. The movement of atoms or ions in liquids is much greater than it is in solids, so the ions are free to carry an electric current. Pure water does not carry an electric current because the oxygen and hydrogen in water do not produce ions.

How Do Ions Combine to Produce Compounds?

Compounds made from ions must have a net charge of zero. That is, the amount of positive charge must be equal to the amount of negative charge. This means that compounds must contain **both** cations and anions. A compound made from ions of potassium and sulfur is an example. Potassium in group 1 of the periodic table forms cations with 1^+ charge, K^+. Sulfur in group 6 forms anions with 2^- charge, S^{2-}. When potassium and sulfur ions combine to form a molecule, they must combine in a ratio so that the net charge on the compound is zero. If one potassium ion combined with one sulfide ion, the net charge on the compound would be 1^-, because the 2^- charge on the sulfide ion is not completely offset by the 1^+ charge on the potassium ion. In order for the compound to have no net charge, two potassium ions must combine with one sulfide ion, K_2S. This rule must be followed when writing formulas for all ionic compounds.

5.4 NAMING COMPOUNDS

Why Do We Need A Systematic Method For Naming Compounds?

At first, chemists who discovered compounds named them whatever they wanted. Often, the names which were chosen provided no information about which elements the compound was made from. It was necessary to memorize the structure for each name. These names are called common names. To avoid memorizing structures for thousands of common names chemists devised a system for naming. Using the system, you should be able to construct a compound's formula from the name, and to name a compound when you are given the formula.

5.5 NAMING COMPOUNDS THAT CONTAIN A METAL AND A NONMETAL

Identifying the group of the metal in the chemical formula is one key to naming these compounds. Remember that metals in groups 1, 2, and aluminum in group 3 form only one type of cation. Metals in other groups can form more than one kind of cation. The naming procedure is different for the two types of metals.

How Can You Name Compounds Where the Metal Forms Only One Cation?

Always name the cation (the metal ion) first, and the anion last. Cation names are easy. They are the same as the name of the metal itself. For example, potassium forms the potassium ion. Anion names are formed by taking the first part of the element name, and adding **ide**. It is not always obvious which part of an element name to keep and which to discard when naming anions. You should memorize the common anion names.

NAMES OF COMMON IONS

Common Anions		**Common Cations**	
Ion Formula	Ion Name	Ion Formula	Ion Name
N^{3-}	nitride	H^+	hydrogen
P^{3-}	phosphide	Li^+	lithium
O^{2-}	oxide	Na^+	sodium
S^{2-}	sulfide	K^+	potassium
F^-	fluoride	Mg^{2+}	magnesium
Cl^-	chloride	Ca^{2+}	calcium
Br^-	bromide	Ba^{2+}	barium
I^-	iodide	Al^{3+}	aluminum
H^-	hydride	Ag^+	silver

How Can You Name Compounds Where the Metal Forms More Than One Cation?

Name the cation first, and the anion last. Cation names are the same as the metals themselves, but you must indicate in the name which cation is present. Copper can form cations with charges of 1^+ or 2^+. To indicate which copper ion is present, use the name of the ion, copper, followed by (I) or (II) to indicate that correct charge. The anions are named as above by using the first part of the element name, and adding **ide**.

Example:

Let's name $CuCl_2$. To determine whether the name of the cation is copper(I) or copper(II) look at the charge on the anion. The charge on a chloride ion is 1^-. There are two of them, so the total negative charge is 2^-. Because compounds must have no net charge, the charge on the copper ion must be 2^+. So the cation name must be copper(II). The correct way to write the name would be copper(II) chloride.

5.6 NAMING BINARY COMPOUNDS THAT CONTAIN ONLY NONMETALS (TYPE III)

How Can You Name Compounds Containing Only Nonmetals?

Look at the formula and give the element which comes first its normal element name. Name the second element as an anion. You will always need to add a prefix to the second element in the name to tell how many atoms are present. Use a prefix in front of the first element in the name only when there is more than one atom of that kind of element. The prefixes are listed below.

PREFIX NAMES

Prefix	**Number of Atoms**
mono	1
di	2
tri	3
tetra	4
penta	5
hexa	6
hepta	7
octa	8

Example:

CS_2 is carbon disulfide. The first element, carbon, is given the full element name. The second element, sulfur, is given the anion name, sulfide. There is one carbon atom, but the mono prefix is never used with the first element in the name, only with the second element. There are two sulfur atoms, so the prefix **di** is used to make the anion name **disulfide**.

5.8 NAMING COMPOUNDS THAT CONTAIN POLYATOMIC IONS

What Are the Polyatomic Ions?

Polyatomic ions are composed of several atoms which act together as a group, and which bear a positive or a negative charge. Compounds which contain polyatomic ions may contain many different atoms. In order for you to distinguish between a compound which contains many different atoms and a compound which contains one or more polyatomic ions, you must be able to recognize and name the polyatomic ions. Each of these ions has a specific name and charge. If you do not learn these names, structures and charges now, you will have trouble throughout the remainder of the course.

NAMES OF POLYATOMIC IONS

Cations		Anions	
Formula	Name	Formula	Name
NH_4^+	ammonium	CO_3^{2-}	carbonate
NO_2^-	nitrite	HCO_3^-	hydrogen carbonate
NO_3^-	nitrate	ClO^-	hypochlorite
SO_3^{2-}	sulfite	ClO_2^-	chlorite
SO_4^{2-}	sulfate	ClO_3^-	chlorate
HSO_4^-	hydrogen sulfate	ClO_4^-	perchlorate
OH^-	hydroxide	$C_2H_3O_2^-$	acetate
CN^-	cyanide	MnO_4^-	permanganate
PO_4^{3-}	phosphate	$Cr_2O_7^{2-}$	dichromate
HPO_4^{2-}	hydrogen phosphate	CrO_4^{2-}	chromate
$H_2PO_4^-$	dihydrogen phosphate	O_2^{2-}	peroxide

How Can You Name Compounds Which Contain Polyatomic Ions?

Name the cation first. Metal cations have the same name as the element name. The names of elements which form more than one type of cation must have a roman numeral to show what the charge on the cation is. If the cation is polyatomic, then you must have already learned the correct cation name. The anion, if monatomic, is named by keeping the first part of the element name and adding **ide**. If the anion is polyatomic, then you must have learned the correct name, and be aware of the charge. If more than one polyatomic ion is required so that the net charge on the molecule is zero, then the polyatomic ion is surrounded by parentheses and a subscript number is used to indicate how many of each ion is present.

Example:

In the formula $Mg(NO_3)_2$, magnesium has a 2^+ charge and the nitrate ion has a 1^- charge. For the net charge on the molecule to be zero, two nitrate ions must be present. The entire nitrate ion is surrounded by parentheses, and the subscript two indicates that there are two of them. The correct name for $Mg(NO_3)_2$ is magnesium nitrate. We name only the cation and the anion and ignore the parentheses.

5.9 NAMING ACIDS

What Are Acids?

Several different kinds of molecules are acids. One kind of molecule will produce H^+ ions when dissolved in water. We want to be able to name the common acids.

How Can You Name Acids Which Do Not Contain Oxygen?

Use the prefix **hydro** as the first part of the acid name. Hydro comes from **hydro**gen ion. The second part of the acid name comes from the name of the remainder of the atoms or groups of atoms in the molecule, and ends in **ic**. You should learn the names of the common acids which do not contain oxygen. For example, HCl is named hydrochloric acid. The **hydro** comes from the presence of hydrogen, **chlor** is derived from the element chlorine, and the suffix **ic** is used on the end.

How Can You Name Acids Which Contain Oxygen?

When the acid contains oxygen, the main part of the acid name comes from the central atom or central anion name. All of these acids have a suffix of either **ic** or **ous**. Use **ic** when the central anion name ends in **ate**. Use **ous** when the central anion name ends in **ite**.

Example:

> The name of H_3PO_4 is phosphoric acid, named after the central anion the phosphate ion. Note that because phosphate ends in **ate**, the acid name ends in **ic**. HNO_2 is named nitrous acid. The central anion is the nitrite ion. You should learn the names of the common acids which contain oxygen.

5.10 WRITING FORMULAS FROM NAMES

How Can You Write a Formula Given the Name?

If you have learned to write the correct name when given a formula, then doing the opposite is straightforward. For example, writing a formula for potassium phosphate requires you to know that the symbol for potassium is K. The formula for the phosphate ion is PO_4^{3-}. The charge on a potassium ion is always 1^+, and the charge on the phosphate ion is always 3^-. To balance the three negative charges on phosphate, three potassium ions with 1^+ charge are needed. The correct formula would be written K_3PO_4.

Remember that when there is more than one of a particular polyatomic ion present you need to surround the ion with parentheses and use a subscript to indicate how many of the ion are present.

Example:

Ammonium carbonate contains two different polyatomic ions, the ammonium ion, NH_4^+ and the carbonate ion, CO_3^{2-}. The charges on the two ions are not identical; the correct formula must have two NH_4^+ ions to balance the charge on the carbonate ion. The correct formula would be written $(NH_4)_2CO_3$. Only the ammonium ion should have parentheses, because two of them are needed to balance the charge on the carbonate ion.

LEARNING REVIEW

1. Some elements exist in nature as diatomic molecules. Which of the elements below will be found as diatomic molecules?

a.	Ar	e.	S
b.	O	f.	N
c.	K	g.	H
d.	F	h.	Cl

2. At room temperature, what is the physical state (solid, liquid, or gas) of each of the elements which naturally form diatomic molecules?

3. Which elements are always found in nature as individual atoms?

a.	carbon	e.	helium
b.	krypton	f.	neon
c.	magnesium	g.	aluminum
d.	chlorine	h.	sulfur

4. Fill in the name of the correct element next to its description at 25 °C.

a. Liquid metal ____________
b. Yellow green gas ____________
c. Colorless gas ____________
d. A 2 carat diamond ____________
e. A reddish brown liquid ____________
f. Dark purple solid ____________

5. Balance the equations for the formation of cations from neutral atoms.

a. Ca $\rightarrow$ ____ + ____
b. K $\rightarrow$ ____ + ____
c. Sr $\rightarrow$ ____ + ____
d. Rb $\rightarrow$ ____ + ____

6. Balance the equations for the reactions of cations with electrons.

a. Mg^{2+} + ____ → ____
b. Li^{+} + ____ → ____
c. $2H^{+}$ + ____ → ____
d. Na^{+} + ____ → ____

7. Fill in the correct number of protons and electrons for either the element or the ion in the table below.

element	protons	electrons	ion	protons	electrons
potassium	19	19			
oxygen				8	10
bromine	35				36
strontium		38		38	
aluminum	13				10

8. Name the ion which is produced from each element below, and tell whether it is a cation or an anion.

a. calcium
b. fluorine
c. oxygen
d. lithium
e. iodine
f. hydrogen

9. You wish to find out whether the compound MgF_2 is composed of ions. What test could you perform to help you make a decision?

10. a. Which diagram represents a solid NaCl crystal?
b. Which diagram represents NaCl dissolved in water?
c. Which form of NaCl, solid or aqueous solution, allows free movement of ions?

11. How many of each ion is needed to form a neutral compound?

a. Ca^{2+} and F^{-}
b. Mg^{2+} and O^{2-}
c. Na^{+} and S^{2-}
d. Li^{+} and I^{-}
e. Sr^{2+} and Cl^{-}
f. K^{+} and P^{3-}
g. Na^{+} and N^{3-}

12. What is wrong with the formulas below? Write the correct formula.

a. $AlCl_2$
b. NaO
c. Mg_2P
d. CaI_3
e. LiN_3
f. KS

13. Name the cations and anions below.

a. Cl^-
b. Mg^{2+}
c. Li^+
d. Ba^{2+}
e. N^{3-}
f. O^{2-}
g. F^-

14. Name the following Type I binary compounds.

a. KF
b. CaS
c. NaI
d. Li_3N
e. HCl
f. Al_2O_3
g. $AgCl$
h. MgF_2

15. The following elements can all form more than one cation. How many cations form and what is the charge on each of them?

a. Cu
b. Fe
c. Sn
d. Hg
e. Pb

16. Name the following Type II binary compounds.

a. $FeCl_3$
b. PbO_2
c. CoI_2
d. SnF_2
e. Fe_2S_3
f. $HgBr$

17. Name each of the compounds below.

a. $CaBr_2$
b. PbS
c. AlP
d. FeS
e. CoO
f. $MgCl_2$

18. Each of the compounds below has an <u>incorrect</u> name. Name each one correctly.

a. KBr potassium (I) bromide
b. Cu_2O cupric oxide
c. PbS_2 lead (IV) sulfide (II)
d. Na_3P sodium (III) phosphide
e. $FeCl_3$ iron chloride

19. Name the following Type III binary compounds.

a. PCl_5
b. CCl_4
c. N_2O_3
d. S_2F_{10}
e. SO_2
f. CO

20. Name the following Type I, Type II, or Type III compounds.

a. KI
b. NO_2
c. $FeCl_2$
d. Al_2O_3
e. Cl_2O_7
f. CaS

21. What are the names of the polyatomic ions below?

a. HCO_3^-
b. OH^-
c. NH_4^+
d. NO_2^-
e. SO_4^{2-}
f. CrO_4^{2-}

22. The compounds below all contain polyatomic ions. Name each one.

a. K_2SO_4
b. $Fe(OH)_3$
c. NH_4NO_3
d. $Al_2(Cr_{27})_3$
e. $Ca(CN)_2$
f. $Mg_3(PO_4)_2$
g. $NaMnO_4$
h. $Cu(ClO_3)_2$
i. $PbCO_3$

23. Check your knowledge of the common acids by naming the acids below.

a. H_2SO_4
b. HCN
c. HBr
d. HNO_3
e. H_2S
f. $HC_2H_3O_2$

24. From their names, write formulas for the compounds below.

a. aluminum chloride
b. cobalt (III) permanganate
c. dinitrogen trioxide
d. sulfur dioxide
e. calcium nitrate
f. silver chloride
g. iron (II) acetate
h. tin (IV) chlorite
i. sodium sulfate
j. lithium hydrogen carbonate
k. mercury (II) dichromate

ANSWERS TO LEARNING REVIEW

1. b. Oxygen is found in nature as O_2 molecules.
   d. Fluorine is found in nature as F_2 molecules.
   f. Nitrogen is found in nature as N_2 molecules.
   g. Hydrogen is found in nature as H_2 molecules.
   h. Chlorine is found in nature as Cl_2 molecules.

2. Most of the diatomic molecules, H_2, N_2, O_2, F_2, Cl_2, are gases at room temperature. Br_2 is a reddish liquid. I_2 is a dark purple solid.

3. a. Carbon is usually found combined with other elements such as hydrogen.
   b. Krypton is always found as individual krypton atoms.
   c. Magnesium is usually found combined with other elements.
   d. Chlorine is found as either Cl_2, or combined with other elements.
   e. Helium is always found as individual helium atoms.
   f. Neon is always found as individual neon atoms.
   g. Aluminum is usually found combined with other elements.
   h. Sulfur is usually found as S_8 molecules or combined with other elements.

4. Only a few of the elements have unique properties, but there are many elements which could be described as colorless gases, or as shiny metals.

a.	Liquid metal	mercury
b.	Yellow-green gas	chlorine
c.	Colorless gas	Many elements fill this description, such as oxygen, hydrogen and others.
d.	A 2 carat diamond	carbon
e.	Reddish brown liquid	bromine
f.	Dark purple solid	iodine

5. Cations have lost one or more electrons and the number of electrons lost always equals the charge on the cation.

a. $Ca \rightarrow Ca^{2+} + 2e^-$
b. $K \rightarrow K^+ + e^-$
c. $Sr \rightarrow Sr^{2+} + 2e^-$
d. $Rb \rightarrow Rb^+ + e^-$

6. Cations will react with electrons to form neutral atoms.

a. $Mg^{2+} + 2e^- \rightarrow Mg$
b. $Li^+ + e^- \rightarrow Li$
c. $2H^+ + 2e^- \rightarrow H_2$
d. $Na^+ + e^- \rightarrow Na$

7. The number of protons does not change when neutral atoms form ions, but the number of electrons either increases or decreases.

element	**protons**	**electrons**	**ion**	**protons**	**electrons**
potassium	19	19	potassium	19	18
oxygen	8	8	oxide	8	10
bromine	35	35	bromide	35	36
strontium	38	38	strontium	38	36
aluminum	13	13	aluminum	13	10

8. Cation names are the same as the element name, but anions have an ide ending.

a.	calcium	cation
b.	fluoride	anion
c.	oxide	anion
d.	lithium	cation
e.	iodide	anion
f.	hydrogen	cation

9. You can place some solid MgF_2 in water. If the solid MgF_2 contains Mg^{2+} ions and F^- ions, when the compound dissolves in water an aqueous solution of Mg^{2+} and F^- ions will form. Test whether or not a solution of MgF_2 will conduct an electrical current by immersing electrodes in the solution. If the solution allows current to flow and a bulb to shine, there is evidence that ions are in solution, free to move around, and able to conduct an electrical current.

10. a. A solid NaCl crystal is an ordered rigid structure, structure i.
    b. Ions are pulled away from the orderly crystal by water molecules, as in structure ii.
    c. Ions in water are free to move around and are not packed close together. Those in a solid are ordered, and packed close together so that each anion is surrounded by cations.

11.
a.	1 calcium and 2 fluoride ions	CaF_2
b.	1 magnesium and 1 oxide ion	MgO
c.	2 sodium and 1 sulfide ion	Na_2S
d.	1 lithium and 1 iodide ion	LiI
e.	1 strontium and 2 chloride ions	$SrCl_2$
f.	3 potassium and 1 phosphide ion	K_3P
g.	3 sodium and 1 nitride ion	Na_3N

12. a. Aluminum forms ions with a 3^+ charge and chlorine forms ions with a 1^- charge. Three chlorine ions will combine with one aluminum ion to form $AlCl_3$.
    b. Sodium forms ions with a 1^+ charge and oxygen forms ions with a 2^- charge. Two sodium ions will combine with one oxide ion to form Na_2O.
    c. Magnesium forms ions with a 2^+ charge and phosphorus forms ions with a 3^- charge. Three magnesium ions will combine with two phosphide ions to form Mg_3P_2.
    d. Calcium forms ions with a 2^+ charge and iodine forms ions with a 1^- charge. One calcium ion will combine with two iodide ions to form CaI_2.
    e. Lithium forms ions with a 1^+ charge and nitrogen forms ions with a 3^- charge. Three lithium ions will combine with one nitride ion to form Li_3N.
    f. Potassium forms ions with a 1^+ charge and sulfur forms ions with a 2^- charge. Two potassium ions will combine with one sulfide ion to form K_2S.

13. The ions in problem 13 are all monatomic ions. The cations all have the same name as the element, while the anions all end in ide.

a.	chloride	e.	nitride
b.	magnesium	f.	oxide
c.	lithium	g.	fluoride
d.	barium		

14. Type I binary compounds form between a metal and a nonmetal.

a.	potassium fluoride	e.	hydrogen chloride
b.	calcium sulfide	f.	aluminum oxide
c.	sodium iodide	g.	silver chloride
d.	lithium nitride	h.	magnesium fluoride

15. a. Cu^+ Cu^{2+}
    b. Fe^{2+} Fe^{3+}
    c. Sn^{2+} Sn^{4+}
    d. Hg^+ Hg^{2+}
    e. Pb^{2+} Pb^{4+}

16. Type II binary compounds form between a metal which forms more than one cation and a nonmetal.

    a. iron(III) chloride
    b. lead(IV) oxide
    c. cobalt(II) iodide
    d. tin(II) iodide
    e. iron(III) sulfide
    f. mercury(I) bromide

17. The compounds are mixed Type I and Type II binary compounds.

    a. calcium bromide
    b. lead(II) sulfide
    c. aluminum phosphide
    d. iron(II) sulfide
    e. cobalt(II) oxide
    f. magnesium chloride

18. a. Potassium only forms cations with 1^+ charge, so potassium(I) bromide should be potassium bromide.
    b. The formula Cu_2O shows copper with a 1^+ charge, which is named the copper(I), or cuprous ion. The correct name for this formula is copper(I) oxide, or cuprous oxide.
    c. The formula PbS_2 tells us that the charge on the lead ion is 4^+, so the first part of the name, lead(IV) is correct. The sulfide ion has a 2^- charge, but we do not use roman numerals after the anion name. So the correct name for this compound is lead(IV) sulfide.
    d. Sodium only forms cations with 1^+ charge, so sodium(III) phosphide should be sodium phosphide.
    e. The formula $FeCl_3$ tell us that iron has a 3^+ charge. Because iron forms cations with more than one charge, the correct name would be iron(III) chloride.

19. Type III binary compounds form between nonmetals. The prefix which indicates 10 atoms is deca. This prefix is used in problem 19d.

    a. phosphorus pentachloride
    b. carbon tetrachloride
    c. dinitrogen trioxide
    d. disulfur decafluoride
    e. sulfur dioxide
    f. carbon monoxide

20. The compounds are a mixture of Type I, Type II and Type III compounds.

a. potassium iodide
b. nitrogen dioxide
c. iron(II) chloride
d. aluminum oxide
e. dichlorine heptoxide
f. calcium sulfide

21. If you have trouble naming these ions, go back and review the names again. You will need these names throughout your chemistry career.

a. bicarbonate
b. hydroxide
c. ammonium
d. nitrite
e. sulfate
f. chromate

22.
a. potassium sulfate
b. iron(III) hydroxide
c. ammonium nitrate
d. aluminum dichromate
e. calcium cyanide
f. magnesium phosphate
g. sodium permanganate
h. copper(II) chlorate
i. lead(II) carbonate

23.
a. sulfuric acid
b. hydrocyanic acid
c. hydrobromic acid
d. nitric acid
e. hydrosulfuric acid
f. acetic acid

24.
a. $AlCl_3$
b. $Co(MnO_4)_3$
c. N_2O_3
d. SO_2
e. $Ca(NO_3)_2$
f. $AgCl$
g. $Fe(C_2H_3O_2)_2$
h. $Sn(ClO_2)_4$
i. Na_2SO_4
j. $LiHCO_3$
k. $HgCr_2O_7$

PRACTICE EXAM

1. Which of the elements below exist in nature as diatomic molecules?

i. hydrogen ii. copper iii. potassium iv. iodine v. nitrogen

a. i and iii
b. i and v
c. ii and iv
d. iii and v
e. i, iv and v

2. Which element forms an anion with a 3^- charge?

a. Al
b. P
c. S
d. Na
e. Mg

3. How many formula/name pairs are incorrect?

i	NO_2^-	nitrate
ii	SO_4^{2-}	sulfate
iii	CrO_4^{2-}	chromate
iv	PO_4^{3-}	phosphite
v	I^-	iodate

a. one
b. two
c. three
d. four
e. five

4. What is the correct name for Al_2S_3?

a. aluminum sulfate
b. aluminum(III) sulfide
c. aluminum sulfate(II)
d. aluminum sulfide(II)
e. aluminum sulfide

5. Which of the formulas below is perchloric acid?

a. HCl
b. HClO
c. $HClO_2$
d. $HClO_3$
e. $HClO_4$

6. What is the correct name for $Sn(Cr_2O_7)_2$?

a. tin(IV) dichromate
b. stannic chromate
c. tin(II) dichromate

d. tin(IV) chromate
e. tin dichromate

7. What is the correct name for PCl_5?

a. monophosphorus pentachloride
b. phosphorous pentachloride
c. potassium pentachloride
d. phosphorus(V) pentachloride
e. phosphorus(V) chloride

8. Which of the ions below does **not** have a correct charge?

a. K^+
b. O^{2-}
c. F^+
d. H^+
e. S^{2-}

9. Which formula/name pair is not correct?

a $PbCl_2$ lead(II) chloride
b. NO nitrogen monoxide
c. $NaMnO_4$ sodium manganate
d. $Ca(NO_3)_2$ calcium nitrate
e. $Mg(CN)_2$ magnesium cyanide

10. Which is the correct formula for ammonium acetate?

a. $(NH_4)_2CO_3$
b. NH_4OH
c. NH_4NO_3
d. $NH_4C_2H_3O_2$
e. $(NH_4)N_3$

PRACTICE EXAM ANSWERS

1. e (5.1)
2. b (5.2)
3. c (5.8)
4. e (5.5)
5. e (5.9)

6. a (5.5 and 5.6)
7. b (5.6)
8. c (5.2)
9. c (5.5 and 5.6)
10. d (5.10)

CHAPTER 6: CHEMICAL REACTIONS: AN INTRODUCTION

INTRODUCTION

Knowing how to write chemical equations and how to interpret what they mean is an important part of chemistry. If you have learned the symbols for the elements, and can write the formulas of compounds from their names, then learning to write chemical equations will be easier. The Answers to Learning Review reviews the logic used when balancing a chemical equation by trial and error. If you are having trouble balancing equations, go over the steps used to balance the equations in the Answers to Learning Review.

AIMS FOR THIS CHAPTER

1. Know the clues which help tell you that a chemical reaction has occurred. (Section 6.1)
2. Given a word description of a reaction, be able to write chemical formulas for reactants and products. (Section 6.2)
3. Be able to balance a reaction so that there are the same numbers of atoms on both sides of the chemical equation. (Section 6.3)

QUICK DEFINITIONS

Chemical reaction — A change during which atoms form new associations with other atoms. (Section 6.2)

Chemical equation — Shows compounds before a chemical reaction takes place on the left, and compounds formed from a chemical reaction on the right. (Section 6.2)

Reactants — Chemicals shown on the left of a chemical equation. They are the chemicals which change to form new chemicals. (Section 6.2)

Products — Chemicals shown on the right of a chemical equation. They are the chemicals which are formed by the reaction. (Section 6.2)

Coefficients — The number in front of a chemical formula that tells how many of each atom or molecule is present. (Section 6.3)

CONTENT REVIEW

6.1 EVIDENCE FOR A CHEMICAL REACTION

What Is the Evidence For a Chemical Reaction?

Any or all of the changes below often indicate that a chemical reaction has taken place.

1. The color changes.
2. A solid forms.
3. Bubbles form.
4. Heat and/or flame is produced, or heat is absorbed.

6.2 CHEMICAL EQUATIONS

The changes which happen during a chemical reaction are shown by writing a chemical equation. The starting materials are called **reactants** and are shown on the left side of the chemical equation. The substances formed in a reaction are called **products** and are shown on the right side of the equation.

The same kinds of atoms must be present before and after a chemical reaction because atoms are neither created nor destroyed during a reaction.

The same number of each kind of atoms must be present before and after a chemical reaction. In other words, the equation must be balanced.

For example, when hydrogen gas and oxygen gas react to form water vapor, the chemical equation must have hydrogen and oxygen atoms on both sides of the chemical equation, and it must have the same number of hydrogen and oxygen atoms on both sides.

Sometimes, the physical state of each of the products and reactants is indicated by a small letter in parentheses after the chemical symbol. Fe(s) indicates that iron is a **solid** reactant. $H_2O(l)$ indicates **liquid** water. $O_2(g)$ indicates **gaseous** oxygen and NaCl(aq) indicates a solution of sodium chloride in water (**aqueous**).

Example:

$$2O_2(g) + H_2(g) \rightarrow 2H_2O(l)$$

This equation reads hydrogen gas reacts with oxygen gas to produce liquid water.

6.3 BALANCING CHEMICAL EQUATIONS

How Can You Balance a Chemical Equation?

1. From the word description of the reaction, write formulas for all of the reactants and all of the products. Separate the reactants and products with an arrow, which means, "reacts to form". If you are having trouble writing the formulas from the word descriptions, you need to review the guidelines in Section 5.10. Put in the physical states of each of the reactants and products if they are given in the word description.
2. Count the number of each kind of atom on both sides of the equation. Usually, the number of one or more atoms is not balanced.
3. Inspect the chemical equation you have written and pick out the most complex chemical formula. Begin balancing the equation with the atoms of this formula. Try to begin balancing equations with atoms other than oxygen and hydrogen. If you save oxygen and hydrogen for last they can often be easily balanced by adjusting the coefficient of water, if it is present in the reaction.

Example:

Under the proper conditions solid potassium chlorate will decompose to form solid potassium chloride and oxygen gas. Write a balanced chemical equation for this reaction.

1. First, write correct chemical formulas for reactants and products.

$$KClO_3(s) \longrightarrow KCl(s) + O_2(g)$$

2. Count the atoms.

$$KClO_3(s) \longrightarrow KCl(s) + O_2(g)$$

1 K, 1 Cl, 3 O atoms   1 K, 1 Cl, 2 O atoms

The number of potassium and chlorine atoms on each side of the equation are already the same, but the left side has three oxygen atoms, while the right side has only two. We will need to balance the equation by adjusting the number of oxygen atoms.

3. Pick out the most complex formula, $KClO_3$. Adjust the numbers of atoms. If we change the coefficient in front of O_2 to 2, we increase the number of oxygen atoms on the right to 4, which still does not balance the three oxygen atoms on the left. If we change the coefficient in front of O_2 to a 3, there will be six oxygen atoms on the right and three on the left. We can now balance the number of oxygen atoms by changing the coefficient of $KCLO_3$ from 1 to 2. Now, there are six oxygen atoms on both sides.

$$2KClO_3(s) \longrightarrow KCl(s) + 3O_2(g)$$

But now the number of K and Cl atoms are not equal on both sides. There are 2 K and Cl atoms on the left, but only one of each on the right. Put a coefficient of 2 in front of KCl on the right. Now, there are equal numbers and kinds of atoms on both sides of the equation.

$$2KClO_3(s) \rightarrow 2KCl(s) + O_2(g)$$

Remember that you can change coefficients when you are balancing an equation, but you can never change the subscript number. For example, to balance an equation for the breakdown of water into oxygen and hydrogen you need two oxygen atoms on the left side of the equation, but you start with H_2O, which contains one oxygen atom.

$$H_2O(l) \rightarrow H_2(g) + O_2(g)$$

The correct way to adjust the number of oxygen atoms is to change the coefficient to $2H_2O$. If you had adjusted the number of oxygen atoms by changing the subscript number to H_2O_2 you would have changed the formula of the reactant from water to hydrogen peroxide, which does not participate in this reaction.

LEARNING REVIEW

1. Which of the following indicates that a chemical reaction has occurred?

   a. Liquid water boils to produce steam.
   b. Burning firewood gives off heat.
   c. Mixing two colorless liquids produces a bright yellow solid.
   d. Solid $NaHCO_3$ dissolves in water.

2. Why is it important that chemical equations be balanced?

3. Count the number of each kind of atom on both sides of the equation and decide which reactions are balanced, and which are not.

   a. $H_2 + Br_2 \rightarrow HBr$
   b. $KClO_3 \rightarrow KCl + O_2$
   c. $2NaOH + CO_2 \rightarrow Na_2CO_3 + H_2O$
   d. $C_2H_5OH + CO_2 \rightarrow 2CO_2 + 3H_2O$
   e. $3Cu + HNO_3 \rightarrow 3Cu(NO_3)_2 + NO + H_2O$
   f. $CaC_2 + 2H_2O \rightarrow C_2H_2 + Ca(OH)_2$

4. Use the following word descriptions to write unbalanced chemical equations. Make sure you include the physical state of reactants and products.

   a. Solid iron metal reacts with oxygen in the atmosphere to form rust, iron(III) oxide.
   b. Solid magnesium metal reacts with aqueous hydrochloric acid to produce hydrogen gas and an aqueous solution of magnesium chloride.
   c. Solid silver oxide decomposes upon heating to produce solid silver metal and oxygen gas.
   d. Aqueous sodium hydroxide reacts with aqueous nitric acid to produce aqueous sodium nitrate and liquid water.

5. Balance these chemical equations. Check your work by counting the number of each kind of atom on both sides of the equation.

   a. $KOH(aq) + H_2S(aq) \rightarrow K_2S(aq) + H_2O(l)$
   b. $HNO_2(aq) \rightarrow N_2O_3(g) + H_2O(aq)$
   c. $NaOH(aq) + H_2SO_4(aq) \rightarrow Na_2SO_4(aq) + H_2O(l)$
   d. $CuSO_4(aq) + Fe(s) \rightarrow FeSO_4(aq) + Cu(s)$
   e. $(NH_4)_2S(aq) + Pb(NO_3)_2(aq) \rightarrow PbS(s) + NH_4NO_3(aq)$
   f. $Al(s) + O_2(g) \rightarrow Al_2O_3(s)$

6. Balance these chemical equations.

   a. $PbO(s) + NH_3(aq) \rightarrow Pb(s) + N_2(g) + H_2O(l)$
   b. $SO_2(g) + O_2(g) \rightarrow SO_3(g)$
   c. $C_4H_{10}(g) + O_2(g) \rightarrow CO_2(g) + H_2O(g)$
   d. $Fe_2O_3(s) + C(s) \rightarrow Fe(s) + CO_2(g)$
   e. $TiCl_4(l) + H_2O(l) \rightarrow TiO_2(s) + HCl(aq)$
   f. $CaC_2(s) + H_2O(l) \rightarrow C_2H_2(g) + Ca(OH)_2(s)$

7. Balance these chemical equations.

   a. $KI(aq) + Br_2(l) \rightarrow KBr(aq) + I_2(s)$
   b. $Al(s) + Cl_2(g) \rightarrow AlCl_3(s)$
   c. $PbO_2(s) \rightarrow PbO(s) + O_2(g)$
   d. $Fe(OH)_3(s) + H_2SO_4(aq) \rightarrow Fe_2(SO_4)_3(s) + H_2O(l)$
   e. $K_3PO_4(aq) + BaCl_2(aq) \rightarrow KCl(aq) + Ba_3(PO_4)_2(s)$
   f. $Ba(ClO_3)_2(s) \rightarrow BaCl_2(s) + O_2(g)$

ANSWERS TO LEARNING REVIEW

1. a. Boiling represents a physical change, not a chemical reaction.
   b. Heat production is an indication of a chemical reaction.
   c. A color change is an indication of a chemical reaction.
   d. No chemical reaction has occurred. Solid $NaHCO_3$ dissolves into ions in water. The ions are so small they cannot be seen but the identity of the compound does not change.

2. When a chemical reaction occurs, atoms are neither created nor destroyed. The atoms are only rearranged to produce new molecules. Therefore it is important that the same kinds and numbers of atoms be present on both sides of a chemical equation.

3. a. $H_2 + Br_2 \rightarrow HBr$

   2 H 2 Br   1 H 1 Br   unbalanced

   b. $KClO_3 \rightarrow KCl + O_2$

   1 K 1 Cl 3 O   1 K 1 Cl 2 O   unbalanced

   c. $2NaOH + CO_2 \rightarrow Na_2CO_3 + H_2O$

   2 Na 2 O 2 H 1 C 2 O   2 Na 1 C 3 O 2 H 1 O

   The total number of atoms of each kind is

   2Na 4 O 2 H 1 C   2 Na 4 O 2 H 1 C   balanced

   d. $C_2H_5OH + 3O_2 \rightarrow 2CO_2 + 3H_2O$

   2 C (5 + 1) H 1 O (3 x 2) O   2 C (2 x 2) O (3 x 2) H 3 O

   The total number of atoms of each kind is

   5 C 6 H 7 O   2 C 6 H 7 O   balanced

   e. $3Cu + HNO_3 \rightarrow 3Cu(NO_3)_2 + NO + H_2O$

   3 Cu 1 H 1 N 3 O   3 Cu (3 x 2) O 2 N 1 N 1 O 2 H 1 O

   The total number of atoms of each kind is

   3 Cu 1 H 1 N 3 O   3 Cu 2 H 3 N 8 O   unbalanced

   f. $CaC_2 + 2H_2O \rightarrow C_2H_2 + Ca(OH)_2$

   1 Ca 2 C (2 x 2) H 2 O   2 C 2 H 1 Ca 2 O 2 H

   The total number of atoms of each kind is

   1 Ca 2 C 4 H 2 O   1 Ca 2 C 4 H 2 O   balanced

4. This problem requires that you be able to write formulas from word descriptions. Do not forget to include the physical state of both products and reactants.

   a. $Fe(s) + O_2(g) \rightarrow Fe_2O_3(s)$
   b. $Mg(s) + HCl(aq) \rightarrow H_2(g) + MgCl_2(aq)$
   c. $Ag_2O(s) \rightarrow Ag(s) + O_2(g)$
   d. $NaOH(aq) + HNO_3(aq) \rightarrow NaNO_3(aq) + H_2O(l)$

5. a. First, find the most complex formula. KOH contains three different kinds of atoms, so begin by adjusting the coefficients of the atoms in KOH. There are 2 potassium atoms on the right and only 1 on the left. Adjust potassium by increasing the coefficient of KOH from 1 to 2.

$$2KOH(aq) + H_2S(aq) \rightarrow K_2S(aq) + H_2O(l)$$

2K 2O 4H 1S   2K 1O 2H 1S

Now, K and S are balanced, but oxygen and hydrogen are not. There are 4 hydrogen and 2 oxygen atoms on the left, but only 2 hydrogen and 1 oxygen atom on the right. If we adjust the coefficient of water to 2, oxygen and hydrogen are the same on each side and the equation is balanced.

$$2KOH(aq) + H_2S(aq) \rightarrow K_2S(aq) + 2H_2O(l)$$

2K 2O 4H 1S   2K 2O 4H 1S

   b. Because HNO_2 is the most complex molecule, begin by adjusting the number of nitrogen atoms on both sides of the equation. There are 2 on the right, but only 1 on the left. Increase the coefficient of HNO_2 to 2. Hydrogen and oxygen are the same on each side, and the equation is balanced.

$$2HNO_2(aq) \rightarrow N_2O_3(aq) + H_2O(aq)$$

2H 2N 4O   2H 2N 4O

   c. Either NaOH or H_2SO_4 is a good place to begin balancing. Let's start by adjusting the number of sodium ions. Put a coefficient of 2 in front of NaOH.

$$2NaOH(aq) + H_2SO_4(aq) \rightarrow Na_2SO_4(aq) + H_2O(aq)$$

2Na 6O 4H 1S   2Na 5O 2H 1S

There is 1 sulfur atom on each side, but the left side now has 4 hydrogen and 6 oxygen atoms, while the right side has only 5 oxygen atoms and 2 hydrogen atoms. The right side can be adjusted by placing a coefficient of 2 in front of H_2O. The equation is now balanced.

$$2NaOH(aq) + H_2SO_4(aq) \rightarrow Na_2SO_4(aq) + 2H_2O(aq)$$

2 Na 6 O 4 H 1 S — 2 Na 6 O 4 H 1 S

d. This equation is balanced. A coefficient of 1 (not written) in front of each reactant and product is sufficient here.

$$CuSO_4(aq) + Fe(s) \rightarrow FeSO_4(aq) + Cu(s)$$

1 Cu 1 S 4 O 1 Fe — 1 Cu 1 S 4 O 1 Fe

e. Begin by adjusting the ammonium ion and the nitrate ion by putting a coefficient of 2 in front of $NH_4NO_3(aq)$. The equation is now balanced.

$$(NH_4)_2S(aq) + Pb(NO_3)_2(aq) \rightarrow PbS(s) + 2NH_4NO_3(aq)$$

4 N 8 H 1 S 1 Pb 6 O — 4 N 8 H 1 S 1 Pb 6 O

f. The number of aluminum atoms and oxygen atoms on the left are less than the right. Begin by increasing the coefficient of aluminum to 2. Aluminum is now balanced on both sides.

$$2Al(s) + O_2(g) \rightarrow Al_2O_3(s)$$

2 Al 2 O — 2 Al 3 O

Aluminum is balanced, but oxygen is not. There are 3 oxygen atoms on the right, and 2 on the left. We cannot use a coefficient on the left to produce 3 oxygen atoms so that the left and right balance. But, we can put a coefficient of 2 in front of Al_2O_3 to make 6 oxygen atoms. A coefficient of 3 on the left adjusts the oxygen atoms on the left to 6.

$$2Al(s) + 3O_2(g) \rightarrow 2Al_2O_3(s)$$

2 Al 6 O — 4 Al 6 O

There are now 4 aluminum atoms on the right, and 2 on the left, so increase the coefficient of aluminum on the left to 4. The equation is now balanced.

$$4Al(s) + 3O_2(g) \rightarrow 2Al_2O_3(s)$$

4 Al 6 O — 4 Al 6 O

6. a. $3PbO(s) + 2NH_3(aq) \rightarrow 3Pb(s) + N_2(g) + 3H_2O(l)$
   b. $2SO_2(g) + O_2(g) \rightarrow 2SO_3(g)$
   c. $2C_4H_{10}(g) + 13O_2(g) \rightarrow 8CO_2(g) + 10H_2O(g)$
   d. $2Fe_2O_3(s) + 3C(s) \rightarrow 4Fe(s) + 3CO_2(g)$
   e. $TiCl_4(l) + 2H_2O(l) \rightarrow TiO_2(s) + 4HCl(aq)$
   f. $CaC_2(s) + 2H_2O(l) \rightarrow C_2H_2(g) + Ca(OH)_2(aq)$

7. a. $2KI(aq) + Br_2(l) \rightarrow 2KBr(aq) + I_2(s)$
   b. $2Al(s) + 3Cl_2(g) \rightarrow 2AlCl_3(s)$
   c. $2PbO_2(s) \rightarrow 2PbO(s) + O_2(g)$
   d. $2Fe(OH)_3(aq) + 3H_2SO_4(aq) \rightarrow Fe_2(SO_4)_3(s) + 6H_2O(l)$
   e. $2K_3PO_4(aq) + 3BaCl_2(aq) \rightarrow 6KCl(aq) + Ba_3(PO_4)_2(s)$
   f. $Ba(ClO_3)_2(s) \rightarrow BaCl_2(s) + 3O_2(g)$

PRACTICE EXAM

1. Which of the following is not evidence of a chemical reaction?

   a. A bowl of soup is warmed from room temperature to the boiling point of water.
   b. A yellow solid forms when two clear yellow liquids are mixed.
   c. After dropping a strip of metal into a colorless liquid, bubbles form on the surface of the metal.
   d. When two solutions are mixed, the temperature of the mixture drops.
   e. When a white solid is dissolved in a colorless liquid, the solution turns blue.

2. Which of the statements about chemical equations is true?

   a. The arrow means "reacts to form".
   b. The products are always shown on the left side.
   c. Often there are atoms present in the products not present in the reactants.
   d. When a reaction occurs, atoms do not recombine with other elements.
   e. The physical state of an element cannot change during a chemical reaction.

3. Which of the following statements about the physical states of reactants and products is not true?

   a. $BaCrO_2(s)$ on the product side indicates the formation of a solid.
   b. $HCl(aq)$ on the reactant side indicates that there is water present.
   c. $Br_2(l)$ indicates that the bromine is dissolved in liquid water.
   d. $CO_2(g)$ on the product side indicates that at least one product is a gas.
   e. Both products and reactants can be present as solids, liquids or gases.

4. Solid calcium carbide, CaC_2, will react with water to produce acetylene gas, C_2H_2, and aqueous calcium hydroxide. What is the correct equation for this reaction?

   a. $Ca(OH)_2(aq) + C_2H_2(g) \rightarrow CaC_2(s) + 2H_2O(l)$
   b. $CaC_2(s) + 2H_2O(g) \rightarrow C_2H_2(g) + Ca(OH)_2(g)$
   c. $CaC_2(s) + 2H_2O(l) \rightarrow C_2H_2(g) + Ca(OH)_2(aq)$
   d. $CaC_2(s) + 2H_2O(aq) \rightarrow C_2H_2(l) + Ca(OH)_2(aq)$
   e. $CaC_2(s) + 2H_2O(l) \rightarrow C_2H_2(s) + Ca(OH)_2(l)$

5. Solid lead(II) oxide reacts with ammonia gas to form solid lead, nitrogen gas and liquid water. What is the correct balanced equation for this reaction?

a. $3PbO_2(s) + 2NH_3(g) \rightarrow 3Pb(s) + N_2(g) + 3_2O(l)$
b. $3PbO(s) + 2NH_4^+(g) \rightarrow 3Pb(s) + N(g) + 3H_2O(l)$
c. $3PbO(s) + 2NH_3(g) \rightarrow 3Pb_2(s) + N_2(g) + 3H_2O(g)$
d. $3PbO(s) + 2NH_3(g) \rightarrow 3Pb(s) + N_2(g) + 3H_2O(l)$
e. $3PbO_2(s) + 2NH_3(l) \rightarrow 3Pb(g) + N_2(l) + 3H_2O(s)$

6. Balance the equation below. What is the sum of the coefficients for the balanced equation?

$$HNO_2 \rightarrow NO + HNO_3 + H_2O$$

a. 4
b. 5
c. 6
d. 7
e. 9

7. What is the correct balanced equation for the reaction of solid phosphorus with solid iodine to produce liquid phosphorus triiodide?

a. $P(s) + 2I_2(s) \rightarrow PI_3(s)$
b. $P_4(s) + 6I_2(s) \rightarrow 4PI_3(l)$
c. $2P(s) + 3I_2(s) \rightarrow P_2I_3(l)$
d. $P_4(s) + 3I_2(s) \rightarrow 2P_2I_3(l)$
e. $3P(s) + I_2(g) \rightarrow P_3I_2(l)$

8. Which equation is **not** balanced?

a. $2MgO \rightarrow 2Mg + O_2$
b. $4Al + 3O_2 \rightarrow 2Al_2O_3$
c. $2Na + H_2O \rightarrow 2NaOH + H_2$
d. $Mg(OH)_2 + 2HCl \rightarrow MgCl_2 + 2H_2O$
e. $2SO_2 + O_2 \rightarrow 2SO_3$

9. Which is the correct balanced equation for the reaction below?

$$K_3PO_4 + BaCl_2 \rightarrow KCl + Ba_3(PO_4)_2$$

a. $K_3PO_4 + 3BaCl_2 \rightarrow 3KCl + Ba_3(PO_4)_2$
b. $4K_3PO_4 + 6BaCl_2 \rightarrow 12KCl + 2Ba_3(PO_4)_2$
c. $K_3PO_4 + BaCl_2 \rightarrow KCl + Ba_3(PO_4)_2$
d. $2K_3PO_4 + 3BaCl_2 \rightarrow 6KCl + Ba_3(PO_4)_2$
e. $2K_3PO_4 + 3BaCl_2 \rightarrow 2KCl + Ba_3(PO_4)_2$

10. Which is the correct balanced equation for the reaction below?

$$C_6H_{14}(l) + O_2(g) \rightarrow CO_2(g) + H_2O(l)$$

a. $C_6H_{14}(l) + 9O_2(g) \rightarrow 6CO_2(g) + 7H_2O(l)$
b. $C_6H_{14}(l) + 9.5O_2(g) \rightarrow 6CO_2(g) + 7H_2O(l)$
c. $2C_6H_{14}(l) + 19O_2(g) \rightarrow 12CO_2(g) + 14H_2O(l)$
d. $C_6H_{14}(l) + 3.5O_2(g) \rightarrow 3CO_2(g) + 7H_2O(l)$
e. $C_6H_{14}(l) + 9O_2(g) \rightarrow 6CO_2(g) + 6H_2O(l)$

PRACTICE EXAM ANSWERS

1. a (6.1)
2. a (6.2)
3. c (6.2)
4. c (6.3)
5. d (6.3)
6. d (6.3)
7. b (6.3)
8. c (6.3)
9. d (6.3)
10. c (6.3)

CHAPTER 7: REACTIONS IN AQUEOUS SOLUTIONS

INTRODUCTION

Water is a good solvent in the chemical laboratory because many compounds are soluble in water. There are many chemical reactions which occur in water and you will study some of them in this chapter. To predict what will happen when two compounds which are dissolved in water are mixed, you will need to learn some rules. These rules will help you predict what kinds of compounds are soluble in water, and what kinds of products you can make.

Water is also an important part of our environment, and many common reactions take place in water. For example, the rusting of iron takes place in water. For these reasons, a whole chapter is devoted to what happens when substances are added to water.

AIMS FOR THIS CHAPTER

1. Know the four common forces which will often cause a reaction to happen. (Section 7.1)
2. Know what happens when an ionic compound is dissolved in water. (Section 7.2)
3. When two water-soluble ionic compounds are mixed, be able to decide what products will form. (Section 7.2)
4. Learn the rules which help predict which ionic salts are soluble in water, and which are not. (Section 7.2)
5. Be able to write the molecular equation, the complete ionic equation, and the net ionic equation for the reactions between ionic compounds dissolved in water. (Section 7.3)
6. Be able to recognize reactions between strong acids which contain H^+ ion and strong bases which contain OH^- ion and which always produce water as a product. (Section 7.4)

QUICK DEFINITIONS

Precipitate — The solid that forms during some chemical reactions, called precipitation reactions. (Section 7.2)

Strong electrolyte — A substance which breaks apart completely into ions when it is dissolved in water. (Section 7.2)

Soluble — A property of substances which dissolve completely in water. (Section 7.2)

Insoluble — A property of substances which do not dissolve in water, or so little of which dissolves that the substance does not seem to disappear. (Section 7.2)

Slightly soluble	The same as insoluble. (Section 7.2)
Molecular equation	A chemical equation which shows the reactants and products as complete molecules. (Section 7.3)
Complete ionic equation	A chemical equation which shows the reactants and products which are strong electrolytes as though they were ions in solution. (Section 7.3)
Net ionic equation	A chemical equation which shows only ions which react to form products, and the products which are formed. (Section 7.3)
Spectator ions	Ions which are present during a chemical reaction, but which do not participate in the reaction. (Section 7.3)
Acid	Defined by Arrhenius as a substance which produces hydrogen ions when dissolved in water. (Section 7.4)
Strong acids	Acids which completely break apart into ions when dissolved in water. (Section 7.4)
Base	Defined by Arrhenius as a substance which produces hydroxide ions when dissolved in water. (Section 7.4)
Strong bases	Bases which completely break apart into ions when dissolved in water. (Section 7.4)
Salt	An ionic compound which dissolves in water. (Section 7.4)

CONTENT REVIEW

7.1 PREDICTING WHETHER A REACTION WILL OCCUR

When You Mix Two Substances Together, How Can You Tell Whether They Will React?

When you mix two substances together, if any one of four common results or processes can happen, then a reaction will probably occur. The four results which indicate that a reaction will take place are the formation of a solid, the formation of water, the formation of a gas, and the transfer of electrons. For example, when $AgNO_3$ is mixed with Na_2SO_4, solid Ag_2SO_4 could form. A reaction is likely to occur.

7.2 REACTIONS IN WHICH A SOLID FORMS

How Can You Tell Whether a Solid Will Form When Two Solutions Are Mixed?

When we dissolve a water-soluble ionic compound in water, the compound breaks apart into individual ions. The ions are free to move around in the water solution. If you mix together two solutions, each containing a different ionic compound, the two sets of ions can mix freely with one another.

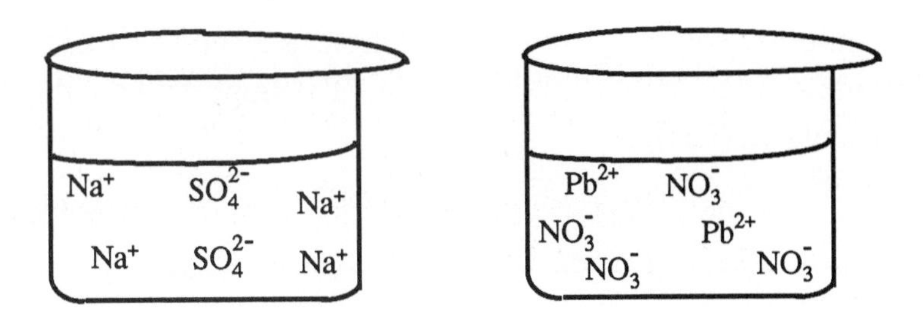

Before the solutions are mixed, Na^+ ions only came into contact with SO_4^{2-} ions. Pb^{2+} ions only came into contact with NO_3^- ions. After the solutions were mixed, each kind of ion can come in contact with three different kinds of ions. When these two solutions containing different ions are mixed together some of the new ion combinations react to form a white solid.

$$Na^+ + SO_4^{2-} + Pb^{2+} + NO_3^- \rightarrow \text{white solid}$$

What combinations of ions are possible? Na^+ ions could combine with SO_4^{2-} ions to form Na_2SO_4. Pb^{2+} ions could combine with NO_3^- ions to form $Pb(NO_3)_2$. Two other combinations are possible. Na^+ ions could combine with NO_3^- ions to form $NaNO_3$ and Pb^{2+} ions could combine with SO_4^{2-} ions to form $PbSO_4$. We already know that Na_2SO_4 and $Pb(NO_3)_2$ do not form solids in water, because our original solutions contained these ions, and neither of them produced a solid when Na_2SO_4 or $Pb(NO_3)_2$ were dissolved in water. If a solid forms, it must be made of Na^+ and NO_3^- ions or of Pb^{2+} and SO_4^{2-} ions.

How do we know which combination of ions results in the white solid? To determine this, you must learn the rules of solubility presented in Table 7.1 of your text and reproduced here.

Simple Rules for Solubility of Ionic Compounds (Salts) in Water
1. Most nitrate (NO_3^-) salts are soluble.
2. Most salts of Na^+, K^+, and NH_4^+ are soluble.
3. Most chloride salts are soluble. Notable exceptions are AgCl, $PbCl_2$, and Hg_2Cl_2.
4. Most sulfate salts are soluble. Notable exceptions are $BaSO_4$, $PbSO_4$, and $CaSO_4$.
5. Most hydroxide salts are only slightly soluble. The important soluble hydroxides are NaOH, KOH, and $Ca(OH)_2$.
6. Most sulfide (S^{2-}), carbonate (CO_3^{2-}), and phosphate (PO_4^{3-}) salts are only slightly soluble.

Using rules 1 and 2 of this table, $NaNO_3$ will be soluble in water. So, the white solid cannot be $NaNO_3$. Rule 4 says that most sulfate salts are soluble, with the exception of $PbSO_4$. The white solid must be $PbSO_4$.

What is the Formula of the Solid?

The following steps will help you determine the formula of the solid which forms during a chemical reaction. First, write all possible formulas for ion pairs, remembering that an anion must pair with a cation. Second, eliminate those ion pair combinations which were present together in your original solutions. These ion pairs are soluble in water; they do not form solids. Third, use the solubility rules to eliminate the remaining ion pairs, and thus identify the solid.

Example:

> Aqueous solutions of $CaCl_2$ and Na_2CO_3 are mixed. A white solid forms. What is the insoluble product of the reaction? The possible ion pairs are Ca^{2+} + $2Cl^-$, Ca^{2+} + CO_3^{2-}, Ca^{2+} + $2Na^+$, Na^+ + Cl^-, $2Na^+$ + CO_3^{2-} and Ca^{2+} + $2Na^+$. We know that a cation must pair with an anion, so we can eliminate two of the pairs. The ones which remain are Ca^{2+} + $2Cl^-$, Ca^{2+} + CO_3^{2-}, Na^+ + Cl^-, and $2Na^+$ + CO_3^-. Two of these ion pairs were present in the original aqueous solutions, so they cannot be solid products. The ion pairs which remain are Ca^{2+} + CO_3^{2-} and Na^+ + Cl^-. The solubility rules tell us that NaCl will be soluble in water (rules 1 and 3) and that $CaCO_3$ will not be soluble in water (rule 6). The solid product is $CaCO_3$.

7.3 DESCRIBING REACTIONS IN AQUEOUS SOLUTIONS

What Are the Different Ways You Can Write Chemical Equations?

You can write the molecular formula for each reactant compound, and write molecular formulas for products, ignoring the fact that the reactants consist of individual ions. We can also ignore the fact that the products will include some individual ions, and a new product. Use the symbol (aq) after each molecular formula to indicate that each substance is present in aqueous solution. The symbol (s) tells you which of the molecular formulas is the solid product. This way of writing chemical equations is called the molecular equation. An example is

$$Na_3PO_4(aq) + AgNO_3(aq) \rightarrow NaNO_3(aq) + Ag_3PO_4(s)$$

Molecular formulas are not very accurate portrayals of reactant solutions because soluble ionic compounds actually exist as individual ions when they are in aqueous solution. Molecular formulas do not present an accurate portrayal of the product side of the equation either, because the products contain some individual soluble ions.

Another way to write chemical equations is to write each ion in the reactants separately, and each ion in the product separately. Insoluble ionic products are shown with their molecular formulas. Use the symbol (aq) for aqueous ions, and (s) to indicate the solid product. For example,

$$Na^+(aq) + PO_4^{3-}(aq) + 3Ag^+(aq) + NO_3^-(aq) \rightarrow Na^+(aq) + NO_3^-(aq) + Ag_3PO_4(s)$$

This way of writing chemical equations is called the complete ionic equation. Complete ionic equations are accurate representations of what kinds of particles are present in solution before and after the reaction.

Another way to write chemical equations is to show on the reactant side only those ions which will form a product when mixed together, and on the product side only the molecular formula of the product or products. All the other ions are left out of the equation. This is called the net ionic equation. For example

$$3Ag^+(aq) + PO_4^{3-}(aq) \rightarrow Ag_3PO_4(s)$$

The net ionic equation is useful because it allows you to see easily what is reacting and what is produced. But don't forget that the reactant solution couldn't contain just Ag^+ ions, or just PO_4^{-3} ions. Each reactant solution must contain both cations and anions. Any time there is a cation present, an anion must accompany it; and any time there is an anion, there must also be a cation.

7.4 REACTIONS THAT FORM WATER: ACIDS AND BASES

What Kinds of Ionic Compounds Form Water When Mixed Together?

Some ionic compounds, when dissolved in water, break completely apart into ions where the cation is always the hydrogen ion (H^+). The anion can be one of several. These ionic compounds are the strong acids. They are called acids because one definition of an acid is a substance which produces hydrogen ions. They are called strong because they completely break apart into hydrogen ions, and an anion. We will discuss the properties and definitions of acids in more detail in Chapter 17. For now, you should learn the formulas of the most common strong acids, hydrochloric acid (HCl), sulfuric acid (H_2SO_4) and nitric acid (HNO_3).

Another kind of ionic compound, when dissolved in water, completely breaks apart into a cation and the hydroxide ion (OH^-). The cation can be one of several. These ionic compounds are the strong bases. They are called bases because one definition of a base is a substance which produces hydroxide ions. They are strong bases because they break completely apart into cations and hydroxides when they are dissolved in water. The most common strong bases are NaOH and KOH. More information about the strong bases can be found in Chapter 17.

A solution containing the strong acid HNO_3 dissolved in water contains H^+ ions and NO_3^- ions. Another solution containing the strong base NaOH dissolved in water contains Na^+ ions and OH^- ions. What happens when the two solutions are mixed together? New combinations of ions will come in contact with each other and mix. What are all the possible ion combinations? Keep in mind that an anion must pair with a cation. H^+ ions can pair with NO_3^- ions, and Na^+ ions can pair with OH^- ions. These are the ion pairs present in our original solutions. They are completely soluble in water, and do not form any new products. The other possible ion pairs which can form when the ions mix in the new solution are $NaNO_3$ and HOH (H_2O). Na^+ and NO_3^- ions mix in solution, but don't form a solid product. They form a soluble salt. According to the rules of solubility in Table 7.1, compounds composed of Na^+ ions and NO_3^- ions are usually soluble in water. But when H^+ ions and OH^- ions mix together, a new product is formed, water. A reaction has occured between the H^+ ion and the OH^- ions. There are no longer H^+ ions and OH^- ions floating in solution; instead, the product water has formed.

All of the strong bases contain OH^- ions. All of the strong acids contain H^+ ions. Whenever strong acids and strong bases are mixed together, the product is always water. The cation which accompanied the OH^- ion and the anion which accompanied the H^+ ion remain in solution as a soluble salt. Water and a salt are always the products when a strong acid and a strong base are mixed together.

LEARNING REVIEW

1. Which one of the following does **not** tend to drive a reaction to produce products?

    a. formation of a gas
    b. transfer of electrons
    c. color change
    d. formation of water
    e. formation of a solid

2. Write the formulas for the ions which are formed when these ionic compounds are dissolved in water. How many of each kind of ion are produced for each molecule dissolved?

a. $(NH_4)_2SO_4$
b. KNO_3
c. $Na_2Cr_2O_7$
d. $MgCl_2$
e. Li_3PO_4
f. $Al(NO_3)_3$

3. When predicting a product for the reaction between two ionic compounds, we can always eliminate some of the ion pairs as possible products. Give a reason for eliminating each of the pairs as a product of the reaction below.

$$AgNO_3 + Na_3PO_4 \longrightarrow$$

a. Ag^+, Na^+
b. Na^+, NO_3^-
c. NO_3^-, PO_3^{3-}

4. Use the solubility rules to predict the water solubility of each of the following compounds.

a. Na_2S
b. $PbCl_2$
c. K_2SO_4
d. $(NH_4)_2CrO_4$
e. $Pb(OH)_2$
f. $Ca(NO_3)_2$
g. $Ba_3(PO_4)_2$
h. $ZnCl_2$

5. For each word description, write the balanced molecular equation, and identify the product of the reaction.

a. Aqueous solutions of sodium sulfate and lead(II) nitrate are mixed. One of the products is a white solid.
b. Aqueous solutions of potassium hydroxide and nickel(II) chloride are mixed. One of the products is a green solid.
c. Aqueous solutions of potassium sulfide and zinc nitrate are mixed. A pale yellow solid is produced.
d. Aqueous solutions of silver nitrate and ammonium phosphate are mixed. A white solid is produced.

6. For each of the balanced equations below, write the complete ionic equation.

a. $3CaCl_2(aq) + 2Na_3PO_4(aq) \longrightarrow Ca_3(PO_4)_2(s) + 6NaCl(aq)$
b. $Cu(NO_3)_2(aq) + K_2S(aq) \longrightarrow CuS(s) + 2KNO_3(aq)$

c. $2AgNO_3(aq) + K_2SO_4(aq) \rightarrow Ag_2SO_4(s) + 2KNO_3(aq)$
d. $3NaOH(aq) + FeCl_3(aq) \rightarrow Fe(OH)_3(s) + 3NaCl(aq)$

7. Complete and balance the reactions below and identify the spectator ions.

a. $Ca(NO_3)_2(aq) + K_2SO_4(aq) \rightarrow$
b. $(NH_4)_2CO_3(aq) + CuCl_2(aq) \rightarrow$
c. $NaOH(aq) + Pb(NO_3)_2(aq) \rightarrow$
d. $MgCl_2(aq) + Na_2S(aq) \rightarrow$
e. $Na_2S(aq) + Zn(NO_3)_2(aq) \rightarrow$
f. $CoCl_2(aq) + Ca(OH)_2(aq) \rightarrow$

8. Write the net ionic equation for each reaction.

a. $K_2CO_3(aq) + CaCl_2(aq) \rightarrow CaCO_3(s) + 2KCl(aq)$
b. $3H_2S(aq) + 2AlCl_3(aq) \rightarrow Al_2S_3(s) + 6HCl(aq)$
c. $Na_2SO_4(aq) + Mn(NO_3)_2(s) \rightarrow 2NaNO_3(aq) + MnSO_4(s)$
d. $Pb(NO_3)_2(aq) + (NH_4)_2S(aq) \rightarrow 2NH_4NO_3(aq) + PbS(s)$
e. $2LiCl(aq) + 2HgNO_3(aq) \rightarrow Hg_2Cl_2(s) + 2LiNO_3(aq)$
f. $2NaOH(aq) + MgCl_2(aq) \rightarrow Mg(OH)_2(s) + 2NaCl(aq)$

9. What salts in aqueous solutions could you mix together to produce the solids below?

a. $Zn(OH)_2$
b. $Ba_3(PO_4)_2$
c. $PbCl_2$
d. $CaSO_4$
e. $CoCO_3$
f. Ag_2SO_4

10. Which of the substances below are strong acids, which are strong bases, and which are neither of these?

a. HNO_3
b. $C_2H_4O_2$
c. H_2SO_4
d. HCl
e. $NaCl$
f. K_2SO_4

11. Write complete ionic equations for the reactions below.

a. Sodium hydroxide reacts with sulfuric acid.
b. Hydrochloric acid reacts with potassium hydroxide.
c. Nitric acid reacts with sodium hydroxide.

12. Write net ionic equations for each reaction in Problem 11.

13. Which of the reactions below are acid/base reactions?

a. $K_2SO_4(aq) + Pb(NO_3)(aq) \rightarrow PbSO_4(s) + 2KNO_3(aq)$
b. $KOH(aq) + HNO_3(aq) \rightarrow KNO_3(aq) + H_2O(l)$
c. $H_2SO_4(aq) + 2NaOH(aq) \rightarrow Na_2SO_4(aq) + 2H_2O(l)$
d. $Na_2CO_3(aq) + CoCl_2(aq) \rightarrow CoCO_3(s) + 2NaCl(aq)$

ANSWERS TO LEARNING REVIEW

1. Only **c**, color change, does not tend to make a reaction occur.

2.
a. 2 NH_4^+ 1 SO_4^{2-}
b. 1 K^+ 1 NO_3^-
c. 2 Na^+ 1 $Cr_2O_7^{2-}$
d. 1 Mg^{2+} 2 Cl^-
e. 3 Li^+ 1 PO_4^{3-}
f. 1 Al^{3+} 3 NO_3^-

3.
a. Both Ag^+ and Na^+ are cations. An anion and a cation are needed to form a neutral product.
b. $NaNO_3$ is soluble in water, and so it exists in solution as Na^+ ions and NO_3^- ions.
c. Both NO_3^- and PO_4^{3-} are anions. An anion and a cation are needed to form a neutral product.

4.
a. water soluble (rule 2)
b. not soluble (rule 3)
c. water soluble (rule 2)
d. water soluble (rule 2)
e. not soluble (rule 5)
f. water soluble (rule 1)
g. not soluble (rule 6)
h. water soluble (rule 3)

5.
a. $Na_2SO_4(aq) + Pb(NO_3)_2(aq) \rightarrow PbSO_4(s) + 2NaNO_3(aq)$
The product is lead(II) sulfate.

b. $NiCl_2(aq) + 2KOH(aq) \rightarrow Ni(OH)_2(s) + 2KCl(aq)$
The product is nickel(II) hydroxide.

c. $K_2S(aq) + Zn(NO_3)_2(aq) \rightarrow ZnS(s) + 2KNO_3(aq)$
The product is zinc sulfide.

d. $3AgNO_3(aq) + (NH_4)_3PO_4(aq) \rightarrow Ag_3PO_4(s) + 3NH_4NO_3(aq)$
The product is silver phosphate.

6.
a. $3Ca^{2+} + 6Cl^- + 6Na^+ + 2PO_4^{3-} \rightarrow Ca_3(PO_4)_2(s) + 6Na^+ + 6Cl^-$
b. $Cu^{2+} + 2NO_3^- + 2K^+ + S^{2-} \rightarrow CuS_{(s)} + 2K^+ + 2NO_3^-$

c. $2Ag^{+} + 2NO_3^{-} + 2K^{+} + SO_4^{2-} \rightarrow Ag_2SO_{4(s)} + 2K^{+} + 2NO_3^{-}$

d. $3Na^{+} + 3OH^{-} + Fe^{3+} + 3Cl^{-} \rightarrow Fe(OH)_3 + 3Na^{+} + 3Cl^{-}$

7. a. $Ca(NO_3)_2(aq) + K_2SO_4(aq) \rightarrow CaSO_4(s) + 2KNO_3(aq)$
   K^{+} and NO_3^{-} are the spectator ions

   b. $(NH_4)_2CO_3(aq) + CuCl_2(aq) \rightarrow CuCO_3(s) + 2NH_4Cl(aq)$
   NH_4^{+} and Cl^{-} are the spectator ions

   c. $2NaOH(aq) + Pb(NO_3)_2(aq) \rightarrow Pb(OH)_2(s) + 2NaNO_3(aq)$
   Na^{+} and NO_3^{-} are the spectator ions

   d. $MgCl_2(aq) + Na_2S(aq) \rightarrow MgS(s) + 2NaCl(aq)$
   Na^{+} and Cl^{-} are the spectator ions

   e. $Na_2S(aq) + Zn(NO_3)_2(aq) \rightarrow ZnS(s) + 2NaNO_3(aq)$
   Na^{+} and NO_3^{-} are the spectator ions

   f. $CoCl_{2(aq)} + Ca(OH)_{2(aq)} \rightarrow Co(OH)_{2(s)} + CaCl_{2(aq)}$
   Ca^{2+} and Cl^{-} are the spectator ions

8. a. $Ca_2^{+} + CO_3^{2-} \rightarrow CaCO_3(s)$
   b. $2Al^{3+} + 3S^{2-} \rightarrow Al_2S_3(s)$
   c. $Mn^{2+} + SO_4^{2-} \rightarrow MnSO_4(s)$
   d. $Pb^{2+} + S^{2-} \rightarrow PbS(s)$
   e. $2Hg^{+} + 2Cl^{-} \rightarrow Hg_2Cl_2(s)$
   f. $Mg^{2+} + 2OH^{-} \rightarrow Mg(OH)_2(s)$

9. It is possible to produce the solids below from several different soluble salts, so your answer could be correct and not the same as the answer below. If your answer does not match the one below, use the solubility rules to help you determine whether the aqueous salt solutions you chose would be soluble in water, and whether an exchange of anions would produce the desired insoluble salt.

   a. $Zn(NO_3)_2$ and $NaOH$
   b. $Ba(NO_3)_2$ and K_3PO_4
   c. $Pb(NO_3)_2$ and $NaCl$
   d. $CaCl_2$ and $(NH_4)_2SO_4$
   e. $Co(NO_3)_2$ and Na_2CO_3
   f. $AgNO_3$ and K_2SO_4

10. a. HNO_3 is a strong acid.
 b. $C_2H_4O_2$ is a weak acid, so the correct answer is neither of these.
 c. H_2SO_4 is a strong acid.
 d. HCl is a strong acid.
 e. NaCl is a salt produced when HCl and NaOH react, so it is neither a strong acid nor a strong base.
 f. K_2SO_4 is a salt produced when H_2SO_4 and KOH react.

11. a. $2Na^+ + 2OH^- + 2H^+ + SO_4^{2-} \rightarrow 2Na^+ + SO_4^{2-} + 2H_2O(l)$
 b. $K^+ + OH^- + H^+ + Cl^- \rightarrow K^+ + Cl^- + H_2O(l)$
 c. $Na^+ + OH^- + H^+ + NO_3^- \rightarrow Na^+ + NO_3^- + H_2O(l)$

12. a. $H^+ + OH^- \rightarrow H_2O(l)$
 b. $H^+ + OH^- \rightarrow H_2O(l)$
 c. $H^+ + OH^- \rightarrow H_2O(l)$

13. a. This is a precipitation reaction.
 b. This is an acid base reaction. The products are water and the salt KNO_3.
 c. This is an acid base reaction. The products are water and the salt Na_2SO_4.
 d. This is a precipitation reaction.

PRACTICE EXAM

1. When the materials below are mixed together, in which pair would a reaction **not** occur?

 a. $HNO_3 + NaOH \rightarrow$
 b. $Na_2SO_4 + CaCl_2 \rightarrow$
 c. $Pb(NO_3)_2 + NH_4Cl \rightarrow$
 d. $HCl + KOH \rightarrow$
 e. $K_2SO_4 + NaNO_3 \rightarrow$

2. Which of the following is a strong electrolyte?

 a. $Zn(OH)_2$
 b. H_2O
 c. $Ca(NO_3)_2$
 d. $PbCl_2$
 e. $Ba_3(PO_4)_2$

3. Which pair of ions will react to form a solid product when aqueous solutions of copper(II) sulfate and sodium nitrate are mixed?

$$CuSO_4 + NaNO_3 \rightarrow$$

a. SO_4^{2-} and NO_3^-
b. Cu^{2+} and $2NO_3^-$
c. $2Na^+$ and SO_4^{2-}
d. Cu^{2+} and $2Na^+$
e. No product would form when solutions containing these ions are mixed.

4. Which of the ionic compounds below is soluble in water?

a. $CuCO_3$
b. Na_2CrO_4
c. $Pb(OH)_2$
d. $CaSO_4$
e. HgS

5. Which equation correctly represents the reaction between aqueous solutions of copper(I) nitrate and potassium carbonate?

a. $2CuNO_3(aq) + K_2CO_3(aq) \rightarrow Cu_2CO_3(s) + 2KNO_3(aq)$
b. $Cu(NO_3)_2(aq) + K_2CO(aq) \rightarrow CuCO(s) + 2KNO_3(s)$
c. $Cu(NO_3)_2(aq) + K_2CO_3(aq) \rightarrow CuCO_3(s) + 2KNO_3(aq)$
d. $2CuNO_3(aq) + K_2CO(aq) \rightarrow Cu_2CO(s) + 2KNO_3(aq)$
e. $2CuNO_3(aq) + K_2CO_3(aq) \rightarrow CuCO_3(aq) + 2KNO_3(s)$

6. Which is the correct complete ionic equation for the reaction between aqueous solutions of lead(IV) nitrate and sodium hydroxide?

a. $Pb^{4+} + 4NO_3^- + 4Na^+ + 4OH^- \rightarrow 4NaNO_3(s) + Pb^{4+} + 4OH^-$
b. $Pb^{4+} + 4NO_3^- + 4Na^+ + 4OH^- \rightarrow Pb(OH)_4(s) + 4Na^+ + 4NO_3^-$
c. $Pb(NO_3)_4(aq) + 4Na^+ + 4OH^- \rightarrow Pb(OH)_4(s) + 4NaNO_3(aq)$
d. $Pb(NO_3)_4(aq) + 4Na^+ + 4OH^- \rightarrow Pb^{4+} + 4OH^- + 4Na^+ + 4NO_3^-$
e. $Pb^{4+} + 4NO_3^- + 4Na^+ + 4OH^- \rightarrow Pb^{4+} + 4NO_3^- + 4NaOH(s)$

7. Which is the correct net ionic equation for the reaction between aqueous solutions of nickel(II) chloride and ammonium phosphate?

a. $3Ni^{2+} + 2PO_4^{3-} \rightarrow Ni_3(PO_4)_2(s)$

b. $3NH_4^+ + PO_4^{3-} \rightarrow (NH_4)_3PO_4(s)$
c. $Ni^{2+} + 2NH_4^+ \rightarrow Ni^{2+} + 2NH_4^+$
d. $Ni^{2+} + 2Cl^- \rightarrow NiCl_2(s)$
e. $NH_4^+ + Cl^- \rightarrow NH_4Cl(s)$

8. In the reaction below, which are the spectator ions?

$$K_2SO_4 + Ba(NO_3)_2 \rightarrow$$

a. K^+, NO_3^-, SO_4^{2-}
b. Ba^{2+}, NO_3^-
c. K^+, NO_3^-
d. Ba^{2+}, K^+, SO_4^{2-}
e. SO_4^{2-}, NO_3^-

9. Which is neither a strong acid nor a strong base?

a. HNO_3
b. KOH
c. H_2SO_4
d. HCl
e. KCl

10. Which represents the correct net ionic equation for the reaction between a strong acid and a strong base?

a. $HCl(aq) + NaOH(aq) \rightarrow NaCl(aq) + H_2O(l)$
b. $HCl(aq) + NaOH(aq) \rightarrow Na^+ + OH^- + H_2O(l)$
c. $H^+ + OH^- \rightarrow H_2O(l)$
d. $H^+ + Cl^- \rightarrow HCl(aq)$
e. $Na^+ + Cl^- \rightarrow NaCl(aq)$

PRACTICE EXAM ANSWERS

1. e (7.2)
2. c (7.2)
3. e (7.2)
4. b (7.2)
5. a (7.2)
6. b (7.3)
7. a (7.3)

8. c (7.3)
9. e (7.4)
10. c (7.4)

CHAPTER 8: CLASSIFYING CHEMICAL REACTIONS

INTRODUCTION

There are many different chemical reactions. When you first look at the equation for a reaction, it often looks completely new and unfamiliar. After you learn the material in this chapter, you will be able to classify many of the new reactions you come across into one or more basic categories. Categorizing a reaction is important. By fitting a reaction into a specific familiar category, you automatically know some things about that reaction, even if you have never seen that specific reaction before.

AIMS FOR THIS CHAPTER

1. Be able to recognize chemical reactions where the reactants transfer electrons to produce products. (Section 8.1)
2. Be able to show which reactant is losing electrons and which is gaining electrons, and how many electrons are transferred. (Section 8.1)
3. Be able to recognize electron transfer reactions between nonmetals by the presence of oxygen gas as a reactant or as a product. (Section 8.1)
4. Be able to classify reactions into one or more categories. (Sections 8.2 and 8.3)

QUICK DEFINITIONS

Oxidation-reduction reaction — A reaction which involves the transfer of electrons from one substance to another. (Section 8.1)

Precipitation reaction — A reaction where a solid product is produced when two anions in solution are exchanged. (Section 8.2)

Double displacement reaction — The same thing as a precipitation reaction. (Section 8.2)

Acid-base reaction — The reaction between a strong acid and a strong base. (Section 8.2)

Single replacement reaction — An anion is exchanged between one of the reactants and another reactant. (Section 8.2)

Combustion reaction — A reaction which includes oxygen gas as a reactant and produces heat energy. A combustion reaction is a type of oxidation-reduction reaction. (Section 8.3)

Synthesis reaction — A reaction in which the reactants are simpler than the product. This type of reaction is also called a combination reaction. Synthesis reactions are a type of oxidation-reduction reaction. (Section 8.3)

Decomposition reaction — A reactant compound is broken down into elements or smaller compounds. A decomposition reaction is a type of oxidation-reduction reaction. (Section 8.3)

CONTENT REVIEW

8.1 REACTIONS OF METALS WITH NONMETALS

What Kind of a Reaction Takes Place Between Metals and Nonmetals?

In Chapter 7 we saw what happens when solutions containing two ionic compounds react. Now we want to see what happens when a metal and a nonmetal react.

When a metal and a nonmetal react, the product is often an ionic compound. For example, when solid potassium metal reacts with liquid bromine, the product is the ionic compound, potassium bromide.

$$2K(s) + Br_2(l) \longrightarrow 2KBr(s)$$

By ionic, we mean that the product potassium bromide is actually composed of potassium ions (K^+) and bromide ions (Br^-), although we write the formula as KBr. The reactant potassium atoms and the bromine molecule are both electrically neutral. The number of protons equals the number of electrons. How were the neutral species potassium and bromine converted to ions? An electron is transferred from two potassium atoms to each atom in the bromine molecule. Let's look closer at this transfer.

The potassium atom has nineteen protons and nineteen electrons. The number of positive charges is counterbalanced by an equal number of negative charges. Likewise, each bromine atom has thirty-five protons and thirty-five electrons. Bromine atoms are also electrically neutral. When the reaction between two potassium atoms and a bromine molecule occurs, one electron from each potassium atom is transferred to each bromine atom. Potassium atoms have become potassium ions, because the atoms are no longer electrically neutral. Each atom is missing one electron.

$$2K \longrightarrow 2K^+ + 2e^-$$

This means that there is an excess of positive charge of 1^+ for each potassium ion. We write the formula of the ion K^+.

Each bromine atom in the bromine molecule becomes a bromide ion, because the atoms are no longer electrically neutral. Each bromine atom has gained one electron. The bromine ion has a 1^- charge, and is written Br^-.

$$Br_2 + 2e^- \rightarrow 2\,Br^-$$

The reaction between potassium and bromine is an example of an electron transfer reaction. When an electron transfer reaction occurs, one substance always loses one or more electrons, and another substance always gains them. Losing and gaining electrons always happen together in the same reaction. Reactions where electrons are transferred are also called **oxidation-reduction** reactions.

How Can We Tell How Electrons Are Transferred In a More Complex Reaction?

$$Cu(s) + 2AgNO_3(aq) \rightarrow 2Ag(s) + Cu(NO_3)_2(aq)$$

The reaction above is an electron transfer reaction. Let's see how electrons are transferred in this example. Copper can form two different cations, Cu^+ and Cu^{2+}. The product $Cu(NO_3)_2$ contains two NO_3^- anions, each of which has a 1^- charge, for a total negative charge of 2^-. For $Cu(NO_3)_2$ to be neutral the copper cation must be copper(II) or Cu^{2+}. This means that each atom of Cu loses two electrons to form the Cu^{2+} cation.

$$Cu \rightarrow Cu^{2+} + 2e^-$$

If copper is losing electrons, some other substance in this reaction must gain electrons. Before the electron transfer occurs, silver is present as the cation Ag^+. After the transfer, silver is present as neutral silver atoms. The silver cation has gained electrons to become elemental silver.

$$2Ag^+ + 2e^- \rightarrow Ag$$

We can show the overall reaction more clearly by showing only reactants and products which are gaining or losing electrons.

$$Cu(s) + 2Ag^+ + 2e^- \rightarrow Cu^{2+} + 2Ag(s) + 2e^-$$

8.2 WAYS TO CLASSIFY REACTIONS

Being able to classify reactions can help you to understand reactions you have never seen before. When you look at the equation for a new reaction, if you are able to determine what class of reaction it is, all the information you know about that class of reaction is true about the new reaction as well. One way to classify reactions is by the driving forces, the things that happen which tend to make a reaction occur. They are the formation of a solid, the formation of water and the transfer of electrons.

What Kinds of Reactions Cause Formation of a Solid?

When two aqueous solutions are mixed and a solid forms, the reaction is called a **precipitation reaction.** For example,

$$2AgNO_3(aq) + Na_2SO_4(aq) \rightarrow Ag_2SO_4(s) + 2NaNO_3(aq)$$

produces the solid Ag_2SO_4. Ionic solids produced in chemical reactions are called **precipitates.** When precipitation reactions occur, the two anions exchange places. This can be symbolized as

$$AB + CD \rightarrow AD + CB$$

Because of this double exchange, these reactions are also called **double displacement reactions.** To identify this kind of reaction, look for an ionic solid present as a product, but not present as a reactant.

What Kinds of Reactions Cause Production of Water?

So far, you have studied one type of reaction which produces water, the reaction between strong acids and strong bases. For example,

$$NaOH(aq) + HNO_3(aq) \rightarrow NaNO_3(aq) + H_2O(l)$$

When any strong acid reacts with any strong base, the product is always water. These are called **acid-base reactions.** To identify this kind of reaction, look for water as a product, and a strong acid and a strong base as reactants.

What Kinds of Reactions Are Electron Transfer Reactions?

A third driving force we have encountered is electron transfer. An example reaction is

$$2K(s) + Br_2(l) \rightarrow 2KBr(s)$$

where each K loses one electron to become K^+ and each bromine atom gains an electron to become Br^-. Electron transfer reactions are called **oxidation-reduction reactions.** Some electron transfer reactions can be identified by comparing reactants and products. If a reactant is present as a neutral element or as a diatomic molecule, but the product contains ions of that element, then an electron transfer has occurred. Also, if ions are present in the reactants, but the product of that element is present either as the neutral element or as a diatomic molecule, then electron transfer has taken place.

8.3 OTHER WAYS TO CLASSIFY A REACTION

There are many categories of reaction besides precipitation, acid-base, and oxidation-reduction. Fortunately, most of the other categories are special cases of one of the three main types.

Combustion Reactions

Some reactions have oxygen gas as a reactant, and produce heat as a product. These are the **combustion reactions**. An example is

$$C_2H_5OH(l) + 3O_2(g) \rightarrow 2CO_2(g) + 3H_2O(g)$$

the combustion of ethyl alcohol, a fuel additive present in gasohol. This reaction appears to be a new type, one we have not classified before. Many reactions which involve oxygen gas as a reactant are examples of oxidation-reduction reactions, although at this point in the course, it is hard to show why this is true. Combustion reactions are a kind of oxidation-reduction reaction.

Synthesis Reactions

This name is a little misleading because every reaction synthesizes (makes) something. **Synthesis reactions** usually mean making a more complex molecule from simpler ones. It often means making a compound from individual elements. For example,

$$H_2(g) + Cl_2(g) \rightarrow 2HCl(g)$$

molecular hydrogen gas and molecular chlorine gas react to produce the compound hydrogen chloride. If we look closely, hydrogen atoms lose an electron

$$H_2 \rightarrow 2H^+ + 2e^-$$

and chlorine atoms gain an electron.

$$Cl_2 + 2e^- \rightarrow 2Cl^-$$

Synthesis reactions are a type of oxidation-reduction reaction.

Decomposition Reactions

Decomposition reactions usually refer to the breaking down of a compound into simpler (often elemental) components.

$$H_2O_2(l) \rightarrow H_2(g) + O_2(g)$$

Hydrogen peroxide can be broken down into molecular hydrogen and molecular oxygen.

How Can You Determine the Classification For a Reaction?

The summary below can help you classify the reactions we have covered so far. Let's try an example.

$$2KClO_3(s) \rightarrow 2KCl(s) + 3O_2(g)$$

Potassium chlorate reacts to produce potassium chloride and oxygen gas.

1. When two water soluble ionic compounds react to produce a solid ionic compound, the reaction is probably a precipitation reaction.

   Solid potassium chloride is formed as a product, but the reactant is a solid too, not in an aqueous solution. There is only one reactant, and two water soluble reactants are needed in a precipitation reaction. This is not a precipitation reaction.

2. If water is a product, and one of the reactants is an acid and the other reactant is a base, then the reaction is an acid-base reaction.

   In this reaction water is not formed, so the reaction is not an acid-base reaction.

3. If the reaction has oxygen as a reactant, then the reaction is most likely a combustion reaction. Combustion reactions are a sub-class of oxidation-reduction reactions.

   In this reaction, there is no oxygen gas on the reactant side, so this is not a combustion reaction.

4. If the reaction has elements or small molecules as reactants and produces larger compounds, then the reaction is a synthesis reaction. Synthesis reactions are a sub-class of oxidation-reduction reaction.

   In this reaction larger compounds are not products, so this is not a synthesis reaction.

5. If the reaction has a large compound as reactant and the products are smaller compounds or elements, then the reaction is a decomposition reaction.

   In this reaction, the reactant is larger than the products. One of the products is molecular oxygen gas; the other one is the small ionic compound, potassium chloride. This is an example of a decomposition reaction.

LEARNING REVIEW

1. How many electrons do the elements below either gain or lose? For example, potassium atoms lose one electron.

$$K \rightarrow K^+ + e^-$$

a. Br_2
b. Mg
c. H_2
d. Al
e. O_2
f. S

2. Show how the ions below can gain or lose electrons to form atoms or molecules. For example, a sodium ion gains one electron to form an atom of sodium.

$$Na^+ + e^- \rightarrow Na$$

a. $2Cl^-$
b. K^+
c. P^{3-}
d. Ca^{2+}
e. $2I^-$
f. Al^{3+}

3. For each reaction below, write equations showing the gain and loss of electrons.

a. $Cu(s) + 2AgNO_3(aq) \rightarrow 2Ag(s) + Cu(NO_3)_2(aq)$
b. $2HCl(aq) + Zn(s) \rightarrow H_2(g) + ZnCl_2(aq)$
c. $2NaBr(aq) + Cl_2(g) \rightarrow 2NaCl(aq) + Br_2(g)$
d. $2Hg(l) + O_2(g) \rightarrow 2HgO(s)$

4. Classify the reactions below as a precipitation reaction, an acid-base reaction, or an oxidation-reduction reaction.

a. $2NaCl(s) + Br_2(l) \rightarrow 2NaBr(s) + Cl_2(g)$
b. $Na_2SO_4(aq) + Pb(NO_3)_2(aq) \rightarrow PbSO_4(s) + 2NaNO_3(aq)$
c. $2NaOH(aq) + H_2SO_4(aq) \rightarrow 2H_2O(l) + Na_2SO_4(aq)$
d. $2AgNO_3(aq) + Fe(s) \rightarrow Fe(NO_3)_2(aq) + 2Ag(s)$
e. $2KOH(aq) + ZnCl_2(aq) \rightarrow Zn(OH)_2(s) + 2KCl(aq)$

5. Classify the reactions below as combustion, synthesis or decomposition reactions.

a. $N_2(g) + 3H_2(g) \rightarrow 2NH_3(g)$
b. $C_7H_{16}(g) + 11O_2(g) \rightarrow 7CO_2(g) + 8H_2O(g)$
c. $16Cu(s) + S_8(s) \rightarrow 8Cu_2S(s)$
d. $2NaNO_3(s) \rightarrow 2NaNO_2(s) + O_2(g)$
e. $SO_3(g) + H_2O(l) \rightarrow H_2SO_4(l)$

6. Write balanced equations for each of the word descriptions. Classify each reaction as precipitation, oxidation-reduction, or acid-base.

   a. Ethyl alcohol, a gasoline additive, burns in the presence of oxygen gas to produce carbon dioxide and water vapor.
   b. Aqueous solutions of ammonium sulfide and lead nitrate are mixed to produce solid lead sulfide and aqueous ammonium nitrate.
   c. Aluminum metal reacts with oxygen gas to produce solid aluminum oxide.
   d. Sodium metal reacts with liquid water to produce aqueous sodium hydroxide and hydrogen gas.
   e. Aqueous solutions of potassium hydroxide and nitric acid are mixed to produce aqueous potassium nitrate and liquid water.
   f. Aqueous solutions of sodium phosphate and silver nitrate are mixed to produce solid silver phosphate and aqueous sodium nitrate.

ANSWERS TO LEARNING REVIEW

1. When atoms or molecules lose electrons, a positively charged cation is produced. When electrons are gained, then a negatively charged anion is produced. You can show how electrons are gained or lost by adding electrons to either the right side or the left side of an equation.

   a. $Br_2 + 2e^- \rightarrow 2Br^-$
   b. $Mg \rightarrow Mg^{2+} + 2e^-$
   c. $H_2 \rightarrow 2H^+ + 2e^-$
   d. $Al \rightarrow Al^{3+} + 3e^-$
   e. $O_2 + 4e^- \rightarrow 2O^{2-}$
   f. $S + 2e^- \rightarrow S^{2-}$

2. Ions can either gain or lose electrons to become neutral atoms or molecules. You can show whether the ions must lose or gain electrons by adding electrons to either the right or the left side of an equation.

   a. $2Cl^- \rightarrow Cl_2 + 2e^-$
   b. $K^+ + e^- \rightarrow K$
   c. $4P^{3-} \rightarrow P_4 + 12e^-$
   d. $Ca^{2+} + 2e^- \rightarrow Ca$
   e. $2I^- \rightarrow I_2 + 2e^-$
   f. $Al^{3+} + 3e^- \rightarrow Al$

3. When presented with a reaction where electrons are transferred, it is possible to extract the parts of the reaction where electrons are lost and where electrons are gained, and to write

each part separately. Notice that the number of electrons lost is equal to the number gained.

a. $Cu \rightarrow Cu^{2+} + 2e^-$ — 2 electrons are lost
$2Ag^+ + 2e^- \rightarrow 2Ag$ — 2 electrons are gained

b. $Zn \rightarrow Zn^{2+} + 2e^-$ — 2 electrons are lost
$2H^+ + 2e^- \rightarrow H_2$ — 2 electrons are gained

c. $2Br^- \rightarrow Br_2 + 2e^-$ — 2 electrons are lost
$Cl_2 + 2e^- \rightarrow 2Cl^-$ — 2 electrons are gained

d. $2Hg \rightarrow 2Hg^{2+} + 4e^-$ — 4 electrons are lost
$O_2 + 4e^- \rightarrow 2O^{2-}$ — 4 electrons are gained

4. a. In this reaction, two chloride ions lose electrons to become a chlorine molecule, and a bromine molecule gains two electrons to become two bromide ions. This is an oxidation-reduction reaction. Because the chloride ion paired with sodium is exchanged for a bromide ion, this kind of reaction is often called a replacement reaction.
   b. Two aqueous solutions containing ionic compounds react and one of the products is the ionic solid $PbSO_4$. This is a precipitation reaction.
   c. The base NaOH reacts with the acid H_2SO_4 to produce water. This is an acid-base reaction.
   d. In this reaction, two Ag^+ ions gain two electrons to become two atoms of silver, and an atom of iron loses two electrons to become an Fe^{2+} ion. This is an oxidation-reduction reaction. Because the silver ion paired with the nitrate ion is exchanged for an iron ion, this kind of reaction is called a replacement reaction.
   e. Two aqueous solutions containing ionic compounds react and one of the products is the ionic solid $Zn(OH)_2$. This is an example of a precipitation reaction.

5. a. Molecular nitrogen and molecular hydrogen react to produce a larger molecule, ammonia. This is a synthesis reaction.
   b. A molecule which is composed of carbon and hydrogen reacts with oxygen gas. The products are carbon dioxide and water. Reactions which have oxygen as a reactant are members of a sub-class of oxidation-reduction reaction called combustion reactions.
   c. Elemental copper reacts with elemental sulfur. A compound containing both elements is the product. This is a synthesis reaction.
   d. Solid sodium nitrate is converted to two simpler molecules, sodium nitrite and

molecular oxygen. This is an example of a decomposition reaction.

e. In this reaction, two small molecules combine to produce one larger molecule. This is an example of a synthesis reaction.

6. a. $C_2H_5OH(l) + 3O_2(g) \rightarrow 2CO_2(g) + 3H_2O(g)$
A molecule reacts with oxygen gas. Because this reaction has oxygen as a reactant, it is an oxidation-reduction reaction.

b. $(NH_4)_2S(aq) + Pb(NO_3)_2(aq) \rightarrow PbS(s) + 2NH_4NO_3(aq)$
Two aqueous solutions are mixed to produce a solid product. This is a precipitation reaction.

c. $4Al(s) + 3O_2(g) \rightarrow 2Al_2O_3(s)$
Elemental aluminum loses 3 electrons to become Al^{3+} and molecular oxygen gains 2 electrons to become O^{2-}. This is an oxidation-reduction reaction.

d. $2Na(s) + 2H_2O(l) \rightarrow 2NaOH(aq) + H_2(g)$
Sodium metal loses an electron to become Na^+ and two hydrogen ions gain an electron to become hydrogen gas. This is an oxidation-reduction reaction.

e. $KOH(aq) + HNO_3(aq) \rightarrow KNO_3(aq) + H_2O(l)$
Aqueous solutions of the base KOH and the acid HNO_3 are mixed to produce liquid water. So this is an acid-base reaction.

f. $Na_3PO_4(aq) + 3AgNO_3(aq) \rightarrow Ag_3PO_4(s) + 3NaNO_3(aq)$
Two aqueous solutions are mixed. The product is a solid, Ag_3PO_4. So this is a precipitation reaction.

PRACTICE EXAM

1. How many electrons are transferred in the oxidation-reduction reaction below?

$$Zn(s) + 2AgNO_3(aq) \rightarrow Zn(NO_3)_2(aq) + 2Ag(s)$$

a. 1
b. 2
c. 4
d. 6
e. 7

2. Which reaction type is **not** an oxidation-reduction reaction?

a. combustion
b. precipitation
c. synthesis
d. decomposition
e. reaction of a metal with a nonmetal

3. Which statement is correct?

a. The formation of water always occurs in a precipitation reaction.
b. The formation of a small molecule or an atom indicates a synthesis reaction.
c. The reaction of a metal with a nonmetal is an electron transfer reaction.
d. A double displacement reaction is an electron transfer reaction.
e. A decomposition reaction is a sub-class of acid-base reactions.

4. When a metal reacts with a nonmetal, which statement is **not** true?

a. The number of electrons lost and gained by each side must be equal.
b. Electrons are both lost and gained.
c. Metal atoms usually gain electrons.
d. While electrons are transferred between atoms, protons are not.
e. At least one product is an ionic compound.

5. Which of the reactions below are oxidation-reduction reactions?

i. $H_2SO_4(aq) + 2KOH(aq) \rightarrow K_2SO_4(aq) + 2H_2O(l)$
ii. $3K_2SO_4(aq) + 2Fe(NO_3)_3(aq) \rightarrow 6KNO_3(aq) + Fe_2(SO_4)_3(s)$
iii. $2C_2H_6(g) + 7O_2(g) \rightarrow 4CO_2(g) + 6H_2O(g)$

a. i, ii
b. ii, iii
c. i, iii
d. iii only
e. none of these

6. In which kind of reaction is an ionic solid always a product?

a. double displacement
b. decomposition
c. synthesis
d. oxidation-reduction
e. acid-base

7. Classify the reaction below.

$$2CuO(s) \rightarrow 2Cu(s) + O_2(g)$$

a. precipitation
b. combustion
c. synthesis
d. decomposition
e. acid-base

8. Classify the reaction below.

$$K_2O(s) + H_2O(l) \rightarrow 2KOH(aq)$$

a. precipitation
b. decomposition
c. synthesis
d. acid-base
e. combustion

9. Classify the reaction below.

$$AgNO_3(aq) + HCl(aq) \rightarrow HNO_3(aq) + AgCl(s)$$

a. precipitation
b. decomposition
c. synthesis
d. acid-base
e. combustion

10. Classify the reaction below.

$$NaOH(aq) + HNO_3(aq) \rightarrow NaNO_3(aq) + H_2O(l)$$

a. precipitation
b. decomposition
c. synthesis
d. acid-base
e. combustion

PRACTICE EXAM ANSWERS

1. b (8.1)
2. b (8.3)
3. c (8.2)
4. c (8.1)
5. d (8.2)
6. a (8.2 and 8.3)
7. d (8.3)
8. c (8.3)
9. a (8.3)
10. d (8.3)

CHAPTER 9: CHEMICAL COMPOSITION

INTRODUCTION

Before beginning a project of any kind, it is always important to know the quantity of material needed to finish the project. It is usually possible to count the number of individual items you will need. In chemistry, it is difficult to count the number of atoms or molecules needed, because the individual particles are too small, and there are too many of them. This chapter will show you how you can count the number of particles by weighing them.

AIMS FOR THIS CHAPTER

1. Understand the concept of counting by weighing. (Section 9.1)
2. Be able to count atoms by weighing. (Section 9.2)
3. Be able to define the term mole, and use your definition to calculate the number of moles of a substance, and the number of units of a substance. (Section 9.3)
4. Be able to calculate the molar mass of any molecule. (Section 9.4)
5. Given the formula of a molecule, be able to calculate the percent composition. (Section 9.5)
6. Understand the distinction between an empirical formula and a molecular formula. (Section 9.6)
7. Given data about the mass of each atom in a formula, be able to calculate the empirical formula. Make sure the empirical formula contains the smallest integers possible. (Section 9.7)
8. Given the percent composition of a molecule, calculate the empirical formula. (Section 9.7)
9. Be able to determine the molecular formula, given the percent composition and the molar mass. (Section 9.8)

QUICK DEFINITIONS

Atomic mass unit — The amu is a unit of mass used to count atoms by weighing. Each amu is 1.66×10^{-24} g. (Section 9.2)

Atomic weight — An average atomic mass in units of either amu or grams of all the naturally occurring isotopes of an element. (Section 9.2)

Mole — A mole of anything is equal to 6.022×10^{23} units. The number 6.022×10^{23} is equal to the number of atoms in 12.01 g of carbon. (Section 9.3)

Avogadro's number — The number of units in a mole of anything, 6.022×10^{23}. (Section 9.3)

Molar mass — Obtained by adding together the average mass of each individual atom in a molecule. (Section 9.4)

Formula weight — Sometimes used instead of molar mass when discussing ionic compounds. (Section 9.4)

Mass percent — The mass of a particular element in a compound, divided by the mass of the whole molecule, multiplied by 100. (Section 9.5)

Percent composition — The mass percent values for each element in a molecule. (Section 9.5)

Empirical formula — A formula with the smallest possible whole numbers to indicate the relative number of each kind of element. (Section 9.6)

Molecular formula — Gives the actual number of each kind of element present in a molecule. The molecular formula is a multiple of the empirical formula. (Section 9.6)

CONTENT REVIEW

9.1 COUNTING BY WEIGHING

What Does It Mean To Count By Weighing?

Sometimes it is hard to determine the number of items in a sample, either because there are too many of them to count individually, or because they are too small to count individually. When either of these situations occurs, it is possible to achieve an accurate count by weighing.

To count by weighing you need to know the average mass of one item. One way to get an average mass is to weigh several individual items, and then divide the total mass of all the items weighed by the number of items.

$$\frac{\text{total mass}}{\text{number of items}} = \text{average mass}$$

The average mass of an item can be used in a conversion factor to help you convert from mass to numbers of items. For example, if the average mass of a nail is 8.5 g, then we know that each 8.5 g of nails represents a count of one nail.

$$\frac{1 \text{ nail}}{8.5 \text{ g}}$$

If we weigh out 748 g of nails, we can calculate how many nails this is.

$$748 \not{g} \times \frac{1 \text{ nail}}{8.5 \not{g}} = 88 \text{ nails}$$

9.2 ATOMIC MASSES: COUNTING ATOMS BY WEIGHING

How Can You Count Atoms By Weighing?

If you know the average mass of an atom, you can count atoms by weighing just as you can nails. It is easy to determine the average mass of a nail. Weigh several of them and determine the average mass. Determining the average mass of atoms is not easy. Chemists have, however, been able to determine average masses of atoms. Because atoms are so small, the average mass is often given in units of **atomic mass units** (amu) instead of grams.

$$1 \text{ amu} = 1.66 \times 10^{-24} \text{ g}$$

The average mass of a sodium atom is 22.99 amu. We can use this as a conversion factor.

$$\frac{1 \text{ sodium atom}}{22.99 \text{ amu}}$$

If a pile of sodium atoms weighs 2368 amu, how many sodium atoms is this?

$$2368 \not{\text{amu}} \times \frac{1 \text{ sodium atom}}{22.99 \not{\text{amu}}} = 103.0 \text{ sodium atoms}$$

9.3 THE MOLE

Why Can We Use Grams to Describe the Mass of Atoms Instead of amu?

The average mass of a sodium atom is 22.99 amu and the average mass of a krypton atom is 83.80 amu. This means that 22.99 amu of sodium and 83.80 amu of krypton contain the same number of atoms. The ratio

$$\frac{\text{number of sodium atoms in 22.99 amu}}{\text{number of krypton atoms in 83.80 amu}} = 1$$

Any masses of sodium and krypton whose ratio is the same as $\frac{22.99 \text{ amu}}{83.80 \text{ amu}}$ contain the same number of atoms. For example, 22.99 g of sodium and 83.80 g of krypton contain the same numbers of atoms. The masses in grams of any two elements which are equal to the ratio of average masses of those elements in amu contain the same number of atoms. The mass in grams of an element which is equal to the average mass of an atom of that element in amu is called the **atomic weight** of an element.

Where Does Avogadro's Number Come From?

The number of atoms found in a mass in grams equal to the mass in amu of an average atom is the same for all the elements. For example, the average mass of a silicon atom is 28.09 amu and the average mass of a sulfur atom is 32.06 amu. 28.09 g silicon and 32.06 g sulfur contain the same number of atoms. This number has been given a precise definition as the number equal to the number of atoms in 12.01 g of carbon. Chemists have determined this number to be 6.022×10^{23} and is called **Avogadro's number**. Any quantity of material which contains 6.022×10^{23} units is said to contain a **mole** of units. A mole of basketballs means that there are 6.022×10^{23} basketballs. A mole of silicon atoms contains 6.022×10^{23} silicon atoms, and a mole of silicon atoms weighs 28.09 g.

How Can Avogadro's Number Be Used In Calculations?

Avogadro's number can be used as a conversion factor to help calculate the number of moles, the number of atoms, or the mass in grams. For example, we can calculate the number of atoms and the number of moles present in 53.5 g of iron.

$$53.5\ \not{g}\ \text{Fe} \times \frac{1\ \text{mol Fe}}{55.85\ \not{g}\ \text{Fe}} = 0.958\ \text{mol Fe}$$

We solved this part of the problem by using a conversion factor between moles and grams. A mole of any element contains a mass in grams equal to the average weight in amu of an atom of that element. The average atomic weight of Fe in amu is 55.85 amu. A mole of Fe weighs (has a mass of) 55.85 g. These average weights are available inside the front cover of your textbook. We can also calculate the number of atoms present in 53.5 g Fe.

$$0.958\ \text{m}\not{\text{o}}\text{l Fe} \times \frac{6.022 \times 10^{23}\ \text{atoms}}{1\ \text{m}\not{\text{o}}\text{l}} = 5.77 \times 10^{23}$$

The answer is just a little under 1 mole, which is sensible, since 53.5 g Fe is just a little under 55.85 g Fe in a mole of Fe.

9.4 MOLAR MASS

What is Molar Mass and How Can We Calculate It?

We know that a mole of any element is equal to the atomic weight in grams of that element. For example, the atomic weight of sodium is 22.99, so a mole of sodium atoms is 22.99 g of sodium. How many grams are in a mole of Na_2O, which contains two atoms of sodium and one atom of oxygen? The mass of sodium oxide which contains one mole of sodium oxide is equal to the mass of each individual atom added together.

$$(2 \times 22.99 \text{ g Na}) + 16.00 \text{ g O} = 61.98 \text{ g}$$

The mass in grams of one mole of a compound is called the **molar mass**. The molar mass of Na_2O is 61.98 g. The molar mass is always calculated by adding together the masses of each individual atom in the compound. A mole (6.022×10^{23} molecules) of Na_2O would have a mass of 61.98 g.

How Is the Molar Mass Used In Calculations Involving the Mole?

We can use the molar mass as a conversion factor to convert from mass in grams to moles.

Example:

How many moles of aluminum hydroxide are there in 225.8 g of $Al(OH)_3$? To solve this problem, we first need to know how many grams of $Al(OH)_3$ there are in one mole. To determine this, we need to calculate the molar mass of $Al(OH)_3$.

$$\text{molar mass} = 26.98 \text{ g Al} + (3 \times 16.00 \text{ g O}) + (3 \times 1.008 \text{ g H}) = 78.00 \text{ g Al(OH)}_3.$$

The conversion factor from g to mol is $\frac{1 \text{ mol Al(OH)}_3}{78.00 \text{ g}}$.

$$225.8 \cancel{\text{g}} \text{ Al(OH)}_3 \times \frac{1 \text{ mol Al(OH)}_3}{78.00 \cancel{\text{g}} \text{ Al(OH)}_3} = 2.895 \text{ mol Al(OH)}_3$$

9.5 PERCENT COMPOSITION OF COMPOUNDS

How Do We Calculate the Percent Composition of a Compound?

The chemical formula for acetaldehyde, which is C_3H_7O, tells us how many of each kind of atom is present. It does not tell us how much mass is contributed by each kind of atom in the compound. We can calculate what the contribution of each atom to the total mass is. The fraction of the mass of each element to the total mass of a molecule is called the **mass fraction.**

$$\text{mass fraction for an element} = \frac{\text{mass of element in one mole of compound}}{\text{mass of a mole of that compound}}$$

In acetaldehyde, the mass fraction of carbon is

$$\frac{(3 \times 12.01 \text{ g C})}{(3 \times 12.01 \text{ g C}) + (7 \times 1.008 \text{ g H}) + 16.00 \text{ g O}} = 0.6098$$

Remember that because there are three carbon atoms in acetaldehyde, the weight of a single carbon atom is multiplied by three when calculating the mass fraction of carbon in acetaldehyde. Multiply 0.6098 by 100 to obtain the mass percent, which is equal to 60.98%. This compound has 60.98% of its mass as carbon. When mass percents for all the elements in a compound are calculated, the results are called the percent composition for a compound. The mass percent of hydrogen in acetaldehyde is

$$\frac{(7 \times 1.008 \text{ g H})}{59.09 \text{ g } C_3H_7O} \times 100 = 11.94 \text{ \% H}$$

The mass percent of oxygen in acetaldehyde is

$$\frac{16.00 \text{ g O}}{59.09 \text{ g } C_3H_7O} \times 100 = 27.08 \text{ \% O}$$

So the percent composition for acetaldehyde is 60.97 % carbon, 11.94 % hydrogen, and 27.08 % oxygen.

9.6 FORMULAS OF COMPOUNDS

What Kinds of Chemical Formulas Are There?

Two types of chemical formulas are the empirical formula and the molecular formula. The **empirical formula** tells us the relative numbers of atoms. For example, the empirical formula C_2H_5 says that for every two carbon atoms there are five hydrogen atoms. The empirical formula does not say how many carbon and hydrogen atoms are actually in the compound. In order to know the actual numbers of atoms, we need to know the **molecular formula**. The molecular formula for the compound with an empirical formula of C_2H_5 might be $3(C_2H_5)$. This means that the molecular formula contains three empirical formulas. We could also write this molecular formula as C_6H_{15}.

What Information Is Needed to Determine an Empirical Formula?

If we know the mass of each kind of atom present in a given mass of compound, we can calculate the empirical formula of the compound. Let's see why this is so. An empirical formula gives the relative numbers of each kind of atom. That is, the empirical formula represents a count of the different kinds of atoms. The masses in grams for each kind of atom does not tell us how many atoms are present, but since we can count by weighing, we can use mass to convert to the number of atoms or the number of moles of atoms.

Example:

In 10.0 g of a compound which contains carbon, hydrogen and chlorine, there is 3.783 g carbon, 0.6349 g hydrogen, and 5.583 g chlorine. If we know the atomic weights for each of these three elements, we can count the numbers of atoms. Recall that the atomic weights for all elements are given in the front inside cover of your textbook.

$$3.783\ \cancel{g}\ C \times \frac{1\ mol\ C}{12.01\ \cancel{g}\ C} = 0.3150\ mol\ C$$

$$0.6349\ \cancel{g}\ H \times \frac{1\ mol\ H}{1.008\ \cancel{g}\ H} = 0.6299\ mol\ H$$

$$5.583\ \cancel{g}\ Cl \times \frac{1\ mol\ Cl}{35.45\ \cancel{g}\ Cl} = 0.1575\ mol\ Cl$$

We now know how many moles of each kind of atom we have in this compound, but the numbers are difficult to interpret because compounds contain whole numbers of atoms, not the fractions of moles which we have just calculated. However, just by looking at the calculated numbers of moles we can see some relationships between the kinds of atoms in this compound. Chlorine is present in the smallest quantity in this compound. There is twice as much carbon as there is chlorine, $\frac{0.3150\ mol\ C}{0.1575\ mol\ Cl} = 2.000$, and four times as much hydrogen as there is chlorine, $\frac{0.6288\ mol\ H}{0.1575\ mol\ Cl} = 3.999$. So, we can represent the empirical formula as C_2H_4Cl. In the next section we will extend our ability to determine the formulas of compounds.

9.7 EMPIRICAL FORMULA CALCULATIONS

How Can You Calculate the Empirical Formula of a Compound If You Know the Mass of Each Element in the Formula?

If we know the mass of each element present in a compound, we can calculate the empirical formula of the compound. For example, 7.560 g of a compound was found to contain 4.028 g carbon, 1.183 g hydrogen, 2.349 g nitrogen. What is its empirical formula? We know the mass of each element in the sample, but not the number of atoms of each element. It is possible to convert from mass to number of moles, though, by using a conversion factor.

$$\cancel{g}\ \text{of an element} \times \frac{1\ \text{mol}}{\text{atomic mass in}\ \cancel{g}} = \text{mol of an element}$$

The number of moles of carbon, hydrogen and nitrogen are calculated below.

$$4.028\ \cancel{g}\ C \times \frac{1\ mol\ C}{12.01\ \cancel{g}\ C} = 0.3354\ mol\ C$$

$$1.183 \not{g}\, H \times \frac{1 \text{ mol H}}{1.008 \not{g}\, H} = 1.174 \text{ mol H}$$

$$2.349 \not{g}\, N \times \frac{1 \text{ mol N}}{14.01 \not{g}\, N} = 0.1677 \text{ mol N}$$

How can the numbers of moles of each type of atom tell us what the formula of the compound is? The number of moles is one way to express how many atoms are present. The compound in this example contains 0.3354 moles carbon and 0.1677 moles of nitrogen. We can say that the compound contains more carbon than it does nitrogen. How much more? The ratio of moles of carbon to nitrogen is $\frac{0.3354 \text{ mol C}}{0.1677 \text{ mol N}} = 2$. There is twice as much carbon as nitrogen. We can begin to write the formula as $C_2H_?N$. How many hydrogen atoms should appear in the formula? We can compare the number of moles of hydrogen to the number of moles of nitrogen, $\frac{1.174 \text{ mol H}}{0.1677 \text{ mol N}} = 7$. There are seven times as many hydrogen atoms as there are nitrogen atoms. We can write the formula as C_2H_7N. The formula we have determined from the masses of three different atoms is a relative formula. It tells us that there are seven times as many hydrogen atoms as nitrogen atoms, and two times as many carbon atoms as nitrogen atoms. A formula which tells the relative numbers of atoms is called an empirical formula.

9.8 MOLECULAR FORMULA CALCULATIONS

How Can You Calculate the Molecular Formula For a Compound If You Know Its Empirical Formula and the Molar Mass Of the Compound?

In order to calculate the empirical formula for a compound, all you need to know is the mass contribution of each of the elements in the compound. This is not enough information to allow you to calculate a molecular formula. In addition to the empirical formula, you need to know the molar mass of the compound.

Example:

If 6.000 g of a compound contains 1.040 g of hydrogen and 4.960 g of carbon and has a molar mass of 58.12, what is the molecular formula? To calculate the molecular formula, it is first necessary to know the relative numbers of hydrogen and carbon atoms in the compound; in other words, you need to calculate the empirical formula first.

$$1.040 \not{g}\, H \times \frac{1 \text{ mol H}}{1.008 \not{g}\, H} = 1.032 \text{ mol H}$$

$$4.960 \not{g}\, C \times \frac{1 \text{ mol C}}{12.01 \not{g}\, C} = 0.4130 \text{ mol C}$$

From these numbers we can see that there are more moles of hydrogen than carbon in this compound because there are 1.032 mol hydrogen compared with only 0.4130 mol of carbon. We need to know the relative numbers of atoms in whole numbers, not in parts of moles. If we divide the number of moles of each element by the moles of the least abundant element, we can get rid of the fractions of moles.

$$\frac{0.4130 \text{ mol C}}{0.4130} = 1 \text{ mol C} \qquad \frac{1.032 \text{ mol H}}{0.4130} = 2.5 \text{ mol H}$$

The empirical formula we have calculated is $CH_{2.5}$. Since half atoms are not possible, we need to convert this empirical formula to all whole numbers by multiplying the entire formula by the smallest number which will make all of the numbers whole numbers. If we multiply the carbon atom by two and the hydrogen atoms by two, the resulting numbers are all whole numbers. C_2H_5 is the correct empirical formula. How can we use this information to calculate the molecular formula? We need to know how many times to multiply the number of atoms in the empirical formula to get the correct molecular formula. The actual molecular formula could be C_2H_5 or $2(C_2H_5)$ or $3(C_2H_5)$ or $4(C_2H_5)$ or any other number of empirical formulas. Our only other piece of information is the molar mass. Let's compare the molar mass of the actual compound with the molar mass of each of the possible empirical formulas. When the molar mass of an empirical formula matches that of the molecular formula, we have found the correct molecular formula. The molar mass of C_2H_5 is 29.06. C_2H_5 cannot be the correct molecular formula, because it does not have the correct molar mass. The molar mass of $2(C_2H_5)$ is 58.12. Because the molar mass of this empirical formula matches that of the molecular formula, this empirical formula is the same as the molecular formula. We can write the molecular formula as C_4H_{10}.

LEARNING REVIEW

1. A hardware store employee determined that the average mass of a certain size nail was 2.35 g.

   a. How many nails are there in 1057.5 g nails?
   b. If a customer needs 1500 nails, what mass of nails should the employee weigh out?

2. Ten individual screws have masses of 10.23 g, 10.19 g, 10.24 g, 10.23 g, 10.26 g, 10.23 g, 10.28 g, 10.30 g, 10.25 g, and 10.26 g. What is the average mass of a screw?

3. The average mass of a hydrogen atom is 1.008 amu. How many hydrogen atoms are there in a sample which has a mass of 25,527.6 amu?

4. The average mass of a sodium atom is 22.99 amu. What is the mass in amu of a sample of sodium atoms which contains 3.29×10^3 sodium atoms?

5. A sample with a mass of 4.100×10^5 amu is 25.00% carbon, and 75.00% hydrogen by mass. How many atoms of carbon and hydrogen are in the sample? The average mass of a carbon atom is 12.01 amu and of a hydrogen atom is 1.008 amu.

6. The average mass of a nitrogen atom is 14.01 amu and the average mass of a sodium atom is 22.99 amu. If a sample of sodium atoms has a mass of 68.97 g, what mass (in grams) of nitrogen contains the same number of atoms as are found in the sample of sodium?

7. The average mass of a neon atom is 20.18 amu.

   a. How many grams of neon are found in a mole of neon?
   b. How many atoms of neon are in a mole of neon?

8. What is the value of Avogadro's number and how is it defined?

9. Use the average mass values found inside the front cover of your textbook to solve the problems below.

   a. A sample of magnesium metal has a mass of 18.9 g. How many moles are in the sample?
   b. A helium balloon contains 5.38×10^{22} helium atoms. How many grams of helium are in the balloon?
   c. A piece of iron was found to contain 3.25 mol Fe. How many grams are in the sample?
   d. A sample of liquid bromine contains 65.00 g Br atoms. How many bromine atoms are in the sample?
   e. A sample of zinc contains 0.78 mol Zn. How many zinc atoms are in this sample?
   f. A sample of potassium contains 9.41×10^{25} potassium atoms. How many moles of potassium are in the sample?

10. What is meant by the term "molar mass"?

11. Calculate the molar mass of the following substances.

   a. Fe_2O_3
   b. NH_3
   c. C_2H_5OH
   d. CO_2
   e. N_2O_5
   f. CCl_3H

12. Calculate the molar mass of these ionic compounds.

a. HCl
b. $MgBr_2$
c. $Pb(OH)_2$
d. $Cu(NO_3)_2$
e. KCl
f. Na_2SO_4

13. Acetone, which has a formula of C_3H_6O, is used as a solvent in some fingernail polish removers. How many moles of acetone are in 5.00 g of acetone?

14. How many grams of potassium sulfate are in 0.623 mol potassium sulfate?

15. Calculate the mass fraction of nitrogen in N_2O_5.

16. Calculate the mass percent of each element in the following substances.

a. CH_3NH_2
b. $Zn(OH)_2$
c. H_2SO_4

17. Explain the difference between the empirical formula and the molecular formula of a compound.

18. The molecular formula of the gas acetylene is C_2H_2. What is the empirical formula?

19. When 2.500 g of an oxide of mercury, Hg_xO_y, is decomposed into the elements by heating, 2.405 g of mercury are produced. Calculate the empirical formula for this compound.

20. A compound was analyzed and found to contain only carbon, hydrogen and chlorine. A 6.380 g sample of the compound contained 2.927 g carbon and 0.5729 g hydrogen. What is the empirical formula of the compound?

21. The compound benzamide has the following percent composition. What is the empirical formula?

$$C = 69.40\% \quad H = 5.825\% \quad N = 11.57\% \quad O = 13.21\%$$

22. The empirical formula for a compound used in the past as a green paint pigment is $C_2H_3As_3Cu_2O_8$. The molar mass is 1013.71 g. What is the molecular formula?

23. A sugar which is broken down by the body to produce energy has the following percent composition.

$$C = 39.99\% \quad H = 6.713\% \quad O = 53.29\%$$

The molar mass is 210.18 g. What is the molecular formula?

ANSWERS TO LEARNING REVIEW

1. a. This problem relies on the principle of counting by weighing. The question "how many nails?" can be answered because we are given the average mass of 1 nail.

$$\frac{1 \text{ nail}}{2.35 \text{ g}} \times 1057.5 \text{ g} = 450. \text{ nails}$$

   b. If we know the mass of 1 nail equals 2.35 g, then the mass of 1500. nails is a multiple of 2.35 g.

$$\frac{2.35 \text{ g}}{1 \text{ nail}} \times 1500. \text{ nails} = 3530 \text{ g}$$

2. Average mass can be determined by adding the masses of each individual screw, then dividing by the number of screws measured.

$$10.23 \text{ g} + 10.19 \text{ g} + 10.24 \text{ g} + 10.23 \text{ g} + 10.26 \text{ g} + 10.23 \text{ g} + 10.28 \text{ g} + 10.30 \text{ g} + 10.25 \text{ g} + 10.26 \text{ g} = 102.47 \text{ g}$$

The total mass of all 10 nails is 102.47 g.

$$\frac{102.47 \text{ g}}{10 \text{ nails}} = 10.25 \text{ g/nail}$$

The average mass of a nail is 10.25 g.

3. This problem is an example of counting by weighing. We are given the average mass of 1 hydrogen atom, and asked for the number of hydrogen atoms in some other mass of hydrogen.

$$\frac{1 \text{ hydrogen atom}}{1.008 \text{ amu}} \times 25{,}527.6 \text{ amu} = 25{,}330 \text{ hydrogen atoms}$$

4. If we know the average mass of an atom, we can calculate the mass of any quantity of atoms.

$$\frac{22.99 \text{ amu}}{1 \text{ sodium atom}} \times 3.29 \times 10^3 \text{ sodium atoms} = 75{,}600 \text{ amu}$$

5. The total mass of the sample is 4.100×10^5 amu. Of this mass, 25.00 % comes from carbon atoms. So the mass contributed by carbon is

$$4.100 \times 10^5 \text{ amu} \times 0.2500 = 1.025 \times 10^5 \text{ amu}$$

The mass contributed by hydrogen is the original mass minus the mass contributed by carbon

$$4.100 \times 10^5 \text{ amu} - 1.025 \times 10^5 \text{ amu} = 3.075 \times 10^5 \text{ amu}$$

Now that we know the total mass of each kind of atom, we can use the average mass of one atom to count the number of atoms present.

$$\frac{1 \text{ hydrogen atom}}{1.008 \text{ amu}} \times 3.075 \times 10^5 \text{ amu} = 3.051 \times 10^{23} \text{ hydrogen atoms}$$

$$\frac{1 \text{ carbon atom}}{12.01 \text{ amu}} \times 1.025 \times 10^5 \text{ amu} = 8.535 \times 10^3 \text{ carbon atoms}$$

6. To solve this problem, you need to recognize that one nitrogen atom and one sodium atom have a ratio of $\frac{14.01 \text{ amu}}{22.99 \text{ amu}}$. If we have masses in grams of nitrogen and sodium in the same ratio as the atomic masses, then we will also have the same numbers of nitrogen and sodium atoms.

$$\frac{14.01 \text{ amu N}}{22.99 \text{ amu Na}} = \frac{? \text{ g N}}{68.97 \text{ g Na}}$$

We can solve for the number of nitrogen atoms in this equation because we know three of the quantities, and only lack the fourth. Isolate the unknown quantity, grams of nitrogen, on one side of the equation

$$\text{g N} = \frac{14.01 \text{ amu N}}{22.99 \text{ amu Na}} \times 68.97 \text{ g Na}$$

$$\text{g N} = 42.03 \text{ g}$$

So 68.97 g sodium and 42.03 g nitrogen contain the same number of atoms.

7. a. A mole of any element always contains a mass in grams equal to the average atomic mass of that element. So there are 20.18 g Ne in 1 mol Ne.

   b. A mole of atoms of any element always contains 6.022×10^{23} atoms.

8. Avogadro's number is the number equal to the number of atoms in 12.01 grams of carbon. Chemists have accurately determined this number to be 6.022×10^{23} atoms.

9. These problems use conversions between moles and grams, or between moles and number of atoms. For each element, you must write a different conversion factor for moles to grams depending upon the average mass for that element.

a. $$18.9 \text{ g Mg} \times \frac{1 \text{ mol Mg}}{24.31 \text{ g Mg}} = 0.777 \text{ mol Mg}$$

b. This problem requires first determining the moles of He, and then converting moles to grams.

$$5.38 \times 10^{22} \text{ He atoms} \times \frac{1 \text{ mol He}}{6.022 \times 10^{23} \text{ He atoms}} \times \frac{4.00 \text{ g He}}{1 \text{ mol He}} = 0.357 \text{ g He}$$

c. $$3.25 \text{ mol Fe} \times \frac{55.85 \text{ g Fe}}{1 \text{ mol Fe}} = 182 \text{ g Fe}$$

d. This is a two step problem, requiring that you first calculate the number of moles of Br, then the number of Br atoms.

$$65.00 \text{ g Br} \times \frac{1 \text{ mol Br}}{79.90 \text{ g Br}} \times \frac{6.022 \times 10^{23} \text{ Br atoms}}{1 \text{ mol Br}} = 4.899 \times 10^{23} \text{ Br atoms}$$

e. $$0.78 \text{ mol Zn} \times \frac{6.022 \times 10^{23} \text{ Zn atoms}}{1 \text{ mol Zn}} = 4.7 \times 10^{23} \text{ Zn atoms}$$

f. $$9.41 \times 10^{25} \text{ K atoms} \times \frac{1 \text{ mol}}{6.022 \times 10^{23} \text{ K atoms}} = 156 \text{ mol K}$$

10. Molar mass is the number of grams found in 1 mole of a substance. The molar mass is calculated by adding together the masses of each atom in the substance.

11. a. Fe_2O_3 contains 2 Fe atoms and 3 O atoms.
$(2 \times 55.85 \text{ g Fe}) + (3 \times 16.00 \text{ g O}) = 159.7 \text{ g}$

b. NH_3 contains 1 nitrogen atom and 3 hydrogen atoms.
$(1 \times 14.01 \text{ g N}) + (3 \times 1.008 \text{ g H}) = 17.03 \text{ g}$

c. C_2H_5OH contains 2 C atoms, 6 H atoms and 1 O atom.
$(2 \times 12.01 \text{ g C}) + (6 \times 1.008 \text{ g H}) + (1 \times 16.00 \text{ g O}) = 46.07 \text{ g}$

d. CO_2 contains 1 C atom and 2 O atoms.
$(1 \times 12.01 \text{ g C}) + (2 \times 16.00 \text{ g O}) = 44.01 \text{ g}$

e. N_2O_5 contains 2 N atoms and 5 O atoms
$(2 \times 14.01 \text{ g N}) + (5 \times 16.00 \text{ g O}) = 108.0 \text{ g}$

f. CCl_3H contains 1 C atom, 3 Cl atoms, and 1 H atom
$(1 \times 12.01 \text{ g C}) + (3 \times 35.45 \text{ g Cl}) + (1 \times 1.008 \text{ g H}) = 119.4 \text{ g}$

12. a. $1.008 \text{ g H} + 35.45 \text{ g Cl} = 36.46 \text{ g}$
b. $24.31 \text{ g Mg} + (2 \times 79.90 \text{ g Br}) = 184.1 \text{ g}$
c. $207.19 \text{ g Pb} + (2 \times 16.00 \text{ g O}) + (2 \times 1.008 \text{ H}) = 241.2 \text{ g}$
d. $63.55 \text{ g Cu} + (2 \times 14.01 \text{ g N}) + (6 \times 16.00 \text{ g O}) = 187.6 \text{ g}$
e. $39.10 \text{ g K} + 35.45 \text{ g Cl} = 74.55 \text{ g}$
f. $(2 \times 22.99 \text{ g Na}) + 32.06 \text{ g S} + (4 \times 16 \text{ g O}) = 142.0 \text{ g}$

13. To solve this problem, we need to know how many grams of acetone are in one mole of acetone. The number of grams of acetone equal to one mole of acetone is the molar mass.

$$\text{molar mass acetone} = (3 \times 12.01 \text{ g}) + (6 \times 1.008 \text{ g}) + 16.00 \text{ g} = 58.08 \text{ g}$$

$$5.00 \text{ g acetone} \times \frac{1 \text{ mol acetone}}{58.08 \text{ g acetone}} = 0.0861 \text{ mol acetone}$$

14.

$$\text{molar mass } K_2SO_4 = (2 \times 39.10 \text{ g}) + 32.06 \text{ g} + (4 \times 16.00 \text{ g}) = 174.3 \text{ g}$$

$$0.623 \text{ mol } K_2SO_4 \times \frac{174.3 \text{ g } K_2SO_4}{1 \text{ mol } K_2SO_4} = 109 \text{ g } K_2SO_4$$

15. Mass fraction is equal to the mass of the desired element, in this case nitrogen, divided by the molar mass.

$$\frac{28.02 \text{ g N}}{108.0 \text{ g total}} = 0.2594$$

16. a. molar mass of CH_3NH_2 is $12.01 \text{ g C} + (5 \times 1.008 \text{ g H}) + 14.01 \text{ g N} = 31.06 \text{ g total}$

$$\frac{12.01 \text{ g C}}{31.06 \text{ g total}} \times 100 = 38.67\% \text{ C}$$

$$\frac{5.040 \text{ g H}}{31.06 \text{ g total}} \times 100 = 16.23\% \text{ H}$$

$$\frac{14.01\text{ g N}}{31.06\text{ g total}} \times 100 = 45.11\%\text{ N}$$

b. molar mass of $Zn(OH)_2$ is 65.38 g Zn + (2 x 16.00 g O) + (2 x 1.008 g H) = 99.40 g total

$$\frac{65.38\text{ g Zn}}{99.40\text{ g total}} \times 100 = 65.77\%\text{ Zn}$$

$$\frac{32.00\text{ g O}}{99.40\text{ g total}} \times 100 = 32.19\%\text{ O}$$

$$\frac{2.016\text{ g H}}{99.40\text{ g total}} \times 100 = 2.028\%\text{ H}$$

c. molar mass of H_2SO_4 is (2 x 1.008 g H) + 32.06 g S + (4 x 16.00 g O) = 98.08 g total

$$\frac{2.016\text{ g H}}{98.08\text{ g total}} \times 100 = 2.055\%\text{ H}$$

$$\frac{32.06\text{ g S}}{98.08\text{ g total}} \times 100 = 32.69\%\text{ S}$$

$$\frac{64.00\text{ g O}}{98.08\text{ g total}} \times 100 = 65.25\%\text{ O}$$

17. The empirical formula gives only the relative number of atoms, that is, a ratio of each kind of atom. The molecular formula tells exactly how many of each kind of atom is present in the molecule.

18. For every 2 atoms of carbon in acetylene there are 2 atoms of hydrogen. The ratio of carbon atoms to hydrogen atoms is 1:1. So, the empirical formula of acetylene is CH.

19.

Hg_xO_y	$\rightarrow$	x Hg	+	y O
2.500 g		2.405 g		? g

Since we know that the mercury and the oxygen combined weighed 2.500 g before the reaction took place, and that the mass of the mercury is 2.405 g, then the mass of oxygen must be 2.500 - 2.405 = 0.095 g. We can now convert grams of mercury and grams of oxygen to moles, using the atomic masses of these elements.

$$2.405\text{ g Hg} \times \frac{1\text{ mol Hg}}{200.59\text{ g}} = 0.01199\text{ mol Hg}$$

$$0.095 \not{g}\,O \times \frac{1 \text{ mol O}}{16.00 \not{g}\,O} = 0.0059 \text{ mol O}$$

The ratio of mercury atoms to oxygen atoms is 2.03 to 1. So, there are twice as many Hg atoms as O atoms, and the empirical formula is Hg_2O.

20. When 6.380 g of a compound which contained only carbon, hydrogen and chlorine was analyzed, it was found to contain 2.927 g carbon and 0.5729 g hydrogen. The mass of chlorine must be equal to the total mass minus the mass of carbon plus hydrogen.

$$\text{mass of chlorine} = 6.380 - (2.927 \text{ g C} + 0.5729 \text{ g H}) = 2.880 \text{ g}$$

The moles of each kind of atom are determined from the average atomic mass.

$$2.927 \not{g} \times \frac{1 \text{ mol C}}{12.01 \not{g}\,C} = 0.2437 \text{ mol C}$$

$$0.5729 \not{g} \times \frac{1 \text{ mol H}}{1.008 \not{g}\,H} = 0.5684 \text{ mol H}$$

$$2.880 \not{g} \times \frac{1 \text{ mol Cl}}{35.45 \not{g}\,Cl} = 0.08124 \text{ mol Cl}$$

Express the mole ratios in whole numbers by dividing each number of moles by the smallest number of moles.

$$\frac{0.2437 \text{ mol C}}{0.08124} = 3.000 \text{ mol C}$$

$$\frac{0.5684 \text{ mol H}}{0.08124} = 7.00 \text{ mol H}$$

$$\frac{0.08124 \text{ mol Cl}}{0.08124} = 1.000 \text{ mol Cl}$$

The empirical formula is C_3H_7Cl.

21. This problem provides only percent composition data for the compound benzamide. It does not provide an analysis in grams for each of the elements present. We need to know how many grams of each element are present in a sample of benzamide so we can calculate the moles of each element. We can convert percent composition data to grams of each element. Assume that we have 100.0 g of benzamide. Of that sample, 69.40 % is carbon. So for a 100.0 g sample, 69.40 g are carbon, 5.825 g are hydrogen, 11.57 g are nitrogen

and 13.21 g are oxygen. We can now calculate the number of moles of each element.

$$69.40 \text{ g C} \times \frac{1 \text{ mol C}}{12.01 \text{ g C}} = 5.779 \text{ mol C}$$

$$5.825 \text{ g H} \times \frac{1 \text{ mol H}}{1.008 \text{ g H}} = 5.779 \text{ mol H}$$

$$11.57 \text{ g N} \times \frac{1 \text{ mol N}}{14.01 \text{ g N}} = 0.8258 \text{ mol N}$$

$$13.21 \text{ g O} \times \frac{1 \text{ mol O}}{16.00 \text{ g O}} = 0.8256 \text{ mol O}$$

Now, divide each number of moles by the smallest number of moles to convert the number of moles to whole numbers.

$$\frac{5.779 \text{ mol C}}{0.8256} = 7.000 \text{ mol C}$$

$$\frac{5.779 \text{ mol H}}{0.8256} = 7.000 \text{ mol H}$$

$$\frac{0.8258 \text{ mol N}}{0.8256} = 1.000 \text{ mol N}$$

$$\frac{0.8256 \text{ mol O}}{0.8256} = 1.000 \text{ mol O}$$

The empirical formula is C_7H_7NO.

22. If you are given both the molar mass and the empirical formula, determining the molecular formula is straightforward. If we multiply all of the atoms in the empirical formula by some number, we will have a correct molecular formula. So the molecular formula is a multiple of the empirical formula. We can determine what this multiple is by comparing the molar mass of the molecular formula with the molar mass of the empirical formula.

$$\text{molar mass empirical formula} = (2 \times 12.01 \text{ g C}) + (3 \times 1.008 \text{ g H}) + (3 \times 74.92 \text{ g As}) + (2 \times 63.55 \text{ g Cu}) + (8 \times 16.00 \text{ g O}) = 506.9 \text{ g}$$

The molar mass of the empirical formula is 506.9 g, and we know that the molar mass of the molecular formula is 1013.7 g. There are 2 empirical formulas in the molecular formula.

$$\frac{1013.7 \text{ g in molecular formula}}{506.9 \text{ g in empirical formula}} = 2.000$$

So, the molecular formula is 2 times the empirical formula. The molecular formula is $2(C_2H_3As_3Cu_2O_8)$ or $C_4H_6As_6Cu_4O_{16}$

23. In this problem, we are asked to find the molecular formula given the molar mass and the percent composition. To determine the molecular formula, we must first find the empirical formula.

$$39.99 \text{ g C} \times \frac{1 \text{ mol C}}{12.01 \text{ g C}} = 3.330 \text{ mol C}$$

$$6.713 \text{ g H} \times \frac{1 \text{ mol H}}{1.008 \text{ g H}} = 6.660 \text{ mol H}$$

$$53.29 \text{ g O} \times \frac{1 \text{ mol O}}{16.00 \text{ g O}} = 3.331 \text{ mol O}$$

Divide each molar quantity by the smallest number of moles to convert the number of moles to a whole number.

$$\frac{3.330 \text{ mol C}}{3.330} = 1.000$$

$$\frac{6.660 \text{ mol H}}{3.330} = 2.000$$

$$\frac{3.331 \text{ mol O}}{3.330} = 1.000$$

The empirical formula is CH_2O. The molar mass of the molecule is 210.18 g. So we need to know the molar mass of the empirical formula.

$$\text{molar mass empirical formula} = 12.01 \text{ g C} + (2 \times 1.008 \text{ g H}) + 16.00 \text{ g O} = 30.03 \text{ g}$$

How many empirical formulas are there in one molecular formula? We can tell by dividing the molar mass of the molecular formula by the molar mass of the empirical formula.

$$\frac{210.18 \text{ g in molecular formula}}{30.03 \text{ g in empirical formula}} = 6.999$$

The molecular formula is 7 times the empirical formula.

$$\text{molecular formula} = 7(CH_2O) \text{ or } C_7H_{14}O_7$$

PRACTICE EXAM

1. How many nitrogen atoms are present in a sample of nitrogen which has a mass of 16,812 amu? The mass of an average nitrogen atom is 14.01 amu.

    a. 600.0
    b. 1.200×10^3
    c. 2.355×10^5
    d. 8.333×10^{-4}
    e. 1.6798×10^4

2. How many grams of aluminum have the same number of atoms as 6.05 g of manganese? The average atomic mass of aluminum is 26.98 amu and the average mass of manganese is 54.94 amu.

    a. 2.97 g Al
    b. 0.491 g Al
    c. 2.04 g Al
    d. 12.3 g Al
    e. 0.0812 g Al

3. How many grams of lead are there in 4.32 mol of lead?

    a. 8.95×10^3 g
    b. 207.2 g
    c. 48.0 g
    d. 0.0208 g
    e. 895 g

4. How many atoms of chlorine are there in 0.98 g of chlorine?

    a. 5.9×10^{23} atoms
    b. 1.7×10^{22} atoms
    c. 2.1×10^{25} atoms
    d. 2.2×10^{25} atoms
    e. 2.7×10^3 atoms

5. What is the molar mass of $Ca_3(PO_4)_2$?

a. 246.1 g
b. 261.0 g
c. 341.1 g
d. 310.2 g
e. 230.0 g

6. How many moles are there in 152.3 g of methyl ether, C_2H_6O?

a. 5.065 mol
b. 7016 mol
c. 0.3025 mol
d. 3.712 mol
e. 3.306 mol

7. How many grams of the solvent cyclohexene, C_6H_{10}, are present in 4.32×10^{19} molecules of cyclohexene?

a. 7.17×10^{-5} g
b. 1.14×10^{6} g
c. 5.89×10^{-3} g
d. 8.72×10^{-7} g
e. 1.69×10^{2} g

8. Which is the correct mass percent of nitrogen in acetamide, C_2H_5NO?

a. 32.56 %
b. 23.72 %
c. 25.46 %
d. 29.77 %
e. 31.09 %

9. A 50.5 g sample of a compound made from phosphorus and chlorine is decomposed. Analysis of the products showed that 11.39 g of phosphorus atoms were produced. What is the empirical formula of the compound?

a. PCl_4
b. PCl
c. PCl_5
d. PCl_3
e. P_2Cl

10. A component of protein called serine has a molar mass of 105.10 g. If the percent composition is as follows, what is the molecular formula of serine?

C = 34.95 % H = 6.844 % O = 46.56 % N = 13.59 %

a. $C_9H_{21}O_9N_3$
b. $C_2H_6O_2N$
c. $C_3H_7O_3N$
d. $C_5HO_7N_2$
e. $C_6H_{14}O_6N_2$

PRACTICE EXAM ANSWERS

1. b (9.2)
2. a (9.3)
3. e (9.3)
4. b (9.3)
5. d (9.4)
6. e (9.4)
7. c (9.4)
8. b (9.5)
9. d (9.7)
10. c (9.8)

CHAPTER 10: CHEMICAL QUANTITIES

INTRODUCTION

In this chapter you will perform many chemical calculations, all of which are based on fundamental principles, such as balanced equations. A balanced equation can provide more information than is apparent at first glance. You can use a balanced equation to help answer such questions as "How much is produced?" and "How much would be needed to make?" Only a balanced equation will provide correct answers to these questions.

Often, when two reactants are mixed together, one of them will run out before the other one is all used up. In a situation like this, the amount of product you can make will be limited by the reactant that is used up first. The balanced equation will help you determine which reactant runs out, and how much product you can make.

AIMS FOR THIS CHAPTER

1. Understand the different ways to write a balanced chemical equation: with molecules or with moles. (Section 10.1)
2. Know how to use the mole ratio from a balanced chemical equation as a conversion factor to calculate moles of product. (Section 10.2)
3. Know how to use the mole ratio from a balanced chemical equation and the molar mass as conversion factors to calculate the mass of product produced or reactant required. (Section 10.3)
4. Be able to recognize a stoichiometry problem where you need to determine the limiting reactant before you calculate the quantity of product or reactant required. (Section 10.4)
5. Be able to determine which reactant in a chemical reaction is limiting and go from there to calculate the quantity of product produced or reactant required. (Section 10.4)
6. Know how to use the actual yield and the theoretical yield of a chemical reaction to calculate the percent yield. (Section 10.5)

QUICK DEFINITIONS

Mole — 6.022×10^{23} units. A mole of eggs would be 6.022×10^{23} eggs. (Section 10.1)

Mole ratios — A conversion factor which relates the number of moles of product and reactant to each other. In the equation $2H_2 + O_2 \rightarrow 2H_2O$ the mole ratio for hydrogen and oxygen is 2 mol H_2 for each mol of O_2, or $\frac{2 \text{ mol } H_2}{1 \text{ mol } O_2}$. (Section 10.2)

Stoichiometry The calculation of the relative amounts of reactants or products using a balanced chemical equation. Pronounced stoy ke om etry. (Section 10.3)

Stoichiometric quantities Amounts of reactants, which, when added together, are both used up, leaving no reactant molecules left over. (Section 10.4)

Limiting reactant When quantities of two reactants are mixed together in non-stoichiometric quantities, one of the reactants will run out first. This is the limiting reactant. (Section 10.4)

Theoretical yield The amount of product which could be produced if a chemical reaction were 100% efficient. Theoretical yield is the amount of product you calculate in a stoichiometry problem. (Section 10.5)

Actual yield The actual amount of product which is produced in a real laboratory situation. Actual yields are smaller than theoretical yields. (Section 10.5)

Percent yield The mass of product actually obtained compared to the amount which is theoretically possible, expressed as a percent. Large numbers near 100 percent mean a high yield (close to the theoretical) and small numbers mean a low yield (far from the theoretical). (Section 10.5)

CONTENT REVIEW

10.1 INFORMATION GIVEN BY CHEMICAL EQUATIONS

How Can You Convert From a Word Description To a Chemical Equation?

An example of a word description of a chemical equation is: When magnesium and oxygen react, magnesium oxide is formed. We can use the information contained in this sentence to help write the chemical equation and to help perform calculations. Because magnesium (Mg) and oxygen (O_2) are mentioned as the substances which are going to react, they are called the reactants and are written on the left side of the chemical equation. The reactants are separated from the products by an arrow

$$Mg + O_2 \longrightarrow$$

The word description says that magnesium oxide (MgO) is formed. In other words, magnesium oxide is the product of the reaction. Products are written on the right side of the equation.

$$Mg + O_2 \rightarrow MgO$$

We now have a chemical equation that incorporates all of the reactants and products mentioned in the word description. However, a chemical equation must always have the same numbers of each kind of atom on both sides of the equation because atoms are not created or destroyed during a chemical reaction. The reaction between magnesium and oxygen above does not have the same numbers of atoms on both sides. On the left side (reactant side) there are two oxygen atoms and one magnesium atom. On the right side (product side) there is one oxygen atom and one magnesium atom. The equation is **not balanced**. We need to adjust the number of oxygen atoms on the right side of the equation to match the number on the left. If we put a two in front of the formula for magnesium oxide, the number of oxygen atoms on the right matches the number on the left.

$$Mg + O_2 \rightarrow 2MgO$$

But now, there are two magnesium atoms on the left and one on the right. We can adjust the number on the left to match the number on the right by putting a two in front of the Mg on the left side of the equation.

$$2Mg + O_2 \rightarrow 2MgO$$

Now, the equation is balanced. When converting a word description of a chemical reaction to a chemical equation, **always** balance the equation after you write the formulas for the reactants and products.

What Information Can You Get From a Balanced Chemical Equation?

Let's look at the reaction between magnesium and oxygen.

$$2Mg + O_2 \rightarrow 2MgO$$

This chemical equation says that two atoms of magnesium and one molecule of oxygen react to form two molecules of magnesium oxide. If we multiply both sides of a chemical equation by the same number, we do not change the relative numbers of atoms and molecules. If we multiply both sides of the equation by 4,

$$4(2Mg) + 4(O_2) \rightarrow 4(2MgO)$$

there are now eight atoms of magnesium on the left and eight on the right, and there are eight atoms of oxygen on the left and eight on the right. The equation is still balanced.

$$8Mg + 4O_2 \rightarrow 8MgO$$

We can multiply both sides of the equation by any number, and the equation will still be balanced. If we multiply both sides of the equation by Avogadro's number (6.022×10^{23}),

$$6.022 \times 10^{23}(2Mg) + 6.022 \times 10^{23}(O_2) \rightarrow 6.022 \times 10^{23}(2MgO)$$

the equation is still balanced. We still have the same numbers of oxygen and magnesium atoms on both sides of the equation. From your experience with counting by weighing, you know that Avogadro's number of atoms or molecules is equal to one mole of atoms or molecules. So, the chemical equation can be written

$$2 \text{ mole Mg} + 1 \text{ mole } O_2 \rightarrow 2 \text{ moles MgO}$$

When we write balanced chemical equations like

$$2Mg + O_2 \rightarrow 2MgO$$

we can interpret them to mean two moles of magnesium react with one mole of oxygen to form two moles of magnesium oxide. When we interpret the coefficients in chemical equations to mean moles instead of individual molecules, we can use the number of moles to calculate the mass of reactants and products involved in the reaction.

10.2 MOLE-MOLE RELATIONSHIPS

How Can You Use a Balanced Equation To Calculate the Number of Moles of Product Produced or Reactant Required?

From the balanced equation for the decomposition of potassium chlorate:

$$2KClO_3 \rightarrow 2KCl + 3O_2$$

two moles of potassium chlorate decompose to produce three moles of oxygen. We can write this as a ratio, called the **molar ratio,** $\frac{2 \text{ mol } KClO_3}{3 \text{ mol } O_2}$. What if a problem asked how many moles of oxygen could be formed from the decomposition of 8.4 moles of potassium chlorate? We can use the balanced equation and a molar ratio to help solve this problem. The given quantity in the problem is moles of potassium chlorate, and the desired quantity is moles of oxygen. There are several molar ratios which can be derived from this balanced equation. The equation says that two moles of $KClO_3$ react to form three moles of O_2 $\frac{2 \text{ mol } KClO_3}{3 \text{ mol } O_2}$, two moles of $KClO_3$ react to form two moles of KCl $\frac{2 \text{ mol } KClO_3}{2 \text{ mol KCl}}$, and two moles of KCl are produced for every three moles of O_2 $\frac{2 \text{ mol KCl}}{3 \text{ mol } O_2}$ The molar ratio we need to solve this problem is $\frac{2 \text{ mol } KClO_3}{3 \text{ mol } O_2}$. By using the molar ratio as a conversion factor, we can cancel the moles of potassium chlorate, and are left with moles of oxygen.

$$8.4 \text{ mol KClO}_3 \times \frac{3 \text{ mol O}_2}{2 \text{ mol KClO}_3} = 13 \text{ mol O}_2$$

We have used the molar ratio of product to reactant to help determine how many moles of product can be formed from a given number of moles of reactant.

10.3 MASS CALCULATIONS

How Can You Use a Balanced Equation to Calculate the Mass of Reactants or Products?

Chemists are often required to answer practical questions. How much product can be made from a reaction, or how much reactant does it take to make a certain amount of product? To answer such questions, chemists need to know the actual masses of products and reactants, not just the number of moles. With a balanced chemical equation we can answer questions about masses. Section 10.2 showed you how you can answer questions about moles with a balanced equation.

Example:

If you have 125 g of sodium metal, how much chlorine gas is required to react with all of the sodium to produce sodium chloride? First, write the balanced chemical equation for this reaction.

$$2Na + Cl_2 \rightarrow 2NaCl$$

We have seen that the mole ratio is a convenient conversion factor when we are given a number of moles of one component in an equation and we want to calculate the number of moles of some other component. But in this problem we want to know the number of grams, not the number of moles. If we use the atomic mass of sodium, we can find the conversion factor we need and then calculate the number of moles of sodium in 125 g of sodium. One mole of any element is equal to the atomic weight of that element in grams. So one mole of sodium is equal to 22.99 g of sodium. We can use this statement as a conversion factor in the form $\frac{1 \text{ mol sodium,}}{22.99 \text{ g sodium}}$ to convert grams to moles.

$$125 \text{ g sodium} \times \frac{1 \text{ mol sodium}}{22.99 \text{ g sodium}} = 5.44 \text{ mol sodium}$$

Now we know the number of moles of sodium, but we want to know the number of grams of chlorine. We need to select a mole ratio from the balanced equation which relates sodium and chlorine. For every two moles of sodium one mole of chlorine is required, so the conversion factor would be $\frac{1 \text{ mol chlorine}}{2 \text{ mol sodium}}$. Sodium appears on the bottom of the conversion factor so that moles of sodium cancel, leaving moles of chlorine.

$$5.44 \text{ mol sodium} \times \frac{1 \text{ mol chlorine}}{2 \text{ mol sodium}} = 2.72 \text{ mol chlorine}$$

We now know the number of moles of chlorine required to react with 125 g of sodium, but the problem asks for the number of grams. One mole of any molecule is equal to its molar mass in grams. A chlorine molecule contains two chlorine atoms so the molar mass of a chlorine molecule is two times the atomic mass of an individual chlorine atom. Two times 35.45 g equals 70.90 g. So one mole of chlorine gas has a molar mass of 70.90 g. Use this statement to construct the conversion factor $\frac{70.90 \text{ g chlorine}}{1 \text{ mol chlorine}}$. Moles of chlorine appear on the bottom of the conversion factor so that they cancel, leaving us with grams of chlorine. To convert moles of chlorine to grams

$$2.72 \text{ mol chlorine} \times \frac{70.90 \text{ g chlorine}}{1 \text{ mol chlorine}} = 193 \text{ g chlorine}$$

Let's see how we can solve this entire problem with one conversion string.

$$125 \text{ g sodium} \times \frac{1 \text{ mol sodium}}{22.99 \text{ g sodium}} \times \frac{1 \text{ mol chlorine}}{2 \text{ mol sodium}} \times \frac{70.90 \text{ g chlorine}}{1 \text{ mol chlorine}} = 193 \text{ g chlorine}$$

Problems such as these, which require the calculation of masses of products or reactants, are often called **stoichiometry** problems.

10.4 CALCULATIONS INVOLVING A LIMITING REACTANT

What Is Meant by the Term Limiting Reactant?

The balanced equation for the reaction between sodium metal and chlorine gas is

$$2Na + Cl_2 \rightarrow 2NaCl$$

When two moles of sodium react with one mole of chlorine, all of the reactants are converted to products. That is, after the reaction is finished, there is no more sodium or chlorine, only sodium chloride. If we add quantities of reactants equal to the mole ratio of the reactants, we will use up all of the reactants. In real chemical experiments, we often do not add enough of each of the reactants to use all of them up. Often, one is used up completely but another one is left over. For example, if we added two moles of sodium to four moles of chlorine, would we use up all the reactants, or would one of them be left over? Let's think through what happens when we mix two moles of sodium with four moles of chlorine. According to the mole ratio, two moles of sodium require one mole of chlorine. Once the two moles of sodium react, there is none left over. When four moles of chlorine react with sodium, only one mole is used up, because there is not enough sodium present to use up the rest of the chlorine. There are three moles of chlorine left over. The excess chlorine cannot react, because there is no more sodium left. Because the sodium has run out, the reaction stops, even though there is enough chlorine present to make more sodium chloride. In this reaction, sodium is called the **limiting reactant**. The correct answer is two moles of sodium chloride will form. In this example, it was straightforward to tell which reactant was the limiting one because all of the quantities were given in moles that you could easily compare. When you are given grams of reactants, you will need to convert to moles to determine which reactant is limiting.

How Can You Determine Which Reactant is Limiting?

Example:

If 36 g sodium reacts with 49 g of chlorine, how much sodium chloride can be produced? In this problem, we cannot tell by looking at the mass of the reactants whether or not they are present in quantities equal to the mole ratio, in other words, whether all of the reactants will be used up, or whether one of them is a limiting reactant. Before we can calculate the grams of sodium chloride produced, we will have to determine whether one of the reactants is limiting. To make the comparison we need to convert grams to moles. Let's see how many moles of product could be formed from each reactant, as we did for the example with moles above. First we will find how many moles of sodium chloride could be produced from 36 g of sodium.

$$36\ \not{g}\ \text{sodium} \times \frac{1\ \text{m}\not{o}\text{l sodium}}{22.99\ \not{g}\ \text{sodium}} \times \frac{2\ \text{mol sodium chloride}}{2\ \text{m}\not{o}\text{l sodium}} = 1.6\ \text{mol sodium chloride}$$

So, 36 g of sodium, if it were all used, could form 1.6 mol sodium chloride. Now, let's find the number of moles of sodium chloride which could be produced from 49 g of chlorine.

$$49\ \not{g}\ \text{chlorine} \times \frac{1\ \text{m}\not{o}\text{l chlorine}}{70.90\ \not{g}\ \text{chlorine}} \times \frac{2\ \text{mol sodium chloride}}{1\ \text{m}\not{o}\text{l chlorine}} = 1.4\ \text{mol sodium chloride}$$

If we used up all the sodium, we would form 1.6 mol sodium chloride, and if we used up all the chlorine, we would form 1.4 mol sodium chloride. We cannot form 1.6 mol of sodium chloride because there is only enough chlorine to make 1.4 mol of sodium chloride. Therefore, chlorine is the limiting reactant. It will limit the amount of product we can make. How can we find the mass of sodium chloride? We already know the number of moles we can make. By using the molar mass of sodium chloride as a conversion factor, we can convert moles of sodium chloride to grams of sodium chloride.

$$1.4\ \text{m}\not{o}\text{l sodium chloride} \times \frac{58.44\ \text{g sodium chloride}}{1\ \text{m}\not{o}\text{l sodium chloride}} = 82\ \text{g sodium chloride}$$

When Do You Need to Determine the Limiting Reactant?

When you are given the masses of two reactants and asked how much product you can make, you need to determine which is the limiting reactant before you calculate grams of product.

When you are given the mass of one reactant, but told that the other reactant is present in **excess**, you do not have to calculate the limiting reactant. Because one reactant is present in excess, the other reactant will automatically be the limiting reactant, the one which will run out first.

Example:

When Al reacts with HCl, the products are $AlCl_3$ and H_2.

$$2Al(s) + 6HCl(aq) \longrightarrow 2AlCl_3(aq) + 3H_2(g)$$

Several different kinds of problems can be written using this balanced equation. For some of them, you will need to calculate the limiting reactant. Let's look at several different problems and determine whether or not it is necessary to find the limiting reactant.

a. When 4.0 g Al react with 5.5 g HCl, how many grams of $AlCl_3$ can be produced? Because masses of two reactants are given, you determine which one is limiting before you can find the mass of product.

b. When 0.3 mol Al reacts with 1.0 mol HCl, how many moles of H_2 can be formed? In this problem you are given moles of two reactants. You must determine which reactant is limiting before you can find moles of product.

c. When 2.5 mol Al reacts with excess HCl, how many moles of $AlCl_3$ can be formed? In this problem you are given the moles of one reactant, but the other reactant is present in excess. In this kind of problem you do not need to determine which is the limiting reactant. Al is limiting. It will run out first because HCl is present in excess.

10.5 PERCENT YIELD

What Is Meant By the Terms Theoretical Yield, Actual Yield and Percent Yield?

Theoretical yield is the amount of product which would be formed under ideal conditions. Theoretical yield is the maximum amount of product possible. Any time we calculate the amount of product which can be formed from a chemical reaction, we are actually calculating the theoretical yield. **Actual yield** is the amount of product you can really make in the laboratory. The actual yield is usually less than the theoretical yield because other reactions which also take place in the reaction container can decrease the amount of product formed. **Percent yield** is a comparison between the actual yield and the theoretical yield. Percent yield is the actual yield divided by the theoretical yield, multiplied by 100 percent.

$$\text{percent yield} = \frac{\text{actual yield}}{\text{theoretical yield}} \times 100\ \%$$

Example:

Magnesium metal reacts with hydrochloric acid to produce hydrogen gas and magnesium chloride.

$$Mg(s) + 2HCl(aq) \rightarrow MgCl_2(aq) + H_2(g)$$

When 5.00 g Mg reacts with excess HCl, 0.415 g of H_2 gas can be formed. 0.415 g H_2 is the theoretical yield, the amount we calculate can be formed using the equation below.

$$5.00\ \cancel{g}\ Mg \times \frac{1\ m\cancel{o}l\ Mg}{24.31\ \cancel{g}\ Mg} \times \frac{1\ m\cancel{o}l\ H_2}{1\ m\cancel{o}l\ Mg} \times \frac{2.016\ g\ H_2}{1\ m\cancel{o}l\ H_2} = 0.415\ g\ H_2$$

The actual yield was found in the laboratory to be 0.329 g. What is the percent yield? Percent yield can be calculated by dividing the actual yield by the theoretical yield, and multiplying by 100 %.

$$\text{percent yield} = \frac{0.329\ g\ H_2}{0.415\ g\ H_2} \times 100\ \%$$

$$\text{percent yield} = 79.3\ \%$$

LEARNING REVIEW

1. Rewrite the equation below in terms of moles of reactants and products.

$$6.022 \times 10^{23}\ \text{molecules}\ H_2(g) + 6.022 \times 10^{23}\ \text{molecules}\ I_2(g) \rightarrow 1.204 \times 10^{24}\ \text{molecules}\ HI(g)$$

2. How many moles of hydrogen gas could be produced from 0.8 mol sodium and an excess of water? Solve this problem by writing the equation using moles and by using the mole ratio for sodium and hydrogen.

$$2Na(s) + 2H_2O(l) \rightarrow 2NaOH(aq) + H_2(g)$$

3. How many moles of aluminum oxide could be produced from 0.12 mol Al?

$$4Al(s) + 3O_2(g) \rightarrow 2Al_2O_3(s)$$

4. How many moles of zinc chloride would be formed from the reaction of 1.38 mol Zn with HCl?

$$Zn(s) + 2HCl(aq) \rightarrow ZnCl_2(aq) + H_2(g)$$

5. How many moles of oxygen gas will react with just 0.25 mol N_2 gas?

$$N_2(g) + O_2(g) \rightarrow 2NO(g)$$

6. Solid silver carbonate decomposes to produce silver metal, oxygen gas and carbon dioxide.

   a. Write a balanced chemical equation for this reaction.

   b. What mass of silver will be produced by the decomposition of 6.32 g silver carbonate?

7. Aluminum metal reacts with hydrochloric acid to produce aluminum chloride and hydrogen gas.

   a. Write a balanced equation for this reaction.

   b. How many grams of H_2 gas can be produced from 8.56 g Al?

8. When aqueous solutions of sodium sulfate and lead(II) nitrate are mixed, a solid white precipitate is formed. How much solid lead(II) sulfate could be produced from 12.0 g Na_2SO_4 if $Pb(NO_3)_2$ is in excess?

$$Na_2SO_4(aq) + Pb(NO_3)_2(aq) \rightarrow PbSO_4(s) + 2NaNO_3(aq)$$

9. Sulfuric acid reacts with sodium hydroxide to form sodium sulfate and water. How many grams of sodium hydroxide will neutralize 0.838 g H_2SO_4?

10. Hydrogen gas and chlorine gas will combine to produce gaseous hydrogen chloride. How many molecules of hydrogen chloride can be produced from 20.1 g hydrogen gas and excess chlorine gas?

11. Some lightweight backpacking stoves use kerosene as a fuel. Kerosene is composed of carbon and hydrogen, and although it is a mixture of molecules, we can represent the formula of kerosene as $C_{11}H_{24}$. When a kerosene stove is lit, the fuel reacts with oxygen in the air to produce carbon dioxide gas and water vapor. If it takes 15 g of kerosene to fry a trout for dinner, how many grams of water are produced?

$$C_{11}H_{24}(l) + 17O_2(g) \rightarrow 11CO_2(g) + 12H_2O(g)$$

12. You are trying to prepare 6 copies of a three-page report. If you have on hand 6 copies of pages one and two, and 4 copies of page three,

    a. How many complete reports can you produce?

    b. Which page limits the number of complete reports you can produce?

13. Manganese(IV) oxide reacts with hydrochloric acid to produce chlorine gas, manganese(II) chloride and water.

$$MnO_2(s) + 4HCl(aq) \rightarrow Cl_2(g) + MnCl_2(aq) + 2H_2O(l)$$

    a. When 10.2 g MnO_2 react with 18.3 g HCl, which is the limiting reactant?

    b. What mass of chlorine gas can be produced?

    c. How many molecules of water can be produced?

14. How much barium sulfate can be formed from the reaction between 0.56 mol barium oxide and 0.35 mol sulfur trioxide?

15. The acid-base reaction between phosphoric acid and magnesium hydroxide produces solid magnesium phosphate and liquid water. If 121.0 g of phosphoric acid react with 89.70 g magnesium hydroxide, how many grams of magnesium phosphate will be produced?

16. If 85.6 g of potassium iodide react with 2.41×10^{24} molecules of chlorine gas, how many grams of iodine can be produced?

$$C12(g) + 2KI(s) \rightarrow 2KCl(s) + I_2(s)$$

17. Aqueous sodium iodide reacts with aqueous lead(II) nitrate to produce the yellow precipitate lead(II) iodide and aqueous sodium nitrate.

    a. What is the theoretical yield of lead iodide if 125.5 g of sodium iodide react with 205.6 g of lead nitrate?

    b. If the actual yield from this reaction is 197.5 g lead iodide, what is the percent yield?

ANSWERS TO LEARNING REVIEW

1. 6.022×10^{23} molecules is equivalent to 1 mol of molecules and 1.204×10^{24} molecules is equivalent to $2(6.022 \times 10^{23}$ molecules), so the equation can be rewritten as

$$1 \text{ mol } H_2(g) + 1 \text{ mol } I_2(g) \longrightarrow 2 \text{ mol } HI(g)$$

2. The balanced equation tells us that two moles of sodium react with two moles of water to form two moles of sodium hydroxide and four moles of hydrogen. By using mole ratios determined from the balanced equation, we can calculate the number of moles of reactants required and products produced from 0.8 mol sodium.

$0.8 \text{ mol Na} \times \frac{2 \text{ mol } H_2O}{2 \text{ mol Na}} = 0.8 \text{ mol } H_2O$ 0.8 mol sodium requires 0.8 mol H_2O.

$0.8 \text{ mol Na} \times \frac{2 \text{ mol NaOH}}{2 \text{ mol Na}} = 0.8 \text{ mol NaOH}$ 0.8 mol Na produces 0.8 mol NaOH.

$0.8 \text{ mol Na} \times \frac{1 \text{ mol } H_2}{2 \text{ mol Na}} = 0.4 \text{ mol } H_2$ 0.8 mol Na produces 0.4 mol H_2.

We can write the molar values we have calculated in equation form.

$$0.8 \text{ mol Na(s)} + 0.8 \text{ mol } H_2O(l) \longrightarrow 0.8 \text{ mol NaOH(aq)} + 0.4 \text{ mol } H_2(g)$$

3. First, make sure the equation is balanced. You should always determine whether or not an equation is balanced, and balance it if necessary. To solve this problem, we need to know the mole ratio for aluminum and aluminum oxide. The mole ratio represents the relationship between the mol of substance given in the problem and the mol of the desired substance, and is taken directly from the balanced equation. The mole ratio for aluminum oxide and aluminum is $\frac{2 \text{ mol } Al_2O_3}{4 \text{ mol Al}}$.

$$0.12 \text{ mol Al} \times \frac{2 \text{ mol } Al_2O_3}{4 \text{ mol Al}} = 0.060 \text{ mol } Al_2O_3$$

4. First, make sure the equation is balanced. The mole ratio for zinc and zinc chloride is taken from the balanced equation and is $\frac{1 \text{ mol } ZnCl_2}{1 \text{ mol Zn}}$.

$$1.38 \text{ mol Zn} \times \frac{1 \text{ mol } ZnCl_2}{1 \text{ mol Zn}} = 1.38 \text{ mol } ZnCl_2$$

Because there is a 1:1 mole ratio of $ZnCl_2$ to Zn, the number of moles of zinc present will produce an identical number of moles of zinc chloride.

5. The mole ratio for nitrogen and oxygen in this reaction is $\frac{1 \text{ mol } O_2}{1 \text{ mol } N_2}$. Equal numbers of moles of nitrogen and oxygen react, so 0.25 mol nitrogen reacts with just 0.25 mol oxygen.

6. a. First, write the formulas for reactants and products. Include the physical states. Then, balance the equation.

$$2Ag_2CO_3(s) \rightarrow 4Ag(s) + O_2(g) + 2CO_2(g)$$

b. It is not possible to solve this problem by converting directly from grams of Ag_2CO_3 to grams of Ag. However, the balanced equation tells us the relationship between Ag_2CO_3 and Ag in moles. If we can convert grams of Ag_2CO_3 to moles, we can use the mole ratio to tell us how many moles of Ag are produced. To convert grams of Ag_2CO_3 to moles, you can produce a conversion factor from the equivalence statement which relates number of moles to molar mass. The correct conversion factor is $\frac{1 \text{ mol } Ag_2CO_3}{275.75 \text{ g } Ag_2CO_3}$.

$$6.32 \text{ g } Ag_2CO_3 \times \frac{1 \text{ mol } Ag_2CO_3}{275.75 \text{ g } Ag_2CO_3} = 0.0229 \text{ mol } Ag_2CO_3$$

Now, we can use the mole ratio for Ag_2CO_3 and Ag to calculate the moles of Ag.

$$0.0229 \text{ mol } Ag_2CO_3 \times \frac{4 \text{ mol Ag}}{2 \text{ mol } Ag_2CO_3} = 0.0458 \text{ mol Ag}$$

We now know the moles of Ag, but we want to know the grams of Ag. The conversion factor $\frac{107.87 \text{ g Ag}}{1 \text{ mol Ag}}$, which is derived from the molar mass of silver, will allow us to calculate grams.

$$0.0458 \text{ mol Ag} \times \frac{107.87 \text{ g Ag}}{1 \text{ mol Ag}} = 4.94 \text{ g Ag}$$

If we string together all the parts of this problem, we can see that the overall strategy is to convert grams to moles using the molar mass, then moles to moles using the mole ratio, and moles to mass using the molar mass.

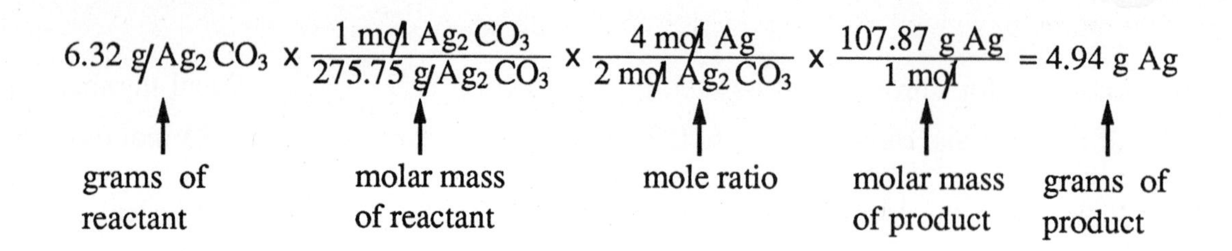

7. a. First write the formulas and physical states for the reactants and products. Then balance the reaction.

$$2Al(s) + 6HCl(aq) \rightarrow 2AlCl_3(aq) + 3H_2(g)$$

b. We cannot directly convert between grams of Al and grams of H_2. However, any time we are given grams, we can always calculate moles using the molar mass. We will first convert grams of Al to moles of Al. The balanced equation tells us the relationship between moles of Al and moles of H_2, so we can determine the moles of H_2. If we know the moles of H_2, we can use the molar mass to calculate the grams of H_2.

$$8.56 \not{g}\ Al \times \frac{1\ m\not{o}l\ Al}{26.98\ \not{g}\ Al} \times \frac{3\ m\not{o}l\ H_2}{2\ m\not{o}l\ Al} \times \frac{2.016\ g\ H_2}{1\ m\not{o}l\ H_2} = 0.959\ g\ H_2$$

grams Al — mol Al — mole ratio — mol H_2 — grams H_2

8. This question provides us with grams of reactant and asks for grams of product. Because we are told that $Pb(NO_3)_2$ is in excess, the limiting reactant must be Na_2SO_4. The amount of precipitate which can be formed is determined by the amount of Na_2SO_4. We cannot convert directly from grams Na_2SO_4 to grams $PbSO_4$. We must first calculate the moles of Na_2SO_4, then use the mole ratio derived from the balanced equation to tell us how many moles of $PbSO_4$ are produced, and finally, we can use the molar mass of $PbSO_4$ to calculate the grams of $PbSO_4$.

$$12.0\ \not{g}\ Na_2SO_4 \times \frac{1\ m\not{o}l\ Na_2SO_4}{142.04\ \not{g}\ Na_2SO_4} \times \frac{1\ m\not{o}l\ PbSO_4}{1\ m\not{o}l\ Na_2SO_4} \times \frac{303.26\ g\ PbSO_4}{1\ m\not{o}l\ PbSO_4} = 25.6\ g\ PbSO_4$$

9. Before attempting to solve this problem, write the balanced equation.

$$H_2SO_4(aq) + 2NaOH(aq) \rightarrow Na_2SO_4(aq) + 2H_2O(l)$$

The problem asks how many grams of sodium hydroxide will neutralize 0.838 g H_2SO_4. This is another way of asking how many grams of sodium hydroxide will react with 0.838 g H_2SO_4. So we can approach this problem as we have several others. First, convert grams of H_2SO_4 to moles of H_2SO_4. Use the mole ratio from the balanced equation to determine the moles of NaOH which will react. Then convert moles of NaOH to grams of NaOH using the molar mass.

$$0.838\ \not{g}\ H_2SO_4 \times \frac{1\ m\not{o}l\ H_2SO_4}{98.08\ \not{g}\ H_2SO_4} \times \frac{2\ m\not{o}l\ NaOH}{1\ m\not{o}l\ H_2SO_4} \times \frac{40.00\ g\ NaOH}{1\ m\not{o}l\ NaOH} = 0.684\ g\ NaOH$$

10. First, write the balanced equation for this reaction.

$$H_2(g) + Cl_2(g) \rightarrow 2HCl(g)$$

This problem gives us grams of hydrogen and asks for molecules of hydrogen chloride. There is no way to convert grams of hydrogen directly to molecules of hydrogen chloride. However, we can convert grams of hydrogen to moles of hydrogen using the molar mass of hydrogen gas. The balanced equation provides a mole ratio so that we can calculate the moles of hydrogen chloride. Converting from moles to molecules can be done because we know that 1 mole of hydrogen chloride equals 6.022×10^{23} molecules of hydrogen chloride.

$$20.1\ \not{g}\ H_2 \times \frac{1\ m\not{o}l\ H_2}{2.016\ \not{g}\ H_2} \times \frac{2\ m\not{o}l\ HCl}{1\ m\not{o}l\ H_2} \times \frac{6.022 \times 10^{23}\ molecules\ HCl}{1\ m\not{o}l\ HCl} = 1.20 \times 10^{25}\ molecules$$

11. We are given grams of kerosene and asked for grams of water vapor. Because we cannot convert directly between grams of kerosene and grams of water, we first convert grams of kerosene to moles of kerosene using the molar mass of kerosene. Then, use the mole ratio of kerosene and water from the balanced equation to determine the moles of water vapor. The molar mass of water will allow us to convert moles of water to grams of water.

$$15\ \not{g}\ kerosene \times \frac{1\ m\not{o}l\ kerosene}{156.30\ \not{g}\ kerosene} \times \frac{12\ m\not{o}l\ H_2O}{1\ m\not{o}l\ kerosene} \times \frac{18.02\ g}{1\ m\not{o}l\ H_2O} = 21\ g\ H_2O$$

12. a. You can only prepare 4 complete copies. Copies 4 and 5 would lack page 3.

b. Page 3 limits the number of complete reports which can be produced.

13. a. By looking at the grams of MnO_2 and the grams of HCl, it is impossible to tell which is the limiting reactant. It is possible to compare moles of reactants because we know the mole ratio of reactants from the balanced equation. So, calculate the number of moles of each reactant. Then determine how many moles of product could be produced from each of the two reactants. The reactant which allows the fewest number of moles of product is the limiting reactant.

$$10.2 \cancel{g}\ MnO_2 \times \frac{1\ m\cancel{o}l\ MnO_2}{86.94\ \cancel{g}\ MnO_2} \times \frac{1\ mol\ Cl_2}{1\ m\cancel{o}l\ MnO_2} = 0.117\ mol\ MnO_2$$

$$18.3 \cancel{g}\ HCl \times \frac{1\ m\cancel{o}l\ HCl}{36.46\ \cancel{g}\ HCl} \times \frac{1\ mol\ Cl_2}{4\ m\cancel{o}l\ HCl} = 0.125\ mol\ Cl_2$$

From 10.2 g MnO_2, 0.117 mol Cl_2 can be produced, and from 18.3 g HCl, 0.125 mol Cl_2 can be produced. So the limiting reactant is MnO_2.

b. We already know that the most chlorine we can make is 0.117 mol. To convert from moles to grams, use the molar mass of a chlorine molecule.

$$0.117\ m\cancel{o}l\ Cl_2 \times \frac{70.90\ g\ Cl_2}{1\ m\cancel{o}l\ Cl_2} = 8.30\ g\ Cl_2$$

c. The limiting reactant is manganese(IV) oxide so we need to calculate the moles of water which can be produced from 10.2 g MnO_2. By using the mole ratio from the balanced equation we can calculate the moles of water. To convert from moles of water to the number of molecules, use Avogadro's number as a conversion factor.

$$10.2 \cancel{g}\ MnO_2 \times \frac{1\ m\cancel{o}l\ MnO_2}{86.94\ \cancel{g}\ MnO_2} \times \frac{2\ m\cancel{o}l\ H_2O}{1\ m\cancel{o}l\ MnO_2} \times \frac{6.022 \times 10^{23}\ molecules\ H_2O}{1\ m\cancel{o}l\ H_2O} =$$

$$1.41 \times 10^{23}\ molecules\ H_2O$$

14. First write the balanced equation for this reaction.

$$BaO(s) + SO_3(g) \rightarrow BaSO_4(s)$$

We are asked to determine how much product can be made when particular molar quantities of two reactants are mixed. We need to determine which reactant is the limiting reactant. Use the mole ratio from the balanced equation to convert from moles of each reactant to moles of $BaSO_4$.

$$0.56 \text{ mol BaO} \times \frac{1 \text{ mol } BaSO_4}{1 \text{ mol BaO}} = 0.56 \text{ mol } BaSO_4$$

$$0.35 \text{ mol } SO_3 \times \frac{1 \text{ mol } BaSO_4}{1 \text{ mol } SO_3} = 0.35 \text{ mol } BaSO_4$$

The limiting reactant is sulfur trioxide. We know how many moles of $BaSO_4$ can be produced when the given quantities of reactants are mixed, so we can now calculate the grams of barium sulfate from the molar mass of barium sulfate.

$$0.35 \text{ mol } BaSO_4 \times \frac{233.4 \text{ g } BaSO_4}{1 \text{ mol } BaSO_4} = 82 \text{ g } BaSO_4$$

15. First, write the balanced equation for this reaction.

$$2H_3PO_4(aq) + 3Mg(OH)_2(s) \rightarrow Mg_3(PO_4)_2(s) + 6H_2O(l)$$

When the mass is given for two reactants, and you are asked to determine the quantity of product which can be produced, you must first determine which reactant is limiting. Determine how much product would be produced from each reactant. The reactant which will produce the least amount of product is the limiting reactant.

$$121.0 \text{ g } H_3PO_4 \times \frac{1 \text{ mol } H_3PO_4}{97.99 \text{ g } H_3PO_4} \times \frac{1 \text{ mol } Mg_3(PO_4)_2}{2 \text{ mol } H_3PO_4} = 0.6174 \text{ mol } Mg_3(PO_4)_2$$

$$89.70 \text{ g } Mg(OH)_2 \times \frac{1 \text{ mol } Mg(OH)_2}{58.33 \text{ g } Mg(OH)_2} \times \frac{1 \text{ mol } Mg_3(PO_4)_2}{3 \text{ mol } Mg(OH)_2} = 0.51 \text{ mol } Mg_3(PO_4)_2$$

In this reaction, the $Mg(OH)_2$ is the limiting reactant. We now know how many moles of $Mg_3(PO_4)_2$ are produced, but we want to know the number of grams. Use the molar mass of $Mg_3(PO_4)_2$ to convert from moles to grams.

$$0.5126 \text{ mol } Mg_3(PO_4)_2 \times \frac{262.87 \text{ g } Mg_3(PO_4)_2}{1 \text{ mol } Mg_3(PO_4)_2} = 134.7 \text{ g } Mg_3(PO_4)_2$$

16. In this problem we are given quantities of two reactants, one expressed in grams and the other in molecules. Before we can calculate grams of product, we need to know which reactant limits the amount of product which can be produced. Convert the grams of KI to moles using the molar mass of KI and calculate the moles of I_2 from the mole ratio.

$$85.6 \text{ g KI} \times \frac{1 \text{ mol KI}}{166.0 \text{ g KI}} \times \frac{1 \text{ mol } I_2}{2 \text{ mol KI}} = 0.258 \text{ mol } I_2$$

The quantity of the other reactant, Cl_2, is given in molecules, not grams. We can convert molecules of Cl_2 to moles of Cl_2 using Avogadro's number. 1 mol Cl_2 = 6.022×10^{23} molecules Cl_2.

$$2.41 \times 10^{24} \text{ molecules } Cl_2 \times \frac{1 \text{ mol } Cl_2}{6.022 \times 10^{23} \text{ molecules } Cl_2} = 4.00 \text{ mol } Cl_2$$

So 2.4×10^{24} molecules of Cl_2 is equivalent to 4.0 moles of Cl_2. Now we can find the moles of I_2 which can be produced from 4.0 moles of Cl_2.

$$4.0 \text{ mol } Cl_2 \times \frac{1 \text{ mol } I_2}{1 \text{ mol } Cl_2} = 4.0 \text{ mol } I_2$$

KI limits the amount of I_2 which can be produced, so it is the limiting reactant. We can calculate the grams of I_2 using the molar mass of I_2.

$$0.258 \text{ mol } I_2 \times \frac{253.8 \text{ g } I_2}{\text{mol } I_2} = 65.5 \text{ g } Cl_2$$

17. a. First, balance the equation.

$$2NaI + Pb(NO_3)_2 \longrightarrow PbI_2 + 2NaNO_3$$

This problem first asks for the theoretical yield of PbI_2 when 2 quantities of reactants are mixed. Before we can calculate the amount of product, we need to know which reactant is limiting. Use the molar masses for each product and the mole ratio for the balanced equation to calculate the moles of PbI_2 which could be produced.

$$125.5 \text{ g NaI} \times \frac{1 \text{ mol NaI}}{149.89 \text{ g NaI}} \times \frac{1 \text{ mol } PbI_2}{2 \text{ mol NaI}} = 0.4186 \text{ mol } PbI_2$$

$$205.6 \not{g}\ Pb(NO_3)_2 \times \frac{1 \text{ m}\not{o}\text{l } Pb(NO_3)_2}{331.22 \not{g}\ Pb(NO_3)_2} \times \frac{1 \text{ mol } PbI_2}{1 \text{ m}\not{o}\text{l } Pb(NO_3)_2} = 0.6207 \text{ mol } PbI_2$$

The limiting reactant is NaI. Now we can answer the question about theoretical yield. Theoretical yield is the amount of product we calculate can be produced, that is, 0.4186 mol PbI_2. In real life, the actual yield might be less than the calculated yield. The theoretical yield of PbI_2 can be calculated from the number of moles of PbI_2, if we know the molar mass.

$$0.4186 \text{ m}\not{o}\text{l } PbI_2 \times \frac{461.00 \text{ g } PbI_2}{1 \text{ m}\not{o}\text{l } PbI_2} = 193.0 \text{ g}$$

b. In part a we calculated the theoretical yield of lead(II) iodide, which is 193.0 g. We are told that the actual yield from this reaction was found to be 164.5 g. The percent yield is equal to the actual yield divided by the theoretical yield, multiplied by 100 percent. So the percent yield of lead(II) iodide is

$$\frac{164.5 \text{ g } PbI_2}{193.0 \text{ g } PbI_2} \times 100\% = 85.23\%$$

PRACTICE EXAM

1. For the reaction of aluminum with iodine to produce aluminum triodide, which of the reactions below does **not** represent a balanced equation?

   a. $2(6.022 \times 10^{23})\ Al + 3(6.022 \times 10^{23})\ I_2 \rightarrow 2(6.022 \times 10^{23})\ AlI_3$
   b. $2 \text{ g } Al + 3 \text{ g } I_2 \rightarrow 2 \text{ g } AlI_3$
   c. $2Al + 3I_2 \rightarrow 2AlI_3$
   d. $2 \text{ mol } Al + 3 \text{ mol } I_2 \rightarrow 2 \text{ mol } AlI_3$
   e. $6 \text{ molecules } Al + 9 \text{ molecules } I_2 \rightarrow 6 \text{ molecules } AlI_3$

2. When lithium reacts with water to produce lithium hydroxide and hydrogen gas, how many moles of hydrogen gas are produced from 2.12 mol lithium?

$$2\ Li + 2H_2O \rightarrow 2LiOH + H_2$$

   a. 0.236 mol H_2
   b. 4.24 mol H_2
   c. 2.12 mol H_2
   d. 0.943 mol H_2
   e. 1.06 mol H_2

3. For the balanced reaction below, which mole ratio does not express a valid relationship between two substances?

$$3Cu(s) + 8HNO_3(aq) \rightarrow 3Cu(NO_3)_2(aq) + 2NO(g) + 4H_2O(l)$$

a. $\frac{3 \text{ mol Cu}}{3 \text{ mol } Cu(NO_3)_2}$

b. $\frac{2 \text{ mol NO}}{4 \text{ mol } H_2O}$

c. $\frac{8 \text{ mol } HNO_3}{2 \text{ mol NO}}$

d. $\frac{8 \text{ mol } HNO_3}{1 \text{ mol Cu}}$

e. All of the mole ratios above express a valid relationship.

4. Solid iron combines with oxygen gas to produce rust, iron(III) oxide. What mass of rust can be formed from a piece of iron with a mass of 10.5 g if oxygen is present in excess?

$$4Fe(s) + 3O_2(g) \rightarrow 2Fe_2O_3(s)$$

a. 60.0 g
b. 1.84 g
c. 30.0 g
d. 15.0 g
e. 5.89×10^{-4} g

5. Solid ammonium dichromate will decompose when heated to produce nitrogen gas, water vapor and chromium(III) oxide. When enough ammonium dichromate is decomposed to produce 0.497 g nitrogen gas, how many grams of water vapor are produced?

a. 1.28 g
b. 0.773 g
c. 3.09 g
d. 0.320 g
e. 2.56 g

6. When potassium chlorate decomposes to solid potassium chloride and oxygen gas, how much potassium chlorate is required to produce 156 g potassium chloride?

a. 142 g
b. 94.9 g
c. 385 g
d. 171 g
e. 256 g

7. When 3.0 mol $TiCl_4$ and 3.0 mol Ti are mixed, how many grams of $TiCl_3$ will be produced?

$$3TiCl_4(l) + Ti(s) \rightarrow 4TiCl_3(s)$$

a. 350 g
b. 460 g
c. 1900 g
d. 620 g
e. 120 g

8. When 5.0 g hydrogen sulfide react with 5.0 g oxygen gas, how many grams of sulfur dioxide can be produced?

$$2H_2S(g) + 3O_2(g) \rightarrow 2SO_2(g) + 2H_2O(g)$$

a. 2.7 g
b. 9.5 g
c. 10. g
d. 13 g
e. 6.7 g

9. When 25 g potassium nitrate is decomposed to potassium nitrite and oxygen gas, the actual yield is 2.6 g oxygen gas. What is the percent yield of oxygen gas?

a. 16 %
b. 65 %
c. 6.6 %
d. 33 %
e. 1.6 %

10. Sodium cyanide can be prepared from sodium carbonate, carbon and nitrogen gas. When 10.0 g carbon were reacted with excess sodium carbonate and nitrogen, the percent yield of NaCN was 67%. What was the actual yield of NaCN in grams?

$$Na_2CO_3(s) + 4C(s) + N_2(g) \rightarrow 2NaCN(s) + 3CO(g)$$

a. 14 g
b. 0.80 g
c. 5.4 g
d. 20. g
e. 27 g

PRACTICE EXAM ANSWERS

1. b (10.1)
2. e (10.2)
3. d (10.2)
4. d (10.3)
5. a (10.3)
6. e (10.3)
7. d (10.4)
8. e (10.4)
9. b (10.5)
10. a (10.5)

CHAPTER 11: MODERN ATOMIC THEORY

INTRODUCTION

It is difficult to get a mental image of atoms because we can't see them. Scientists have produced models which account for the behavior of atoms by making observations about the properties of atoms. So we know quite a bit about these tiny particles which make up matter even though we cannot see them. In this chapter you will learn how chemists believe atoms are structured.

AIMS FOR THIS CHAPTER

1. Learn how electromagnetic radiation is characterized, and what different types of electromagnetic radiation there are. (Section 11.1)
2. Know what happens when a hydrogen atom in an excited state loses energy. (Section 11.2)
3. Be able to cite the evidence for an excited atom possessing discrete quanta of energy. (Section 11.2)
4. Be able to explain what the Bohr model of the atom says about electron movement and know why the Bohr model is not correct. (Section 11.3)
5. Know what the wave mechanical model of the atom tells us about the location of an electron. (Section 11.4)
6. Know the physical shapes, the relative distances from the nucleus, and labels (symbols) for the orbitals of hydrogen through the third principal energy level. (Section 11.5)
7. Know what a probability map of an orbital represents. (Section 11.5)
8. Know that electron spin limits the number of electrons in each orbital to two. (Section 11.6)
9. Know how to write electron configurations and box diagrams for the elements, using the periodic table. (Sections 11.7 and 11.8)
10. Understand the relationship between the arrangement of atoms in the periodic table and the arrangement of electrons in orbitals. (Section 11.8)
11. Understand what atomic trends exist in the periodic table. (Section 11.9)

QUICK DEFINITIONS

Electromagnetic radiation — A form of energy called radiant energy that travels at the speed of light with wave-like behavior. (Section 11.1)

Wavelength — The distance from one wave peak to the next peak, abbreviated lambda, λ. (Section 11.1)

Frequency — Number of waves passing a particular point per second, abbreviated nu, ν. (Section 11.1)

Photons — Separate bundles of electromagnetic energy. (Section 11.1)

Excited state — The state of an electron in an atom with excess energy. (Section 11.2)

Ground state — An atom in its lowest energy state. (Section 11.2)

Quantized — The amount of energy associated with each energy level has a certain value. The energy amounts are fixed for each level. The fixed amount of energy is called a quantum of energy, so the energy values are said to be quantized. (Section 11.2)

Wave mechanical model — An atomic model which describes the distribution of electrons into orbitals. The wave mechanical model differs from the Bohr model, which suggested that electrons moved around the nucleus in circular orbits. (Section 11.4)

Orbital — A space described by the probability of finding an electron within that particular volume. (Section 11.5)

1*s* orbital — The orbital which is closest to the nucleus and of the lowest energy. (Section 11.5)

Principal energy levels — The major energy levels found in atoms. We refer to them as n=1, n=2, n=3, and so on. (Section 11.5)

Sublevels — Subdivisions of the principal energy levels. (Section 11.5)

Pauli Exclusion Principle — An atomic orbital can contain a maximum of two electrons, and they must be of opposite spin. (Section 11.6)

Electron configuration — The arrangement of electrons into principal energy levels and sublevels. We show the electron configuration by using symbols such as $2s^1$. The 2 refers to the principal energy level, the *s* to the type of sublevel, and the superscript 1 tells how many electrons occupy the orbital. The electron configuration for lithium is $1s^2 2s^1$. (Section 11.7)

Orbital diagram Shows the arrangement of electrons using boxes to represent the orbitals. A number on top of the box tells the principal energy level, and a letter tells the type of sublevel. The number of arrows tells how many electrons are in the orbital. An orbital diagram is also called a box diagram. (Section 11.7)

Lanthanide series A group of elements beginning with the first element after lanthanum (cerium) which fill their 4*f* orbitals. The lanthanides are usually shown below the main body of the periodic table. (Section 11.7)

Actinide series A group of elements beginning with the element after actinium (thorium), which fill their 5*f* orbitals. The actinides are usually shown below the main body of the periodic table, and below the lanthanides. (Section 11.7)

Valence electrons Electrons in the principal energy level furthest from the nucleus (the highest energy level). (Section 11.7)

Core electrons All electrons which are not in the highest energy level. These electrons are all closer to the nucleus than the valence electrons. (Section 11.7)

Metal An element with a lustrous appearance, the ability to change shape without breaking, and which conducts heat and electricity. (Section 11.9)

Nonmetal An element which usually has none of the characteristics of metals given above. Some elements such as iodine and carbon have some but not all the characteristics of metals. (Section 11.9)

Ionization energy The amount of energy required to remove an electron from a gaseous atom. (Section 11.9)

CONTENT REVIEW

11.1 ELECTROMAGNETIC RADIATION AND ENERGY

What Is Electromagnetic Radiation?

Electromagnetic radiation is a form of energy. This form of energy has some of the characteristics of waves and some of the characteristics of particles. A wave is characterized by wavelength, frequency, and speed. Wavelength is the distance between two wave peaks or troughs and is symbolized with the Greek letter lambda, λ. Frequency indicates how many waves pass per second and is symbolized with the Greek letter nu, ν. It is hard to imagine how energy can be both a wave and a particle, but experimental evidence justifies this view. This type of energy travels in packets called **photons**.

11.2 THE ENERGY LEVELS OF HYDROGEN

How Does Electromagnetic Radiation Relate To the Study of Atoms?

All atoms can absorb and release extra energy. Atoms which have absorbed energy are in an **excited state**. Atoms in their lowest energy state are in the **ground state**. When atoms release energy, it is in the form of electromagnetic radiation. They emit a photon of light. The color of the light emitted is related to the amount of energy released. A photon of red light has less energy than a photon of blue light. The color of the photon and the amount of energy it contains depend upon the amount of energy the atom releases when it goes from an excited state to the ground state.

When an excited hydrogen atom which loses energy (goes from an excited state to a ground state) is observed to see what colors of light are emitted, it is found that only photons of certain colors are emitted. Since each color of photon is associated with a particular energy value, it means that hydrogen atoms are releasing only specific amounts of energy.

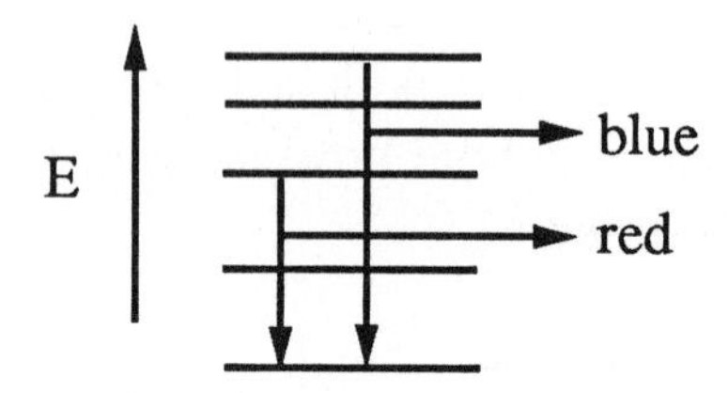

Hydrogen atoms always emit photons with these specific colors. That is, they always release the same amount of energy in the same packets. Because atoms jumping from a higher energy level to a lower one always release the same colored photons each time, their energy levels are quantized, that is, the energy of the photons is always one of the allowed values. We can tell this because photons of certain colors (certain energies) are the only ones we can detect.

11.3 THE BOHR MODEL OF THE ATOM

What Is the Bohr Model Of the Atom?

Niels Bohr looked at the quantized energies produced when a hydrogen atom in an excited state loses energy. From these results, he proposed that the electron surrounding a hydrogen atom moves around the nucleus in circular orbits. When a hydrogen atom absorbed energy, the electron jumped from an orbit nearer the nucleus to an orbit farther away from the nucleus.

Likewise, when a hydrogen atom released energy, the electron jumped from an orbit farther away from the nucleus to one closer to the nucleus, releasing a photon of light of characteristic color.

Although Bohr's model explained why only light of certain colors is emitted when hydrogen atoms jump from an excited state to the ground state, it is basically incorrect. Electrons do not move in circular orbits.

11.4 THE WAVE MECHANICAL MODEL OF THE ATOM

What Is the Wave Mechanical Model Of the Atom?

In thinking about how the electrons are distributed around an atom, two scientists, de Broglie and Schrödinger, theorized that since light can be thought of as both a particle and a wave, maybe an electron could also be considered as a particle and as a wave. When these ideas were treated mathematically, the results were the **wave mechanical model** of the atom. The wave mechanical model describes the behavior of the single electron of hydrogen, and the electrons of other atoms as well.

In the wave mechanical model, electrons are found in locations outside the nucleus called **orbitals**. Orbitals are not the same as circular orbits. Unlike our knowledge of planets in orbits, we cannot know at any one time exactly where an electron is. We can only know where an electron is likely to be. An electron is likely to be somewhere inside the volume of space described by the orbital.

11.5 THE HYDROGEN ORBITALS

What Is Indicated By An Orbital Probability Map?

It can be hard to draw a picture of an orbital, because it represents an area of space where an electron is likely to be. We do not know exactly where the electron is. Theoretically, it is possible to find an electron very far away from the nucleus, even outside of its orbital. Chemists have decided to describe the shape of an orbital based on ninety percent probability. This means that 90 percent of the time, the electron will be found inside the volume of the orbital, and the other 10 percent, the electron will be found somewhere outside the orbital volume. By arbitrarily deciding where to stop including all the places an electron might be, everyone knows what is meant when someone draws an orbital shape. They mean that there is a 90 percent probability that the electron will be found inside that space.

How Do We Organize and Name the Orbitals Of Hydrogen?

When the electron of hydrogen is in the ground state, the electron occupies the **1*s* orbital.**

1*s*

The 1*s* orbital is the orbital closest to the nucleus, and describes a sphere. Remember that 90 percent of the time, the electron will be found somewhere within the sphere, and 10 percent of the time, the electron will be found somewhere outside the sphere.

Hydrogen has other orbitals besides the 1*s*. You will recall that excited hydrogen atoms emit energy when electrons jump from a higher energy level to a lower energy level. Because the 1*s* orbital is the one of lowest energy (closest to the nucleus), there must be other orbitals of higher energy, farther away from the nucleus. Each of the energy levels is called a **principal energy level**. The first principal energy level is indicated with the number 1, the second with the number 2, the third with the number 3, and so on. As the number increases, the distance of the electrons from the nucleus increases. Electrons in the first principal energy level are closer to the nucleus than electrons in the third principal energy level.

Each of the principal energy levels is broken down into one or more sublevels. The first principal energy level has one sublevel. The second principal energy level has two sublevels. The third principal energy level has three sublevels, and so on. We use numbers and symbols to indicate which principal level and which sublevel an electron occupies. The 1*s* orbital of hydrogen means that the electron is in the first principal energy level (indicated by the number 1), and is in the *s* sublevel. The integer always indicates the principal level, and the letter always indicates the sublevel. The 1*s* sublevel has only one orbital, which is spherical in shape.

The second principal energy level has two sublevels, 2*s* and 2*p*. The 2*s* sublevel has one orbital, which is spherical in shape, like the 1*s* orbital. But the 2*s* orbital is larger than the 1*s*.

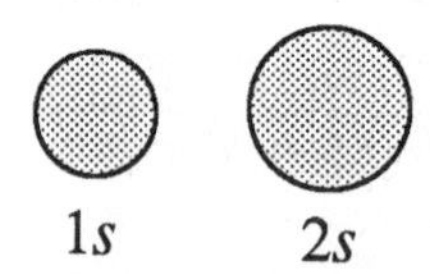

The 2*p* sublevel has three orbitals called $2p_x$, $2p_y$, and $2p_z$. The three 2*p* orbitals have a shape different from the 2*s* orbital. Each of the orbitals is dumbbell shaped, and is oriented along x, y or z axes. Remember that the pictures we draw of the shapes of orbitals tells us that 90 percent of the time, the electron will be found somewhere within that shape, and 10 percent of the time, somewhere outside the shape.

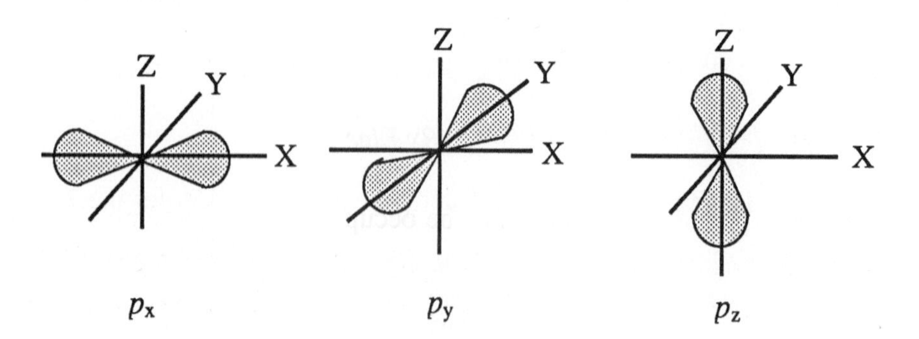

The third principal energy level has three sublevels, $3s$, $3p$, and $3d$. The $3s$ sublevel has one orbital which is spherical in shape. In fact, the s sublevels of all the principal energy levels have orbitals which are spherical in shape. There are three $3p$ orbitals, $3p_x$, $3p_y$, and $3p_z$, each of which is dumbbell shaped along x, y or z axes. The $3d$ sublevel has five orbitals.

As you go from one principal energy level to the next, several facts about the orbitals emerge. All principal energy levels have an s sublevel, which contains one spherical orbital. The spherical orbital increases in size as you move farther away from the nucleus. Each principal energy level adds a new sublevel, with orbital shapes not found in the level below. For example, the second principal energy level has $2p$ orbitals, not found in the first energy level. The third principal energy level has $3d$ orbitals, not found in the second level.

11.6 THE WAVE MECHANICAL MODEL: A SUMMARY

How Many Electrons Occupy Each Orbital?

Electrons are continuously spinning. When two electrons occupy the same orbital, it is necessary that they spin in opposite directions. Because of the need for opposite spin, only two electrons can occupy an orbital. This is called the **Pauli Exclusion Principle.**

How Can the Wave Mechanical Model Be Summarized?

There are seven principal components of the Wave Mechanical model, which are summarized below.

1. Atoms have a series of energy levels called principal energy levels which are described by whole numbers and symbolized by n. Level 1 corresponds to $n = 1$.
2. The energy increases as the value of n increases.
3. Each principal energy level contains one or more types of orbitals, called sublevels.
4. The number of sublevels present in a given principal energy level equals n.
5. The n value is always used to label the orbitals of a given principal level and is followed by a letter that indicates the type (shape) of the orbital.
6. An orbital can be empty or it can contain one or two electrons, but never more than two. If two electrons occupy the same orbital, they must have opposite spins.
7. The shape of an orbital does not indicate the details of electron movement. It indicates the probability of electron distribution for an electron residing in that orbital.

11.7 ELECTRON ARRANGEMENTS IN THE FIRST EIGHTEEN ATOMS

How Can We Show Which Orbitals Are Occupied By Electrons?

There are two ways that electrons are shown to be occupying orbitals. The electrons of lithium distributed into orbitals can be shown by $1s^2 2s^1$. This is called an **electron configuration** for lithium. $1s^2$ indicates that the s sublevel of the first principal energy level contains two electrons.

The superscript number tells the number of electrons in the sublevel. $2s^1$ indicates that the *s* sublevel of the second principal energy level contains only one electron.

The second way to indicate how the electrons of an atom are distributed is to use an **orbital diagram**, or a **box diagram.** A box diagram for lithium would look like

$1s$	$2s$
↑↓	↑

Each box represents one orbital. Remember that an orbital can hold only two electrons, so each box can have a maximum of two arrows, each one indicating an electron with opposite spin.

How Do Atoms With Atomic Numbers Through 18 Fill the Available Orbitals?

In general, atoms fill sublevels closer to the nucleus before they fill sublevels farther away from the nucleus. Hydrogen has its only electron in the first principal energy level, closest to the nucleus. The first eighteen elements fill the first principal level first, followed by the $2s$ and the $2p$, then the $3s$ and the $3p$.

The element with atomic number sixteen is sulfur. Let's look at both an electron configuration and a box diagram for sulfur. $1s^2 2s^2 2p^6 3s^2 3p^4$ is the electron configuration for sulfur. Notice that electrons fill the innermost sublevels first. When writing electron configurations, it is common to lump the electrons occupying orbitals in a kind of sublevel together. For example, all of the electrons occupying the three $2p$ orbitals are indicated with one symbol, $2p^6$, instead of creating a separate symbol for the $2p_x$, $2p_y$, and $2p_z$ orbitals.

The box diagram for sulfur would be

$1s$	$2s$		$2p$		$3s$		$3p$	
↑↓	↑↓	↑↓	↑↓	↑↓	↑↓	↑↓	↑	↑

Notice that each individual $2p$ and $3p$ orbital has a separate box, in contrast to the electron configuration, which lumps all the electrons in a kind of sublevel together. Notice also that the $3p$ sublevel is not completely filled. There are only four electrons in that sublevel. When drawing box diagrams, put one electron in each sublevel box before putting two electrons in any box. For example, the $3p_x$, $3p_y$, and $3p_z$ orbitals would each receive one electron before the $3p_x$ orbital is filled with two electrons. We draw the arrows for $3p_y$ and $3p_z$ pointing in the same direction to indicate that they spin in the same direction.

What Are Valence and Core Electrons?

Valence electrons are the electrons found in the principal energy level farthest from the nucleus. For example, in sulfur, the valence electrons are the $3s$ and $3p$ electrons. **Core electrons** are the ones which are not valence electrons. Core electrons are inside the valence electrons and closer to the nucleus.

11.8 ELECTRON CONFIGURATIONS AND THE PERIODIC TABLE

How Do Electrons Fill Orbitals For Elements With Atomic Numbers Above 18?

Element eighteen, argon, has an electron configuration of $1s^22s^22p^63s^23p^6$. We might expect that the element with atomic number of nineteen, potassium, to begin filling the 3*d* sublevel. But when we look at the location of potassium in the periodic table, we see that it is a member of the alkali metals, along with lithium and sodium. Lithium has its last electron in the 2*s* sublevel, and sodium has its last electron in the 3*s* sublevel. Since groups of elements in the periodic table have similar properties and these properties are determined by the arrangement of valence electrons, it is likely that potassium has its last electron in the 4*s* sublevel, not the 3*d* sublevel. Experimental evidence has shown this to be true. The electron configuration of potassium is $1s^22s^22p^63s^23p^64s^1$. After the 4*s* sublevel is filled by element twenty, calcium, the 3*d* orbitals are filled. In the periodic table, the elements which are filling 3*d* orbitals are the transition metals.

We can make some generalizations from the fourth period of the periodic table about how orbitals in elements in the rest of the table are filled. If a principal energy level has *d* orbitals, the *s* orbital of the next higher principal level (n + 1) always fills before the *d* orbitals of the previous principal energy level. For example, if the 4*p* orbitals are full, the 5*s* orbital is filled first, then the 4*d*.

After the element lanthanum is a series of elements called the **lanthanides**, usually shown below the main body of the periodic table, as in Figure 11.27 of your textbook. These elements are filling their 4*f* orbitals. Another series of elements which occur after the element actinium are called the **actinides**. These elements are filling their 5*f* orbitals. They are also shown below the body of the table for convenience.

It is possible to use the periodic table to tell which sublevels the outermost or valence electrons occupy. For example, elements in group 1, the alkali metals, have one electron in the outermost sublevel, an *s* sublevel. Elements in group 2, the alkaline earth metals, have two electrons in the outermost sublevel, an *s* sublevel. Elements in group 3 have three electrons in their outermost sublevel, a *p* sublevel, and so on. For the representative elements, the group number tells how many electrons are in the outermost sublevel.

11.9 ATOMIC PROPERTIES AND THE PERIODIC TABLE

What Are the Properties of Metals and Nonmetals?

Metals have a lustrous or shiny appearance. They can be beat into thin sheets and pulled into wires (are malleable), and they conduct heat and electricity. Nonmetals generally lack the properties of metals, although some nonmetals have some of the properties of metals. For example, solid iodine is lustrous.

The chemical properties of metals and nonmetals are important to chemists. Metals tend to lose electrons to form positive ions, while nonmetals tend to gain electrons to form negative ions.

How Are Metals and Nonmetals Grouped In the Periodic Table?

Metals are found in the left side and in the center of the periodic table. Most of the elements are metals. Some metals lose electrons easier than other metals. In general, metals at the bottom of the periodic table lose electrons easier than metals at the top. For example, barium loses an electron easier than beryllium does. The outermost electrons in barium are 6*s* electrons and are farther away from the nucleus than the 2*s* electrons of berellium, and so the electrons of barium are not held as tightly.

Nonmetals are found in the upper right corner of the periodic table. Compared to the number of metals, there are fewer nonmetals. Nonmetals vary in their ability to gain electrons from metals. The nonmetals with the greatest pull are found in the upper righthand corner.

Are All Elements Either Metals Or Nonmetals?

Some elements have properties of both the metals and the metalloids. These elements are located between the metals and nonmetals. Figure 11.31 in your textbook shows the divisions between metals, metalloids, and nonmetals.

How Does the Ionization Energy of An Atom Relate To Location In the Periodic Table?

The **ionization energy** of an atom is the amount of energy required to remove an electron from a gaseous atom. For any atom A

$$A(g) \xrightarrow[\text{energy}]{\text{ionization}} A^+(g) + e^-$$

Metals lose electrons easier than nonmetals, and metals at the bottom of the periodic table lose electrons easier than metals at the top of the periodic table. So ionization energy decreases from the top of the periodic table to the bottom and increases from the left side of the periodic table to the right.

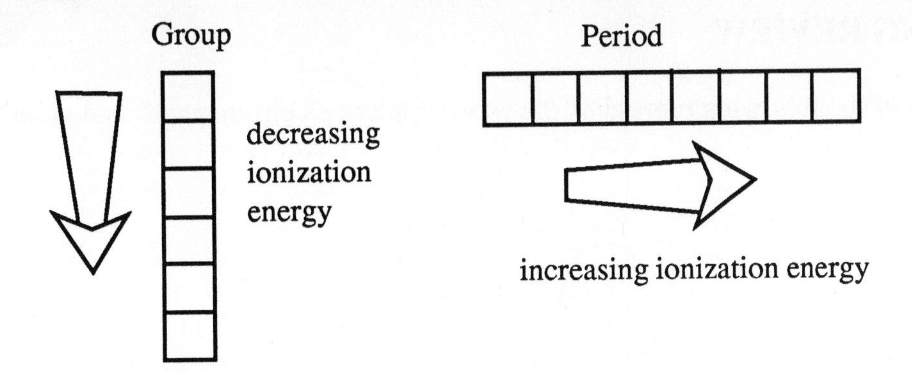

How Does Atomic Size Relate to Location in the Periodic Table?

Atoms are not all the same size. Atoms get larger as we move from the top of the periodic table to the bottom, and they get smaller as we move from left to right.

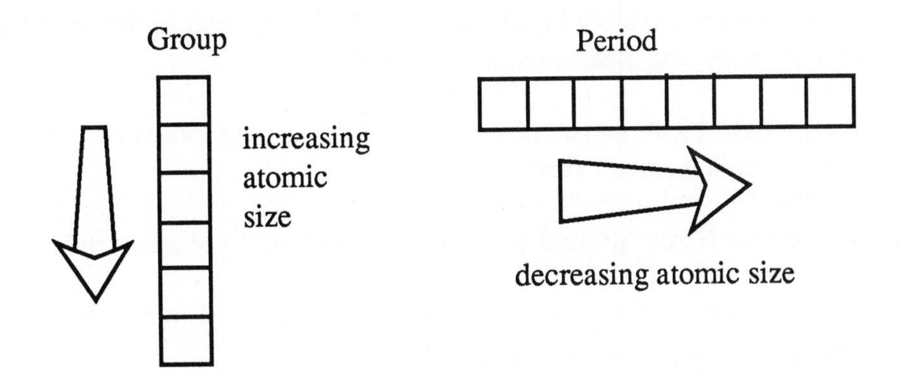

As we move from top to bottom, electrons are added to larger principal energy levels, so atoms become larger. As we move from right to left the number of electrons increases. All atoms in a period have their outermost electrons filling the same principal energy level. Because all the orbitals in a principal energy level would be expected to be the same size, we would expect that the size of an atom in a period would stay the same, not become smaller. As we move from left to right, the number of protons in the nucleus also increases. The increase in positive charge pulls the electrons closer, so as the number of protons increases from left to right, the closer the electrons are pulled, and the smaller the atomic size.

Example:

Which atom, silicon or lead, has a larger ionization energy and a smaller atomic size? Ionization energy decreases from top to bottom of the periodic table, so silicon would have a higher ionization energy than lead. Atomic size increases from top to bottom, so silicon would be a smaller atom than lead.

LEARNING REVIEW

1. Which of the following represents the wavelength of electromagnetic radiation?

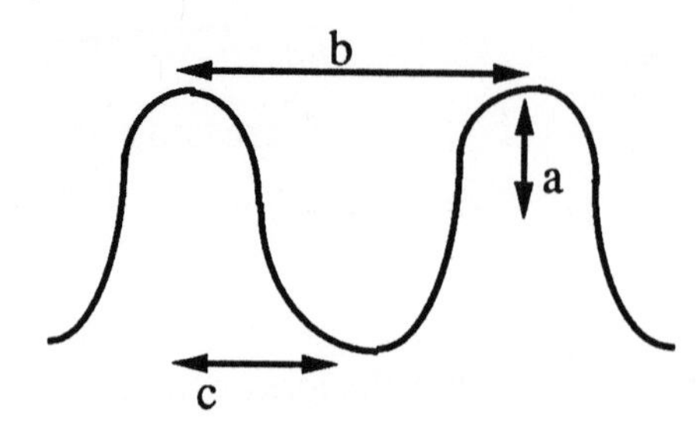

2. How does a microwave oven warm food?

3. Which has the shortest wavelength, ultraviolet light or infrared light? See Figure 11.3 in your textbook for help with electromagnetic radiation.

4. Light can be thought of as waves of energy. There is also evidence that light exists in another form. What is the other form?

5. What is meant by the terms "ground state" and "excited state" of an atom?

6. A sample of helium atoms absorbs energy. Will the photons of light emitted by the helium atoms be found at all wavelengths? Explain your answer.

7. Which energy level represents the ground state?

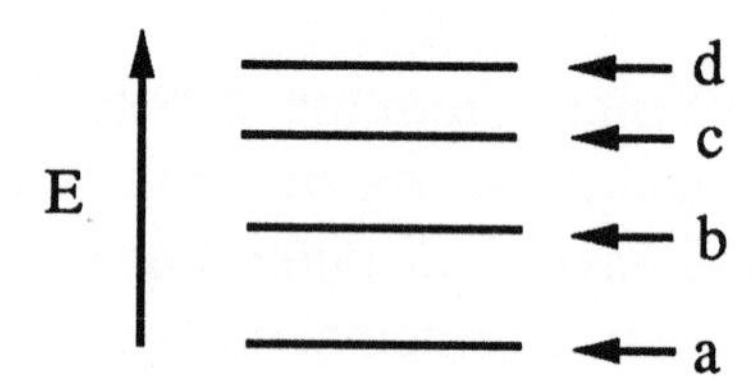

8. Which quantum has the greatest energy?

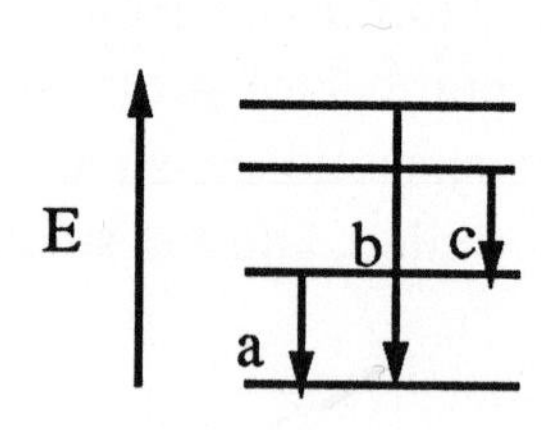

9. What does the Bohr model of the atom say about electron movement?

10. What characteristics of light led de Broglie and Schrödinger to formulate a new model of the atom?

11. Which of the following statements about the wave mechanical model of the atom are true, and which are false?

    a. The probability of finding an electron is the same in any location within an orbital.
    b. The electron will probably spend most of its time close to the nucleus.
    c. The path in which an electron moves is not known exactly.
    d. Electrons travel in circular orbits around the nucleus.

12. Which of the following statements about the orbital below are true, and which are false?

    a. An electron will be found inside the orbital 90 percent of the time.
    b. An electron travels around the surface of the orbital.
    c. An electron cannot be found outside the orbital.

13. Consider the third principal energy level of hydrogen.

    a. How many sublevels are found in this level?
    b. How many orbitals are found in the 3*d* sublevel?
    c. Which shape represents a 3*p* orbital?

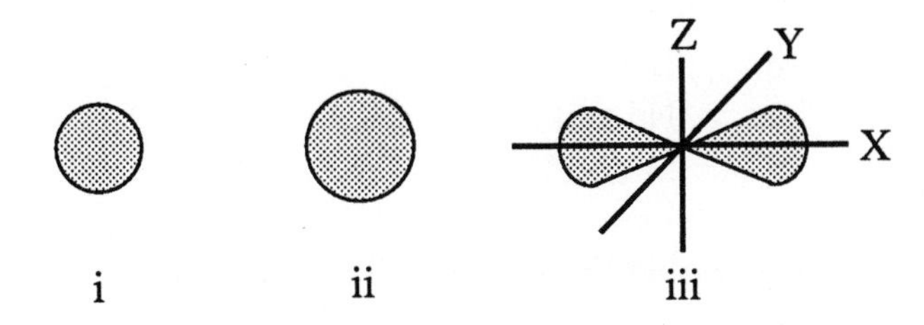

14. How many sublevels do you think would be found in the n=5 principal energy level?

15. What is meant by each part of the orbital symbol, $4p_x^1$?

16. How many electrons can occupy an orbital?

17. How does each column of the periodic table relate to electron configuration?

Use a periodic table such as the one found inside the front cover of your textbook to help answer questions 18, 19, and 20.

18. a. Write the complete electron configuration and the complete orbital diagram for aluminum.
    b. How many valence electrons and how many core electrons does aluminum have?

19. How many valence electrons are found in the elements beryllium, magnesium, calcium and strontium?

20. Write an electron configuration and box diagram for the elements below.

    a. vanadium
    b. copper
    c. bromine
    d. tin

21. How does each row of the periodic table relate to electron configuration?

22. What is characteristic of the electron configuration of the noble gases?

23. Which orbitals are filling in the lanthanide series elements?

24. Decide whether the elements below are representative elements or transition metals.

a.	Ar	c.	N
b.	Fe	d.	Sr

25. Which element in each pair would have a lower ionization energy?

a.	F and C	d.	Li and Rb
b.	O and As	e.	Ne and Rn
c.	Ca and Br	f.	Sr and Br

26. Which element in each pair would have the smaller atomic size?

a.	Ne and Xe	d.	F and Sr
b.	In and I	e.	Ba and Bi
c.	Na and Cs	f.	Cl and Al

ANSWERS TO LEARNING REVIEW

1. Wavelength is the distance from crest to crest, or trough to trough, so the correct answer is b.

2. The kind of electromagnetic radiation generated by the oven, microwave radiation, is of the right frequency to be absorbed by water in food. As the water molecules in food absorb microwave energy, their movement increases. The extra energy is transferred to other molecules in the food when they collide with the water molecules, so heat is transferred from rapidly moving water molecules to other molecules, causing the food to heat up.

3. Infrared light has a wavelength of 10^{-4} meters and ultraviolet light has a wavelength of around 10^{-8} meters, so ultraviolet light has a shorter wavelength than infrared light.

4. There is evidence that light consists of packets of energy called photons.

5. The lowest possible energy state of an atom is the ground state. When an atom has absorbed excess energy it is in an excited state.

6. When helium atoms absorb energy, some of the electrons move from the ground state to an excited state. When helium atoms lose this excess energy they will often emit light. The light is not of just any wavelength. Only certain wavelengths, corresponding to the differences in energy, are allowed.

7. The energy level marked a is the ground state because it is the level with the lowest amount of energy.

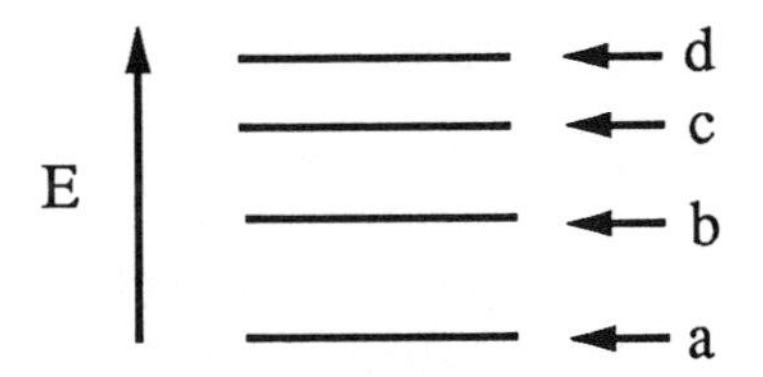

8. The quantum marked by b has the highest amount of energy because this excited state has more energy than the excited states below it.

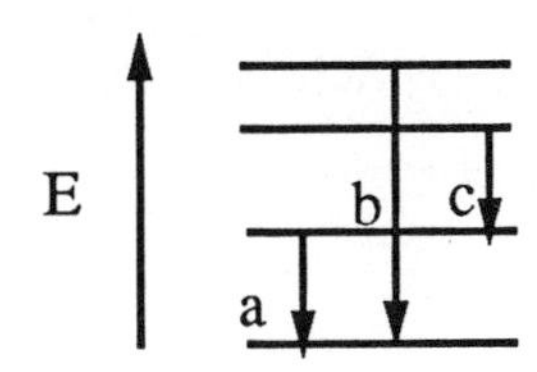

9. Bohr's model said that electrons move in circular orbits around the nucleus. Each circular orbit represents an excited state farther from the nucleus.

10. The fact that light could act as both a wave and as a particle led de Broglie and Schrödinger to suggest that an electron might also exhibit characteristics of both waves and particles. When Schrödinger used these ideas to analyze the problem mathematically, the wave mechanical model of the atom was the result.

11. a. The probability of finding an electron within an orbital is 90 percent, but some locations within the orbital shape are more likely to contain an electron at any given time than are others. So the correct answer is false.
    b. The electron does tend to spend most of its time around the nucleus. This statement is true.

c. The exact path an electron takes is not known. We can only say where it will probably be found. This statement is true.

d. Bohr thought that electrons traveled in circular orbits around the nucleus, but the current model says that electrons are found in orbitals. We do not know their exact paths. This statement is false.

12. a. The orbital shape represents a probability cloud. It is true that an electron will be found within the orbital 90 percent of the time.

b. The orbital marks the area of 90 percent probability. It does not mark the surface on which the electron travels. This statement is false.

c. Ten percent of the time, an electron will be found outside the orbital. This statement is false.

13. a. The third principal energy level, n=3, contains 3 sublevels.

b. There are five orbitals in the 3*d* sublevel.

c. i and ii represent *s* orbitals. iii is a *p* orbital.

14. Each principal energy level has n sublevels, so the fifth principal energy level, n=5, would have 5 sublevels.

15. 4 is the principal energy level, *p* is the sublevel type, x is the specific orbital within the *p* sublevel, and the superscript 1 says that there is one electron in the orbital.

16. Each orbital can hold 2 electrons.

17. The period or row number indicates which *s* and *p* orbitals are being filled for elements in that row. For example, antimony, which is in row 5, has filled its 5*s* orbitals, 4*d* orbitals and is filling 5*p* orbitals.

18. a. $1s^2 2s^2 2p^6 3s^2 3p^1$

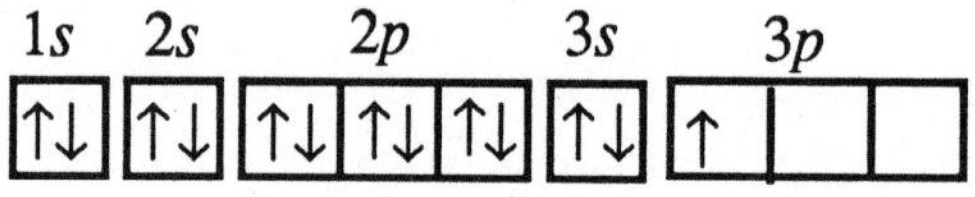

1s	2s	2p			3s	3p		
↑↓	↑↓	↑↓	↑↓	↑↓	↑↓	↑		

b. Aluminum has three valence electrons and ten core electrons.

19. Each of these elements has two valence electrons.

20. a. $1s^2 2s^2 2p^6 3s^2 3p^6 4s^2 3d^3$

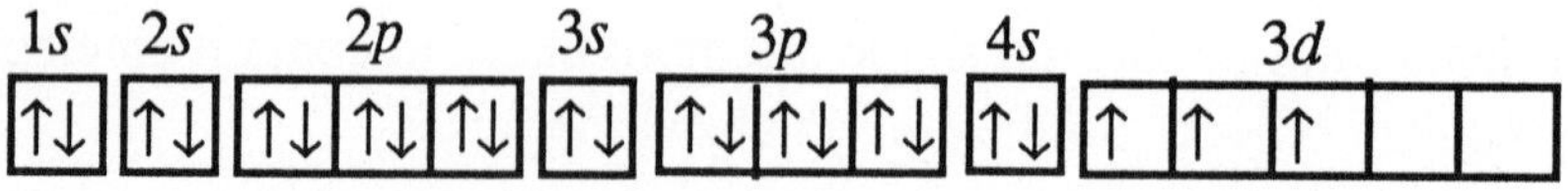

1s	2s	2p			3s	3p			4s	3d				
↑↓	↑↓	↑↓	↑↓	↑↓	↑↓	↑↓	↑↓	↑↓	↑↓	↑	↑	↑		

b. $1s^22s^22p^63s^23p^64s^13d^{10}$ or $[Ar]4s^13d^{10}$

1s	2s	2p	3s	3p	4s	3d
↑↓	↑↓	↑↓ ↑↓ ↑↓	↑↓	↑↓ ↑↓ ↑↓	↑	↑↓ ↑↓ ↑↓ ↑↓ ↑↓

Note that copper does not completely fill 4*s* before filling 3*d*.

c. $1s^22s^22p^63s^23p^64s^23d^{10}4p^5$ or
$[Ar]4s^23d^{10}4p^5$

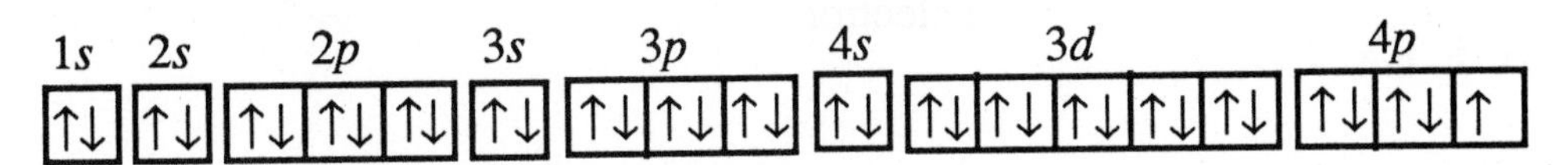

d. $1s^22s^22p^63s^23p^64s^23d^{10}4p^65s^24d^{10}5p^2$

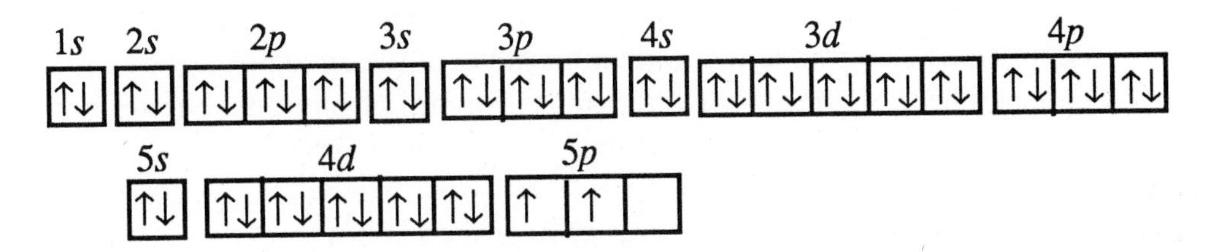

21. The group number on the top of each column of the periodic table is the same as the sum of *ns* + *np* electrons for the highest principal energy level and is equal to the number of valence electrons for that element. For example, sulfur is in group 6 and its electron configuration is $1s^22s^22p^63s^23p^4$. The sum *ns* + *np* for n=3 is 6, which is the same as the group number. Sulfur has 6 valence electrons, $3s^2$ and $3p^4$.

22. The *s* and *p* orbitals for the principal energy level, which is the same as the row number, are full. That is, all these orbitals contain a maximum of 2 electrons for a total of 8 electrons.

23. In the lanthanide series elements, the 4*f* orbitals are filling.

24. a. Representative element (noble gas)
    b. Transition metal
    c. Representative element
    d. Representative element (alkaline earth metal)

25. a. Carbon would have a lower ionization energy than fluorine.
    b. Arsenic would have a lower ionization energy than oxygen.
    c. Calcium would have a lower ionization energy than bromine.
    d. Rubidium would have a lower ionization energy than lithium.

e. Radon would have a lower ionization energy than neon.
f. Strontium would have a lower ionization energy than bromine.

26. a. Neon would have a smaller atomic size than xenon.
 b. Iodine would have a smaller atomic size than indium.
 c. Sodium would have a smaller atomic size than cesium.
 d. Fluorine would have a smaller atomic size than strontium.
 e. Bismuth would have a smaller atomic size than barium.
 f. Chlorine would have a smaller atomic size than aluminum.

PRACTICE EXAM

1. Which statement about electromagnetic radiation is false?

 a. Frequency is the number of waves which pass a point per second.
 b. X rays have a shorter wavelength than microwaves.
 c. All forms of electromagnetic radiation travel at the speed of light.
 d. Light can be thought of as a wave or as a stream of photons, but not both.
 e. Wavelength is the distance from one wave crest to another.

2. Which statement about the energy levels of hydrogen is not true?

 a. The energy levels of a hydrogen atom are quantized.
 b. Photons released by hydrogen atoms contain different amounts of energy and have different colors.
 c. The energy levels of hydrogen resemble the continuous positions along a ramp.
 d. Hydrogen atoms can absorb energy and become excited.
 e. The lowest possible energy state of a hydrogen atom is the ground state.

3. Which of the statements are true about our current model of the atom?
 i. An orbital represents a 90 percent probability map.
 ii. The current model tells us precisely where an electron can be found at any one time.
 iii. Electrons move around the nucleus in circular orbits.

 a. i b. i, ii c. iii d. i, iii e. ii only

4. Which statement concerning the notation $4p_y^1$ is <u>not</u> correct?

a. The shape of the orbital is

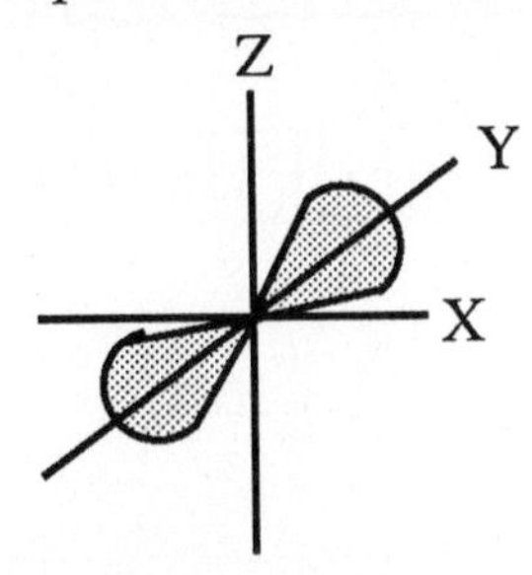

b. The orbital is in the fourth principal energy level.
c. The orbital is smaller than a $5p$ orbital.
d. There are 2 additional orbitals in the $4p$ sublevel.
e. There are 2 electrons present in this orbital.

5. What is the correct electron configuration for sulfur?

a. $1s^2 2s^2 2p^8 3s^2 3p^2$
b. $1s^2 2s^2 2p^6 3s^2 3p^4$
c. $1s^2 2s^2 2p^6 3p^4 4s^2$
d. $1s^2 2p^6 3s^2 3p^6$
e. $1s^2 2s^2 3s^2 2p^8 3p^2$

6. What is the correct box diagram for tungsten?

a.

1s [↑↓] 2s [↑↓] 2p [↑↓][↑↓][↑↓] 3s [↑↓] 3p [↑↓][↑↓][↑↓] 4s [↑↓] 3d [↑↓][↑↓][↑↓][↑↓][↑↓] 4p [↑↓][↑↓][↑↓]

5s [↑↓] 4d [↑↓][↑↓][↑↓][↑↓] 6s [↑↓] 5p [↑↓][↑↓][↑↓] 4f [↑↓][↑↓][↑↓][↑↓][↑↓][↑↓][↑↓] 5d [↑][↑][↑][↑][ ]

b.

1s [↑↓] 2s [↑↓] 2p [↑↓][↑↓][↑↓] 3s [↑↓] 3p [↑↓][↑↓][↑↓] 4s [↑↓] 3d [↑↓][↑↓][↑↓][↑↓][↑↓]

4p [↑↓][↑↓][↑↓] 5s [↑↓] 5p [↑↓][↑↓][↑↓] 4d [↑↓][↑↓][↑↓][↑↓][↑↓] 4f [↑↓][↑↓][↑↓][↑↓][↑↓][↑↓][↑↓] 6s [↑↓]

5d [↑↓][↑][↑][↑][↑]

c.

1s	2s	2p			3s	3p			4s	3d				
↑↓	↑↓	↑↓	↑↓	↑↓	↑↓	↑↓	↑↓	↑↓	↑↓	↑↓	↑↓	↑↓	↑↓	↑↓

4p			5s	4d					5p			6s	5d				
↑↓	↑↓	↑↓	↑↓	↑↓	↑↓	↑↓	↑↓	↑↓	↑↓	↑↓	↑↓	↑↓	↑↓	↑↓	↑↓	↑↓	↑↓

6p			4f						
↑↓	↑↓	↑↓	↑	↑					

d.

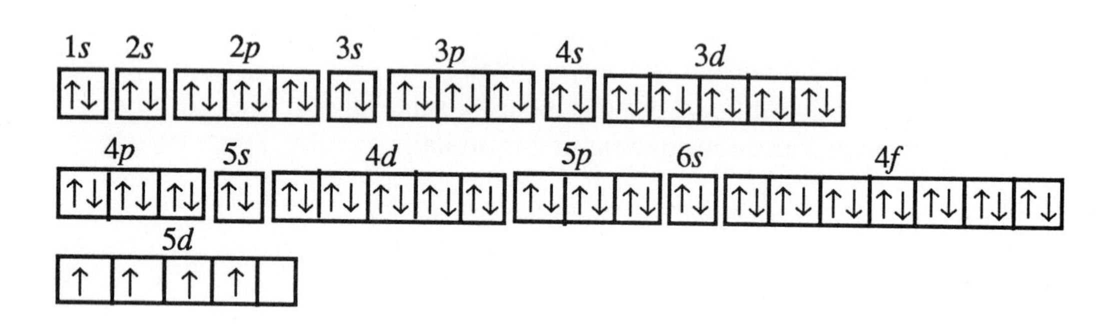

e.

1s	2s	2p			3s	3p			4s	3d				
↑↓	↑↓	↑↓	↑↓	↑↓	↑↓	↑↓	↑↓	↑↓	↑↓	↑↓	↑↓	↑↓	↑↓	↑↓

4p			5s	4d					6s	5p			4f						
↑↓	↑↓	↑↓	↑↓	↑↓	↑↓	↑↓	↑↓	↑↓	↑↓	↑↓	↑↓	↑↓	↑↓	↑↓	↑↓	↑↓	↑↓	↑↓	↑↓

5d				
↑	↑	↑	↑	

7. Which element has its last electron in a 5*f* orbital?

    a. radium
    b. mercury
    c. europium
    d. osmium
    e. neptunium

8. Which element has 7 valence electrons?

    a. tin
    b. iodine
    c. selenium
    d. lead
    e. molybdenum

9. Which element has the lowest ionization energy?

    a. barium
    b. iodine
    c. iron
    d. phosphorus
    e. beryllium

10. Which element has the largest atomic size?

    a. carbon
    b. xenon
    c. barium
    d. sulfur
    e. chlorine

PRACTICE EXAM ANSWERS

1. d (11.1)
2. c (11.2)
3. a (11.3 and 11.4)
4. e (11.5)
5. b (11.7)
6. d (11.8)
7. e (11.8)
8. b (11.7)
9. a (11.9)
10. c (11.9)

CHAPTER 12: CHEMICAL BONDING

INTRODUCTION

There are one hundred and nine elements now known, which seems like a lot of elements, yet the elements can combine to produce a far greater number of different molecules. One of the most important substances in our environment, water, is made from two hydrogen atoms and one oxygen atom which have bonded together. Understanding how and why the elements combine is a fundamental chemical concept that helps chemists predict the structures and properties of new molecules.

AIMS FOR THIS CHAPTER

1. Know the different types of bonds which can form between atoms. (Section 12.1)
2. Know why different kinds of bonds form. (Section 12.1)
3. Be able to describe covalent bonds in terms of electronegativity. (Section 12.2)
4. Know how electronegativity varies from left to right across the periodic table, and from bottom to top. (Section 12.2)
5. Be able to tell which molecules will be polar, based on bond polarity. (Section 12.3)
6. Show how representative metals and nonmetals react to form ions with noble gas electron configurations. (Section 12.4)
7. Predict formulas of ionic compounds based on noble gas configurations of ions. (Section 12.4)
8. Understand the structure of ionic compounds. (Section 12.5)
9. Predict the size of an anion or cation, relative to the size of the neutral atom, and based on location within the group of the periodic table. (Section 12.5)
10. Be able to write Lewis structures for ionic and covalent compounds and molecules with double and triple bonds. (Sections 12.6 and 12.7)
11. Be able to determine the molecular structure, electron arrangement, and bond angles between pairs of atoms for a given molecule. (Section 12.8)
12. Be able to state the premises of the VSEPR model. (Section 12.9)
13. Be able to apply the VSEPR model to molecules with double bonds. (Section 12.10)

QUICK DEFINITIONS

Bond A force between atoms which holds them together as a unit. (Section 12.1)

Bond energy The energy required to break a bond. (Section 12.1)

Ionic bond A bond formed when electrons are transferred from one atom to another. (Section 12.1)

Ionic compound	The kind of bond which forms when a metal reacts with a nonmetal. (Section 12.1)
Covalent bond	A bond formed when electrons are shared between two atoms. (Section 12.1)
Polar covalent bond	A bond formed when electrons are shared unequally between two atoms. One atom has a partial positive charge, and the other atom has a partial negative charge. (Section 12.1)
Electronegativity	A measure of the attraction an atom has for the electrons the atom shares in a covalent bond. (Section 12.2)
Dipole moment	Molecules which have a polar covalent bond and have distinct positive and negative ends. These molecules are said to have a dipole moment, centers of positive and negative charge. (Section 12.3)
Lewis structure	A diagrammatic representation of a molecule showing how valence electrons are arranged among the atoms in the molecule. (Section 12.6)
Duet rule	When hydrogen shares electrons, it tries to fill its valence electron orbital with two (a duet of) electrons. (Section 12.6)
Octet rule	Atoms other than hydrogen try to fill their valence electron orbitals with eight (an octet of) electrons. (Section 12.6)
Bonding pair	A pair of valence electrons shared in a covalent bond. (Section 12.6)
Lone pairs	Pairs of electrons not shared in a covalent bond. (Section 12.6)
Single bond	One pair of electrons being shared between two atoms. (Section 12.7)
Double bond	Two pairs of electrons being shared between two atoms. (Section 12.7)
Triple bond	Three pairs of electrons being shared between two atoms. (Section 12.7)

Resonance A situation which occurs when more than one Lewis structure can be drawn for a molecule. (Section 12.7)

Resonance structures The possible Lewis structures which can be drawn for one molecule. (Section 12.7)

Molecular structure The arrangement in space of atoms in a molecule. Also called the geometric structure. (Section 12.8)

Bond angle An angle described by the location of two or three atoms in space. Bond angles help describe the shape of a molecule. (Section 12.8)

Linear A molecular structure which occurs when there are two pairs of electrons which spread out on either side of a central atom to form bond angles of 180 degrees. (Sections 12.8 and 12.9)

Trigonal planar A molecular structure which occurs when there are three pairs of electrons which spread out flat in space to form bond angles of 120 degrees. (Section 12.8)

Tetrahedral structure A molecular structure which occurs when there are four pairs of electrons which spread out to the four corners of a tetrahedron to form bond angles of 109.5 degrees. (Sections 12.8 and 12.9)

VSEPR model The Valence Shell Electron Pair Repulsion model says that electrons shared in a bond tend to spread out away from other bonding electrons as far as possible. (Section 12.9)

Trigonal pyramid A molecular structure which occurs when there are three shared pairs of electrons and one lone pair of electrons, as in ammonia. (Section 12.9)

Effective pair Two pairs of electrons (a double bond) shared between two atoms that act as a single unit. (Section 12.10)

CONTENT REVIEW

12.1 TYPES OF CHEMICAL BONDS

How Does a Bond Form Between Two Atoms?

A chemical bond can form between any two atoms when the new combination of atoms is more stable than are the individual atoms.

When a metal forms a bond with a nonmetal, electrons are transferred from the metal to the nonmetal, creating oppositely charged ions. The oppositely charged ions are attracted to each other which produces an **ionic bond.** Two nonmetals can also form a bond. When two nonmetals bond, they share electrons to form a **covalent bond.**

Potassium is an alkali metal from Group 1 of the periodic table, and bromine is a halogen (nonmetal) from Group 7. They can form a chemical bond when potassium loses its one valence electron to bromine. The potassium ion with a positive charge is attracted to the bromide ion with a negative charge and the ionic compound KBr is formed.

Carbon and chlorine are both nonmetals on the right end of the periodic table. Reactions between metals and nonmetals result in the formation of ionic bonds. But when two nonmetals combine they form a chemical bond by sharing electrons. Each nucleus attracts the electrons and as a result, they are held between each nucleus.

12.2 ELECTRONEGATIVITY

What Happens When Electrons Are Not Shared Equally?

Sometimes electrons are not shared equally between two bonding atoms. The electrons are attracted to one of the atoms more than the other. **Electronegativity** is a measure of the amount of attraction an atom has for the electrons it shares. Unequal sharing results in the formation of **polar covalent bonds.**

In bonds which form between two atoms of the same kind, the electrons are shared equally between the two atoms. The attraction of each identical nucleus for the electrons is the same. For example, in a molecule of N_2 the bonding electrons are shared equally by both nitrogen atoms.

In bonds which form between two dissimilar atoms such as hydrogen and chlorine, the electrons are shared unequally. The tendency of an atom in a bond to attract the electrons it shares toward itself is often greater for one atom than for the other. The measure of this attraction is called electronegativity. A difference in electronegativy between two atoms results in unequal sharing. Chlorine has an electronegativity value of 3.0 and hydrogen has a value of 2.1. Because chlorine has a value greater than hydrogen, the electrons in the bond are more likely to be found closer to chlorine. A listing of electronegativity values is given in Figure 12.3 of your textbook.

There are some periodic trends in electronegativity. Elements on the upper right of the periodic table have the highest electronegativity values. Electronegativity values generally increase from left to right, and from bottom to top. If you forget the exact electronegativity value for an atom, you should be able to tell which atom in a bond is more electronegative just from the position of the two atoms in the periodic table relative to each other.

In the bond formed between nitrogen and hydrogen, hydrogen has an electronegativity value of 2.1, while nitrogen has a value of 3.5. The electronegativity difference is 1.4 units. Because nitrogen has a larger electronegativity value, the nitrogen atom attracts the electrons more than H does. We say that nitrogen has a partial negative charge because the probability of finding the shared electrons near the nitrogen atom is greater than near the hydrogen atom. In this bond hydrogen has a partial positive charge because the probability of finding the shared electron near it is smaller than near the nitrogen atom.

Both atoms in the carbon-sulfur bond have the same electronegativity value. The shared electrons are attracted equally by both atoms. This is an example of a covalent bond between two dissimilar atoms which is not polar. Neither atom is partially positive nor partially negative.

12.3 BOND POLARITY AND DIPOLE MOMENTS

How Does the Polarity of a Bond Affect the Polarity of a Molecule?

The bond between hydrogen and fluorine in HF is a polar bond. The fluorine atom bears a partial negative charge, and the hydrogen a partial positive charge. Diatomic molecules such as HF which have a center of positive charge and a center of negative charge are said to have a **dipole moment**.

$$\overset{\delta^+}{H}—\overset{\delta^-}{F}$$

Some molecules with more than two atoms also have a dipole moment. Water has two atoms of hydrogen and one atom of oxygen. Each of the bonds between hydrogen and oxygen is a polar bond. The oxygen bears a partial negative charge, while each of the hydrogen atoms bears a partial positive charge. The end of the water molecule with the oxygen atom has a center of negative charge, and the end with the two hydrogen atoms has a center of positive charge.

H
δ^+ O δ^-
H

The dipole moment of a water molecule plays an important role in the properties of water. For example, polar water molecules attract each other. The partial positive end of one water molecule is attracted to the partial negative end of another. So water molecules stick together. A great amount of heat is required to change liquid water to a gas. As a result, water on the earth's surface remains in the liquid state. Polar water molecules also interact with ionic compounds. The partial positive end of water is attracted to negative ions of an ionic compound and dissolves them in solution.

12.4 STABLE ELECTRON CONFIGURATIONS AND CHARGES ON IONS

How Can You Determine How Many Electrons Will Be Lost or Gained When Ions Form?

With a few exceptions atoms tend to form bonds by achieving a noble gas configuration of valence electrons. This means most atoms share or transfer enough electrons to have eight (an octet) in their valence shells. Magnesium, a metal from Group 2 of the periodic table, loses two electrons to a nonmetal.

$$Mg \rightarrow Mg^{2+} + 2e^-$$

By losing two electrons, magnesium achieves the same stable valence electron configuration as neon, which is $1s^22s^22p^6$. Elements tend to gain, lose or share electrons so that their valence electron configuration is the same as the nearest noble gas.

Each chlorine atom in a chlorine molecule gains an electron from a metal to form two chloride ions.

$$2e^- + Cl_2 \rightarrow 2Cl^-$$

A chlorine atom which has an electron configuration of $1^22s^22p^63s^23p^5$ can achieve the valence electron configuration of the noble gas argon, $1s^22s^22p^63s^23p^6$, by gaining 1 electron.

Most representative elements follow the octet rule, but there are exceptions among the nonrepresentative elements.

12.5 IONIC BONDING AND STRUCTURES OF IONIC COMPOUNDS

What Is the Structure of Ionic Compounds?

Potassium chloride forms a regular crystalline structure made of potassium and chloride ions held together by the attractions between the oppositely charged ions. The smaller cations alternate in the structure with larger anions. In the diagram below, the black spheres represent the anions, while the white spheres represent the cations. The diagram shows only a small part of a potassium chloride crystal.

Ionic compounds can be made from a polyatomic ion and a metal, as well as from a metal and a nonmetal. The ionic compound sodium sulfate is composed of two sodium ions, Na^+, and a sulfate ion, SO_4^{2-}. The sulfate ion as a unit has a 2^- charge. Two sodium ions and the sulfate ion are bonded together by attractions between the oppositely charged ions.

What is the Size of an Ion, Compared With an Atom?

Anions are larger than the parent atoms because they have one or more extra electrons, which increases the distance from the nucleus to the outermost electron. Cations are usually smaller than parent atoms because they have lost one or more electrons.

12.6 LEWIS STRUCTURES

How Can You Write Lewis Structures For Simple Molecules?

Lewis structures show the arrangement of the valence electrons in compounds. When writing Lewis structures of ionic compounds, show the valence electron configuration <u>after</u> electrons have been transferred from the cation to the anion. You can also draw Lewis structures for covalent molecules. When you first begin to draw Lewis structures, follow the steps from your textbook until you have gained some experience. These steps are repeated below.

Steps For Writing Lewis Structures

Step 1	Obtain the sum of the valence electrons from all of the atoms. Do not worry about keeping track of which electrons come from which atoms. It is the *total* number of electrons that is important.
Step 2	Use one pair of electrons to form a bond between each pair of bound atoms. For convenience, a line (instead of a pair of dots) is usually used to indicate each pair of bonding electrons.
Step 3	Arrange the remaining electrons to satisfy the duet rule for hydrogen and the octet rule for the second-row elements.

Example:

What is the Lewis structure of lithium iodide?

Lithium iodide is an ionic compound. Lewis structures of ionic compounds are drawn by removing one or more electrons from the metal to produce an ion with a noble gas configuration of valence electrons. A bromine atom has seven valence electrons, while a lithium atom has one. When an ionic bond is formed between lithium and bromine, an electron is removed from lithium and transferred to bromine. Bromine achieves an octet with the same valence electron configuration as the noble gas krypton, while lithium achieves a duet with the same valence electron configuration as helium. When writing Lewis structures of ionic compounds, remember to show the charges present on the ions.

$$Li^{+}\ :\ddot{\underset{..}{Br}}:^{-}$$

We do not put a dash between the lithium and the bromide ions because there are no shared electrons.

Example:

What is the Lewis structure of Br_2?

Let's follow the three steps for writing Lewis structures given on page 380 of your textbook. Step 1. Determine the total number of valence electrons in all atoms of the molecule. In this molecule each bromine atom has seven valence electrons, so the total number of valence electrons is fourteen. Step 2. Place a pair of electrons between each pair of atoms which are bonded together. In this molecule, arrange one pair of electrons between the two bromine atoms. Step 3. If there are electrons remaining, distribute them around the atoms in the molecule so that hydrogen satisfies the duet rule and other atoms satisfy the octet rule. In this molecule the number of valence electrons left is twelve. Arrange these electrons around the bromine atoms as unshared pairs until each bromine atom has an octet.

$$:\ddot{\underset{..}{Br}}-\ddot{\underset{..}{Br}}:$$

Example:

What is the Lewis structure of HCl?

Step 1. There are eight valence electrons in HCl, one contributed by hydrogen and seven by chlorine. Step 2. One electron pair will be shared between hydrogen and chlorine.

$$H-Cl$$

Step 3. The six remaining electrons should be distributed around the atoms until complete duets or octets are achieved. By sharing an electron pair with chlorine, hydrogen has achieved a duet, so the remaining six electrons must be arranged as unshared pairs around the chlorine atom.

$$H-\ddot{\underset{..}{Cl}}:$$

Now, both hydrogen and chlorine are satisfied.

12.7 LEWIS STRUCTURES OF MORE COMPLEX MOLECULES

How Can You Write Lewis Structures For Molecules with Double and Triple Bonds?

Example:

What is the Lewis structure of H_2CO?

For more complicated structures, use the same three rules we have used previously. Step 1. The total number of valence electrons is twelve, two from the hydrogens, four from carbon, and six from oxygen. Step 2. Arrange a pair of electrons between each atom.

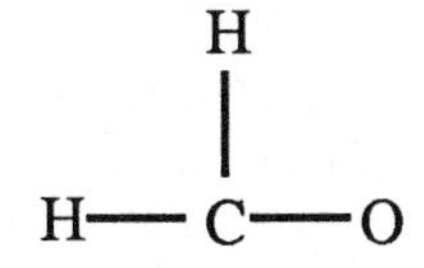

```
      H
      |
  H — C — O
```

Step 3. We have used six electrons between carbon and the two hydrogen atoms and between carbon and oxygen. There are now six electrons left unused. Each hydrogen satisfies the duet rule, so distribute the remaining six electrons around the oxygen atom.

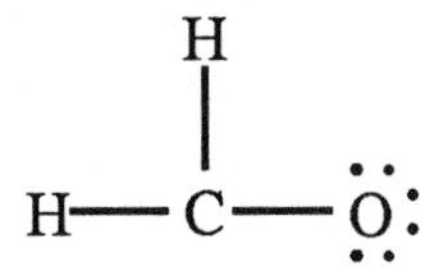

```
      H
      |     ..
  H — C — O :
            ..
```

In this Lewis structure the total number of valence electrons is OK, but carbon does not have an octet. There must be another solution. If carbon and oxygen share two pairs of electrons, then the remaining four electrons can be arranged as unshared electrons around the oxygen. All the atoms are now satisfied according to the octet rule. As you can see, the first solution will not always work. You must find the arrangement which accounts for all the valence electrons, and satisfies the octet or duet rule for all atoms.

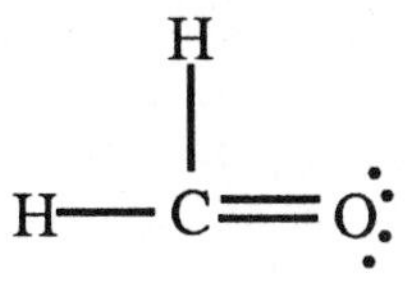

```
      H
      |
  H — C ═ O:
```

12.8 MOLECULAR STRUCTURE

What Is Meant by Molecular Structure?

The **molecular structure** of a molecule is the three-dimensional shape the molecule assumes in space. We can often define molecular structure by looking at the angle formed by atoms in a molecule. This angle is called the **bond angle**. Typical bond angles are 180° for **linear** molecules and 120° for **trigonal planar** molecules. The bond angle for a water molecule is 106°. When there are four atoms in a molecule, the atoms spread out to form a **tetrahedron** around the central atom.

12.9 MOLECULAR STRUCTURE: THE VSEPR MODEL

What Is the VSEPR Model?

The **VSEPR model** says that the structure of a molecule depends on each electron pair spreading out in space as far away from each other as possible. When there are two electron pairs around an atom, each pair spreads out so that there is one pair on one side of the atom and the other pair on the other side of the atom. This produces a **linear structure**. We can illustrate this with some hypothetical atoms, x and y. x is the central atom and is bonded to two y atoms. When the two electron pairs shared between atom x and atom y spread out, they move to opposite sides of the x atom, producing an angle of 180°.

180°

y— x — y

When there are three pairs of electrons around the x atom, the y atoms spread out to form a flat plane, with equal bond angles of 120°. This type of structure is called **trigonal planar**.

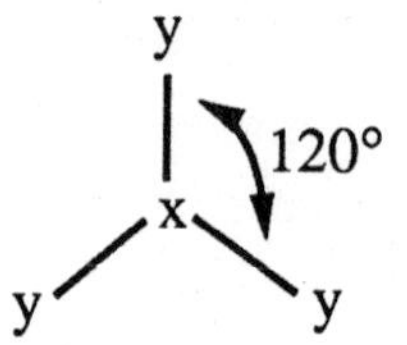

y
120°
x
y y

When there are four electron pairs surrounding the x atom, the y atoms appear to produce bond angles of 90°. But, the electron pairs can spread out even further in three dimensions to form a tetrahedron. The bond angles in this shape are 109.5°, which is greater than 90°. This is called a **tetrahedral arrangement**.

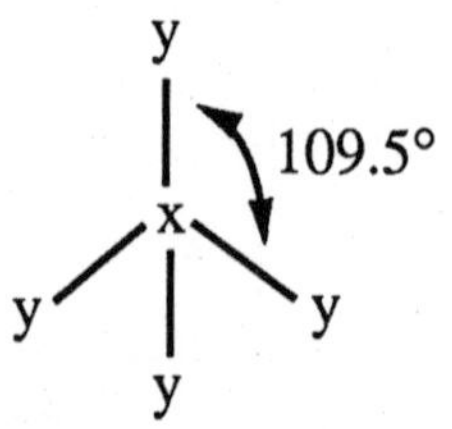

y
109.5°
x
y y
y

How Can the Shape of a Molecule Be Predicted Using the VSEPR Model?

It is possible to predict the shape of a molecule whose shape you do not know. For example, carbon tetrachloride has the formula CCl_4. To determine the molecular structure for CCl_4, we can follow the steps on page 393 of your textbook. Step 1. Draw the Lewis structure of the molecule. Let's draw the Lewis structure for carbon tetrachloride. There are a total of thirty-two valence electrons in CCl_4, four from carbon and seven from each of the four chlorine atoms. Arrange four pairs between the carbon atom and each of the chlorine atoms.

```
      Cl
      |
Cl —  C  — Cl
      |
      Cl
```

This gives carbon an octet. Distribute the remaining electrons around the chlorine atoms to produce

```
       ..
      :Cl:
        |
  ..    |     ..
 :Cl —  C  — Cl:
  ..    |     ..
        |
      :Cl:
       ..
```

Step 2. Count the electron pairs and arrange them as far apart as possible. From the Lewis structure, we can see that there are four pairs of electrons around the central atom, carbon. Step 3. All four pairs of electrons are shared between carbon and chlorine atoms. All four chlorine atoms will occupy a corner of the tetrahedron. Step 4. When four atoms occupy the corners of a tetrahedron, the molecular shape is a tetrahedron and the bond angle is 109.5 °.

```
      Cl
      |  \ 109.5°
      C
    / | \
  Cl  |  Cl
      Cl
```

12.10 MOLECULAR STRUCTURE: MOLECULES WITH DOUBLE BONDS

How Can the Shape of Molecules With Double Bonds be Predicted Using the VSEPR Model?

The VSEPR model can predict the molecular geometry of molecules with double bonds as well as those with only single bonds. We can use the sulfate ion as an example. In each of the two resonance structures sulfur shares six electron pairs with four oxygen atoms. Two of the bonds to oxygen are double bonds. Sulfur shares two pairs of electrons with two of the oxygen atoms to form the double bonds.

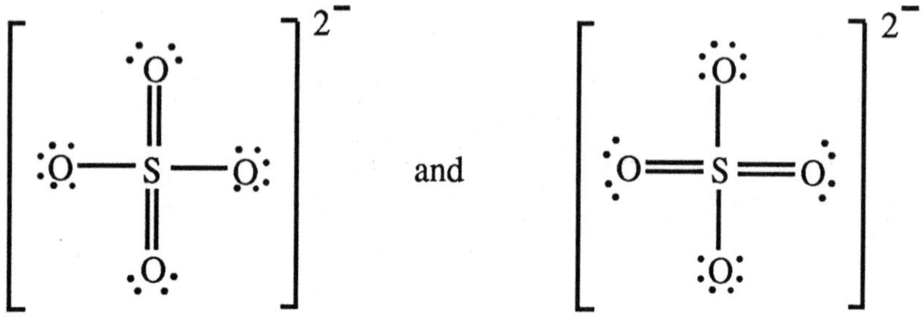

The VSEPR model as applied to molecules with double bonds says that two pairs of electrons act as one region of electron density. The two pairs of electrons become an "effective pair". So when counting the pairs of electrons surrounding an atom a double bond counts as one pair. The sulfate ion has four effective pairs of electrons. The molecular geometry of molecules with four effective pairs of electrons is tetrahedral.

LEARNING REVIEW

1. What are the two kinds of bonds which can form between atoms?

2. What kind of bond forms between

   a. two identical atoms?
   b. a metal and a nonmetal?

3. What is meant by a polar covalent bond?

4. Which of the choices has an ionic bond?

   a. CO
   b. $CaBr_2$
   c. HBr
   d. Cl_2

5. Arrange the following atoms based on electronegativity. Put the most electronegative atom on the right and the least electronegative atom on the left.

P Al Cl Mg

6. Which bond is the most polar? Which bond is the least polar?

   a. P-Cl
   b. H-H
   c. N-H
   d. C-F

7. Write electron configurations for both reactants and products for the reactions below.

   a. $Mg + Cl_2 \rightarrow Mg^{2+} + 2Cl^-$
   b. $2Li + S \rightarrow 2Li^+ + S^{2-}$

8. When two nitrogen atoms combine to form a nitrogen molecule, how many electrons are shared to give each atom a complete octet?

9. How many electrons do each of the atoms below need to gain, lose or share to achieve a noble gas valence electron configuration?

   a. S
   b. Mg
   c. C

10. In which of the atom/ion pairs is the ion **smaller** than the atom?

    a. S/S^{2-}
    b. Ca/Ca^{2+}
    c. Li/Li^+
    d. I/I^-

11. Draw Lewis structures for these ionic compounds.

    a. MgS
    b. Na_2O

12. Write valence electron configurations for the following atoms.

   a. B
   b. Sr
   c. Kr
   d. Cl

13. Draw Lewis structures for the molecules or ions below.

   a. H_2S
   b. ClO^-
   c. HCCH
   d. PO_4^{3-}
   e. HI
   f. PCl_3

14. Some molecules have Lewis structures which violate the octet rule. Draw a probable Lewis structure for BeI_2.

15. Use the VSEPR model to determine the molecular structure of each of the molecules below.

   a. SbF_3
   b. BH_3
   c. SiH_4
   d. 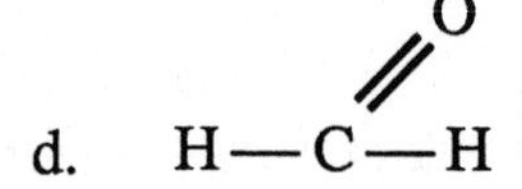

ANSWERS TO LEARNING REVIEW

1. When atoms combine they form either ionic bonds or covalent bonds. In an ionic bond, electrons are completely transferred. In a covalent bond, electrons are shared between atoms. Ionic bonds usually form when a metal and a nonmetal react. Covalent bonds usually form when two nonmetals react.

2. a. When two identical atoms bond, a nonpolar covalent bond forms.
   b. When a metal and a nonmetal bond, an ionic bond forms.

3. Electrons shared between two atoms are not always shared equally. Sometimes, the electrons are attracted to one of the atoms more than the other. The atom in a bond which attracts the electron pair will have an extra electron part of the time and so bear a partial negative charge. The atom which does not strongly attract the electron pair it shares will be electron deficient part of the time and so bears a positive charge. The kind of covalent bond where the electrons are not shared equally is called a polar covalent bond.

4. Ionic bonds are formed when a metal loses an electron to a nonmetal. Among these choices, the only bond between a metal and a nonmetal is the bond formed between calcium and bromine to form calcium bromide. So the correct answer is b.

5. Elements on the right side of the periodic table have higher electronegativity values than do elements on the left side of the periodic table. The atom with the highest electronegativity would be Cl, because it is in the upper righthand corner of the periodic table; then P, then Al, and Mg, on the left side of the periodic table is the lowest.

Mg Al P Cl

6. To determine the polarity of a bond, subtract the electronegativity of the least electronegative atom from the electronegativity of the most electronegative atom. The largest difference is the most polar bond and the smallest difference is the least polar bond.

   a. P is 2.1 while Cl is 3.0. The difference is 0.9.
   b. Both H atoms are 2.1. The difference is 0.0.
   c. N is 3.0 while H is 2.1. The difference is 0.9.
   d. C is 2.5 while F is 4.0. The difference is 1.5.

   The most polar of these bonds is the C-F bond, because the difference in electronegativities is the highest, 1.5. The least polar bond is that which is formed between two hydrogen atoms. The difference in electronegativities is 0, which means that this bond is completely nonpolar.

7. a. $Mg + Cl_2 \rightarrow Mg^{2+} + 2Cl^-$
   $1s^22s^22p^63s^2 + 1s^22s^22p^63s^23p^5 \rightarrow$
   $1s^22s^22p^6 + 2(1s^22s^22p^63s^23p^6)$
   b. $2Li + S \rightarrow 2Li^+ + S^{2-}$
   $2(1s^22s^1) + 1s^22s^22p^63s^23p^4 \rightarrow 2(1s^2) +$
   $1s^22s^22p^63s^23p^6$

8. Nitrogen is in Group 5 of the periodic table, which means that it has 5 valence electrons. Most elements obey the octet rule when they bond with other atoms, so a nitrogen molecule should be composed of two nitrogen atoms each with a complete octet. When

two nitrogen atoms bond, they can only achieve an octet by sharing six electrons, three from each nitrogen atom. So in a nitrogen molecule each nitrogen atom has two valence electrons of its own and shares six with another nitrogen atom.

9. a. Sulfur is in Group 6 of the periodic table so it has six valence electrons. Sulfur needs two more electrons to fill its valence shell. Sulfur often gains two electrons to become a sulfide ion, S^{2-}. The sulfide ion has the same electron configuration as argon.

   b. Magnesium is in Group 2 of the periodic table and has two valence electrons. Magnesium atoms lose two electrons to form the Mg^{2+} ion. The magnesium ion has the same electron configuration as neon.

   c. Carbon is in Group 4 of the periodic table so it has four valence electrons. Each carbon atom will share four valence electrons with other atoms. When a carbon atom shares four electrons with other atoms, it has the same electron configuration as neon.

10. a. The sulfide ion has gained two electrons so it is larger than the sulfur atom.

    b. The calcium ion has lost two electrons so it is smaller than the calcium atom.

    c. The lithium ion has lost one electron so it is smaller than the lithium atom.

    d. The iodide ion has gained one electron so it is larger than the iodine atom.

11. a. In ionic compounds, the electrons removed from the metal are transferred to the nonmetal so that each atom can achieve a noble gas configuration.

$$Mg^{2+} \quad [:\ddot{\underset{..}{S}}:]^{2-}$$

    b. Two sodium atoms each lose one electron to an oxygen atom so that all three atoms achieve a noble gas configuration.

$$Na^{+} \quad [:\ddot{\underset{..}{O}}:]^{2-} \quad Na^{+}$$

12. a. B $2s^2 2p^1$

    b. Sr $5s^2$

    c. Kr $4s^2 4p^6$

    d. Cl $3s^2 3p^5$

13. a. Step 1. Find the total number of valence electrons in all the atoms. There are eight valence electrons in a molecule of H_2S. Six come from sulfur and one each from hydrogen. Step 2. Begin distributing the available valence electrons by putting a pair of electrons between each atom. An electron pair is often symbolized with a line.

H—S—H

Step 3. We have used up four of the available electrons. There are four more to distribute. Each of the two hydrogen atoms is satisfied, so the four extra electrons must appear as unshared pairs on the sulfur atom.

H—$\ddot{\underset{\cdot\cdot}{S}}$—H

Now all atoms satisfy either the octet or the duet rule.

b. You can draw a Lewis structure for ions like the hypochlorite ion by following the same rules you do for a molecule. Step 1. Add together all the valence electrons, plus an extra one for the negative charge. ClO^- would have a total of fourteen valence electrons: seven from chlorine, six from oxygen, and one extra one because of the negative charge. Step 2. Arrange a pair of electrons between the two atoms.

Cl—O

Step 3. There are now twelve valence electrons left to distribute. The leftover valence electrons must be unshared pairs around the chlorine and oxygen atoms. If we distribute the remaining electrons so that the octet rule is obeyed, there are no electrons left over. Put square brackets around the Lewis structure to indicate that this is an ion, and show the negative charge.

$[:\ddot{\underset{\cdot\cdot}{Cl}}—\ddot{\underset{\cdot\cdot}{O}}:]^-$

c. Step 1. The total number of valence electrons for two carbon atoms and two hydrogens is ten, four from each carbon atom and one each from the hydrogens. Step 2. Arrange the electron pairs between atoms.

H—C—C—H

Step 3. There are now four electrons left. If we distribute two around each carbon atom we have

H—$\ddot{C}$—$\ddot{C}$—H

The carbon atoms do not fulfill the octet rule and we have run out of electrons. Some of the atoms must share more than one pair of electrons in order for the octet rule to be satisfied. We cannot share more electrons between the carbon and the hydrogen atoms because hydrogen already satisfies the duet rule. So the electrons must be shared between the two carbon atoms. Let's begin by sharing one more pair of electrons. The result is

$$H—C{=}C—H$$

This leaves us with two electrons left.

$$H—\dot{C}{=}\dot{C}—H$$

Carbon is still not satisfied. Let's share another electron pair between the two carbon atoms.

$$H—C{\equiv}C—H$$

Now, all of the valence electrons are used and both hydrogen and carbon are satisfied.

d. Step 1. The phosphate ion, PO_4^{3-}, has a total of thirty-two valence electrons: twenty-four from the oxygen atoms, five from phosphorus, and three extra electrons because of the 3^- charge on the phosphate ion. Step 2. Distribute pairs of electrons among the atoms.

```
      O
      |
  O — P — O
      |
      O
```

Step 3. There are twenty-four electrons left. Begin arranging the electrons around the atoms to satisfy the octet rule for all the atoms. Three unshared pairs of electrons surround each oxygen atom. The complete Lewis structure is

```
 [       :Ö:        ] 3-
 [        |         ]
 [  :Ö — P — Ö:     ]
 [        |         ]
 [       :Ö:        ]
```

e. Step 1. HI has a total of eight valence electrons. Step 2. Arrange an electron pair between the two atoms.

$$H—I$$

Step 3. Now, hydrogen is satisfied and there are six electrons left. Arrange the remaining electron pairs around the iodine atom.

H—:Ï̤:

Both atoms are satisfied.

f. Step 1. Phosphorus trichloride has a total of twenty-six valence electrons. Step 2. Arrange electron pairs among the atoms.

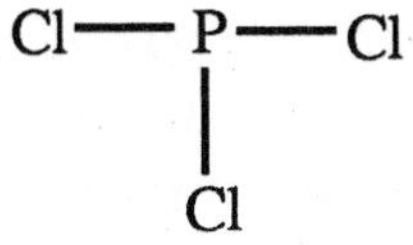

Cl—P—Cl
   |
   Cl

Step 3. There are twenty electrons left. Then, arrange the electrons around each atom, until each has eight electrons. Three unshared pairs surround each chlorine atom, and the phosphorus atom has one unshared pair.

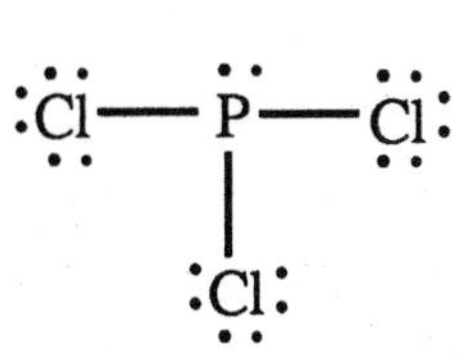

:C̤̈l—P̈—C̤̈l:
   |
  :C̤l:

14. Step 1. Add together the valence electrons contributed by each atom in the compound. Each iodide atom has seven, and berellium has two, for a total of sixteen valence electrons. Step 2. Arrange the electrons in pairs between the atoms.

I—Be—I

Step 3. This leaves twelve electrons to distribute as unshared pairs. The iodide atoms obey the octet rule, but berellium atoms often do not. Berellium is electron deficient and will have only four valence electrons.

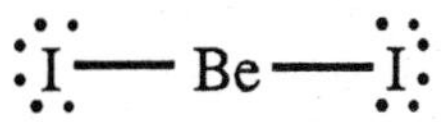

:Ï̤—Be—Ï̤:

15. a. Step 1. Draw the Lewis structure for the molecule. Antimony has five valence electrons and fluorine has seven. Antimony can share an electron pair with each of the three fluorine atoms, so the Lewis structure looks like

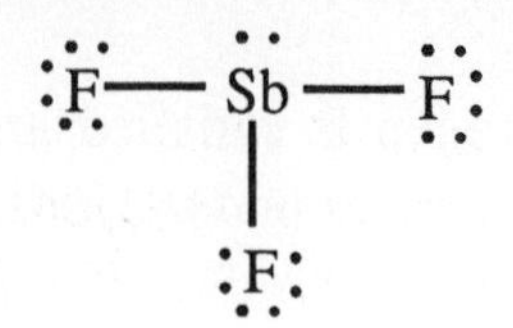

Step 2. Count the electron pairs and arrange them so that they are as far apart as possible. When three pairs of electrons are shared and there is one pair left which is not shared, we can arrange the electrons in a tetrahedron. Step 3. Determine the positions of the atoms using the electron pairs as a guide. The fluorine atoms will occupy the corners of a tetrahedron. The lone pair of electrons on the antimony atom will occupy the fourth corner of the tetrahedron. Step 4. Determine the molecular structure using the positions of the atoms. In this molecule there are three atoms surrounding antimony, and each one is located in the corner of a tetrahedron. Because only three corners of the tetrahedron are occupied by atoms, this molecular shape is a trigonal pyramid.

b. Step 1. The Lewis structure of boron trihydride is

```
      H
      |
  H — B — H
```

Boron has three valence electrons and shares electrons with three hydrogen atoms. Note that boron is an exception to the octet rule.

Step 2. The three pairs of electrons spread themselves out as far as possible to form a trigonal planar shape. Step 3. The three hydrogen atoms occupy the corners of a triangle. Step 4. This molecular structure is called trigonal planar. The bond angle is 120°.

```
     H
     |  ↘ 120°
     B   ↓
    / \
   H   H
```

c. Step 1. Silicon has four valence electrons and each of the hydrogen atoms has one. Silicon can share electron pairs with each of the four hydrogen atoms so the Lewis structure looks like

```
      H
      |
  H — Si — H
      |
      H
```

Step 2. When four pairs of electrons are shared, the VSEPR model predicts that they will spread apart to form a tetrahedron. Step 3. There are four hydrogen atoms bonded to a central silicon atom. The four hydrogen atoms will occupy the four corners of a tetrahedron. Step 4. The molecular structure of this molecule will be a tetrahedron. The bond angle is 109.5°.

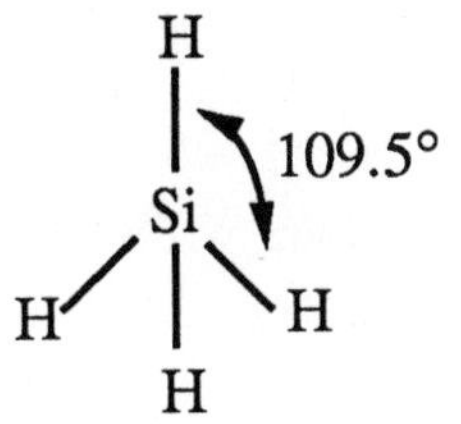

H
Si 109.5°
H H
H

d. Step 1. The Lewis structure for this molecule shows that there is a carbon-oxygen double bond.

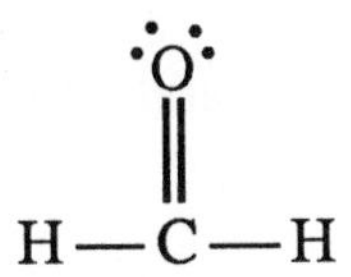

O
H—C—H

Step 2. The three effective pairs of electrons spread themselves out as far as possible to form a trigonal planar shape. Step 3. The two hydrogen atoms and the oxygen atom occupy the corners of a triangle. Step 4. This molecular structure is trigonal planar. The bond angle around the center atom is 120 °.

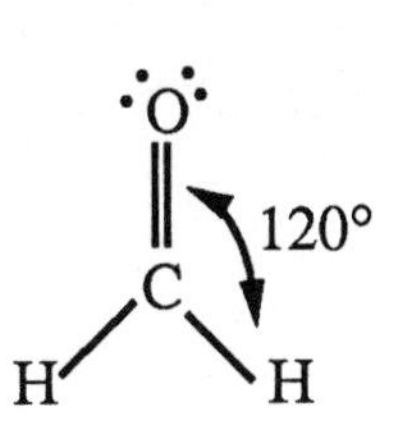

O
C 120°
H H

PRACTICE EXAM

1. Which statement about chemical bonds is **not** true?

   a. A polar bond forms when one atom in a bond strongly attracts the pair of electrons it shares with another atom.
   b. In some molecules, one end will have a partial positive charge, and the other end will have a partial negative charge.
   c. Bond energy is the strength required to break a chemical bond.
   d. A covalent bond usually forms when a metal and a nonmetal bond.
   e. In an ionic bond, electrons are transferred from one atom to another.

2. Which of the atoms below would have the lowest electronegativity value?

   a. H
   b. O
   c. Mg
   d. S
   e. Ba

3. Which of the covalent bonds below will be **most** polar?

   a. H-Cl
   b. O-H
   c. C-O
   d. C-N
   e. S-H

4. Which of the molecules below would exhibit a dipole moment?

   a. $SiCl_4$
   b. CO_2
   c. CH_2Br_2
   d. CH_4
   e. N_2

5. For the ionic compound which forms between potassium and sulfur, which of the ion pairs below shows the correct charges?

   a. K^+, S^{2-}
   b. K^{2+}, S^-
   c. K^{3+}, S^-
   d. K^{2+}, S^{2-}
   e. K^+, S^{3-}

6. Which ion has the largest radius?

   a. bromide
   b. lithium
   c. chloride
   d. iodide
   e. potassium

7. The Lewis structure for which molecule or ion is **not** correct?

a. N_2; $:N{\equiv}N:$

b. MgO, $Mg^{2+} [:\ddot{\underset{..}{O}}:]^{2-}$

c. NH_4^+, $\left[\begin{array}{c} H \\ | \\ H-N-H \\ | \\ H \end{array}\right]^+$

d. H_2CO, $\begin{array}{l} H-C{=}O \\ \quad\;\, | \\ \quad\;\, H \end{array}$

e. HI, $H-\ddot{\underset{..}{I}}:$

8. Which molecule or ion will most likely be paramagnetic?

a. BF_3
b. NH_3
c. XeF_2
d. SO_3^-
e. NO

9. According to the VSEPR model, which molecule will exhibit a linear shape?

a. CH_3-O-H
b. NH_3
c. $SiBr_4$
d. PCl_3
e. CO_2

10. In which of the molecules below would the bond angle around the marked atom be close to 120°?

a.

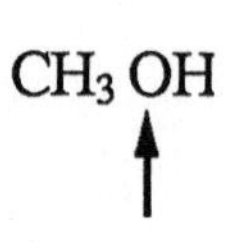

b.

c.

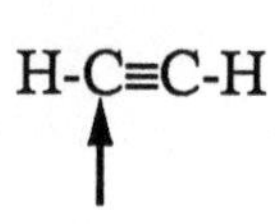

d.

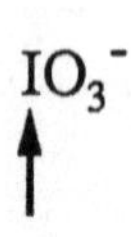

e.

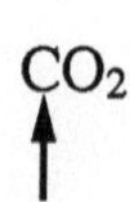

PRACTICE EXAM ANSWERS

1. d (12.1)
2. e (12.2)
3. b (12.2)
4. c (12.3)
5. a (12.4)
6. d (12.5)
7. d (12.6 and 12.7)
8. e (12.7)
9. e (12.8 and 12.9)
10. d (12.9)

CHAPTER 13: GASES

INTRODUCTION

We live in a gaseous environment. The air we breath is a mixture of gases. Yet gases are not as conspicuous as liquids and solids are, and it is easy to overlook their significance. Understanding why gases behave as they do can help us understand everyday occurrences such as "low pressure systems" in our weather and the apparent decrease of the amount of air in car tires in cold weather. This chapter focuses on how gases behave under various circumstances and why they behave as they do.

AIMS FOR THIS CHAPTER

1. Know why a barometer can be used to measure atmospheric pressure. (Section 13.1)
2. Know the common units for measuring pressure, and how to convert between units. (Section 13.1)
3. Know how to use Boyle's law to calculate the new volume or new pressure of a gas. (Section 13.2)
4. Know how to use Charles's law to calculate the new volume or new temperature of a gas. (Section 13.3)
5. Know how to use Avogadro's law to calculate the new volume or the new number of moles of a gas. (Section 13.4)
6. Know how to use the ideal gas law to solve many gas law problems. (Section 13.5)
7. Know how to use Dalton's law of partial pressures to calculate the total pressure of a mixture of gases, or the partial pressure of one gas in a mixture. (Section 13.6)
8. Know the difference between a law and a model. (Section 13.7).
9. Know the parts of the Kinetic Molecular Theory, and how the Kinetic Molecular Theory explains gas behavior. (Sections 13.8 and 13.9)
10. Be able to use the ideal gas law or the molar volume of an ideal gas to solve stoichiometric problems. (Section 13.10)

QUICK DEFINITIONS

Barometer — An instrument which is used to measure atmospheric pressure. (Section 13.1)

mm Hg — A unit of pressure. 760.0 mm Hg equals 1.000 atm. (Section 13.1)

Torr — A unit of pressure named after Evangelista Torricelli. 760.0 torr equals 1.000 atm. (Section 13.1)

Standard atmosphere

A common unit of pressure. 1.000 atm equals 760.0 mm Hg. (Section 13.1)

Pascal

The SI unit of pressure, abbreviated Pa. 1.000 atm equals 101,325 Pa. (Section 13.1)

Boyle's law

Relates pressure and volume and can be written mathematically as $P_1V_1 = P_2V_2$. (Section 13.2)

Extrapolation

Extending a line on a graph where we do not have any data points, based on the trend of the data points we do have. (Section 13.3)

Absolute zero

The temperature below which matter cannot be cooled. This temperature is -273 °C and 0 K. (Section 13.3)

Charles's law

Relates volume and temperature and can be written mathematically as $\frac{V_1}{T_1} = \frac{V_2}{T_2}$.

Avogadro's law

Relates volume and the number of moles and can be written mathematically as $\frac{V_1}{n_1} = \frac{V_2}{n_2}$.

Universal gas constant

When all the constants from Boyle's law, Charles's law and Avogadro's law are combined, the result is the universal gas constant. The gas constant is abbreviated R and has a value of 0.08206 L atm/K mol. (Section 13.5)

Ideal gas law

Combines temperature, pressure, volume, and number of moles of a gas. Can be written mathematically as $PV = nRT$. (Section 13.5)

Ideal gas

A gas for which the ideal gas law has been experimentally shown to hold true. At low pressures and high temperatures most gases follow the ideal gas law. (Section 13.5)

Combined gas law

Combines temperature, pressure and volume, but not the number of moles. Can be expressed mathematically as $\frac{P_1V_1}{T_1} = \frac{P_2V_2}{T_2}$. (Section 13.5)

Partial pressure	The pressure a gas in a mixture of gases would exert on the container if it were the only gas in the container. (Section 13.6)
Dalton's law of partial pressures	The total pressure a mixture of gases exerts is the sum of the partial pressures for all the gases. (Section 13.6)
Kinetic molecular theory	A theory which attempts to explain the behavior of ideal gases by making assumptions about the nature of the gas particles, and the relationships between the particles. (Section 13.8)
Standard temperature and pressure	Standard conditions of 0 °C and 1 atmosphere pressure. Abbreviated STP. (Section 13.10)

CONTENT REVIEW

13.1 PRESSURE

How Can Atmospheric Pressure Be Measured?

Pressure is the force a gas exerts on an object. **Atmospheric pressure** is the force the atmosphere, a mixture of gases, exerts on the earth's surface. The pressure results from gravity pulling the mass of the atmosphere toward the earth's center.

Atmospheric pressure can be measured with a **barometer**. One kind of barometer, invented by Evangelista Torricelli, is made from a long glass tube filled with mercury. The tube is inverted in a dish also containing mercury. Some of the mercury runs out, but most of it stays in the tube. All of the mercury does not run out of the tube because the atmospheric pressure pushes down on the dish of mercury, causing the mercury to remain in the tube. The amount of mercury remaining in the tube depends upon the atmospheric pressure. When the weather changes, the atmospheric pressure often changes as well. Stormy weather is often accompanied by low pressure, which means less pressure on the dish of mercury, and more mercury runs out of the tube. Clear, cool weather often means high pressure, which means more pressure on the dish of mercury, and mercury is pushed higher up in the glass tube. Measuring the height of the column of mercury in the tube gives an indication of atmospheric pressure.

What Are the Common Units of Atmospheric Pressure?

Because the height of mercury in a glass column is a common way of measuring atmospheric pressure, the **mm Hg** is a common unit for reporting pressure. 1 mm Hg is equal to 1 torr, named in honor of Torricelli. Another common unit is the **standard atmosphere**, usually abbreviated atm. One standard atmosphere is exactly 1.000 atm, which is equal to 760.0 mm Hg and 760.0 torr.

The SI unit of pressure is the Pascal, abbreviated Pa. This unit is used less often because the size of a Pascal is small compared with 1 atmosphere. 1.000 atm equals 101,325 Pa.

Some everyday pressure measurements are given in pounds per square inch, psi. 1 standard atmosphere = 14.69 psi.

It is important that you be able to convert among the units of pressure.

Example:

If the compressed air in a cylinder is under 12,700 torr of pressure, what is the pressure in the cylinder expressed in atm, in psi, and in Pa? In order to answer this kind of problem, you need to know the conversions between the various units of pressure. If you can convert the given units to atm, only one additional conversion is needed to produce an answer in any other desired unit. To convert from atm to torr, use the expression 1.000 atm = 760.0 torr We can use the expression as a conversion factor to cancel the unwanted units, torr. The units which remain are atmospheres.

$$12{,}700\ \text{torr} \times \frac{1.000\ \text{atm}}{760.0\ \text{torr}} = 16.7\ \text{atm}$$

Now that we know the number of atmospheres, we can calculate psi. The conversion factor we need is 1.000 atm equals 14.69 torr.

$$16.7\ \text{atm} \times \frac{14.69\ \text{psi}}{1.000\ \text{atm}} = 245\ \text{psi}$$

The expression which relates psi and atm was used to construct the conversion factor which eliminated the unwanted units, atm, giving an answer in the desired units, psi. We can also convert from atmospheres to pascals. We need to use the conversion factor which relates atmospheres and pascals. 1.000 atm equals 101,325 Pa.

$$16.7\ \text{atm} \times \frac{101{,}325\ \text{Pa}}{1\ \text{atm}} = 1.69 \times 10^6\ \text{Pa}$$

Notice that all of the pressure values are expressed to three significant figures because the initial measurement of 12,700 torr contained three significant figures.

13.2 VOLUME AND PRESSURE: BOYLE'S LAW

What Is Boyle's Law and How Can It Be Used?

Robert Boyle studied the relationship between the pressure and the volume of a gas. What happens to the pressure of a gas when the volume changes? Boyle discovered that there was an **inverse relationship** between pressure and volume. When the pressure increased, the volume of the gas went down, and when the pressure decreased, the volume of the gas went up.

Boyle performed many experiments in which he measured the pressure and volume of a gas. When he multiplied the pressure of the gas by the volume of the gas when the temperature and the amount of gas was held constant, he noticed that the product of volume and pressure was always the same. In other words, pressure times volume produces a constant when the temperature and amount of gas are not changed. From this observation we are able to write Boyle's law mathematically as

$$P_1V_1 = \text{constant} = P_2V_2$$

or, as

$$P_1V_1 = P_2V_2$$

where P_1 is the initial pressure, V_1 is the initial volume, P_2 is the new pressure and V_2 is the new volume. This equation is useful because if we know the pressure and volume of a gas (initial conditions) and we change either the pressure or the volume (final conditions), we can calculate the value we do not know.

Example:

If 5.2 L of oxygen gas at 4.2 atm pressure is transferred into a container which holds 7.6 L, what is the new pressure of the oxygen gas? Before you begin solving a gas problem, it is always a good idea to sort the information given to you. In this case, we know both initial conditions; $V_1 = 5.2$ L and $P_1 = 4.2$ atm. We know the final volume, V_2, to be 7.6 L. The quantity we do not know and must find is the final pressure (P_2). Boyle's law can be written as

$$P_1V_1 = P_2V_2$$

We need to rearrange this equation so that P_2 is isolated on one side. To do this, we divide both sides by V_2.

$$P_1 \times \frac{V_1}{V_2} = \frac{P_2 \cancel{V_2}}{\cancel{V_2}}$$

$$P_2 = P_1 \times \frac{V_1}{V_2}$$

Now, substitute known values into the equation.

$$P_2 = 4.2 \text{ atm} \times \frac{5.2 \cancel{\text{L}}}{7.6 \cancel{\text{L}}}$$

$$P_2 = 2.9 \text{ atm}$$

Notice that liters appear in the top and bottom of the equation, and therefore cancel. We are left with the units atm, which is what we wanted. To check your answer, remember that when the volume gets larger, the pressure gets smaller. In this case, the volume increases from 5.2 L to 7.6 L. We started with a pressure of 4.2 atm, and the calculated value for final pressure was 2.9 atm. The pressure decreased, as it should.

13.3 VOLUME AND TEMPERATURE: CHARLES'S LAW

What Is Charles's Law and How Can It Be Used?

Jacques Charles studied the relationship between the volume of a gas and its temperature. What happens to the volume of a gas when the temperature changes? Charles discovered that when the temperature of a gas went up, the volume went up, and when the temperature went down, the volume went down. When an increase or decrease in one property causes a corresponding increase or decrease in another property, there is a **linear relationship** between the two properties.

If a graph of temperature versus volume for several different gases is made, we can follow the path of the line back until we reach a temperature where the volume of the gas should be zero. Each different gas reaches zero volume at the same temperature, -273 °C. This is **absolute zero**, or 0 K, the lowest possible temperature. At this temperature, the volume of a gas should be zero. Chemists have never reached this temperature in the laboratory.

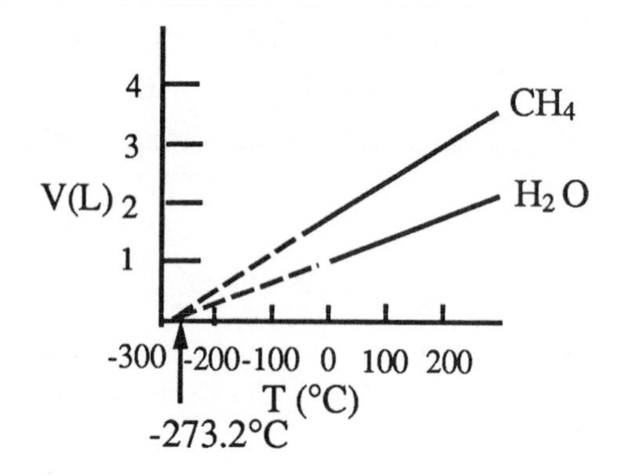

There is a linear relationship between volume and temperature, volume equals temperature times a constant when the pressure is not changed. We can write Charles's law as

$$V = bT \text{ or, by rearranging, } \frac{V}{T} = b$$

This relationship holds true for any volume of gas, as long as the pressure does not change. When working Charles's law problems, the temperature **must** be in kelvins and the volume **must** be in liters. If other units are given, they must be converted to K and L before the problem is solved. We can also write Charles's law as

$$\frac{V_1}{T_1} = \text{constant} = \frac{V_2}{T_2}$$

or, as

$$\frac{V_1}{T_1} = \frac{V_2}{T_2}$$

If we know the volume (V_1) of a gas for one temperature (T_1), and change either the temperature (T_2), or the volume, (V_2), we can calculate the value we do not know.

Example:

If the temperature of 0.54 L of N_2 gas at 383 K is raised by 60. K, what is the new volume? Before you use Charles's law to solve this problem, analyze what information you are given, and what you are asked to find. We know that the initial conditions are 0.54 L and 383 K. Initial conditions are given in liters and kelvins, so no conversions are necessary. Final conditions are an increase in temperature of 60. K. Because the initial temperature was 383 K, the final temperature is 383 plus 60., which equals 443 K. We do not know the final volume. To use Charles's law to solve for V_2, we need to rearrange $\frac{V_1}{T_1} = \frac{V_2}{T_2}$ so that V_2 is isolated on one side of the equation. Multiply both sides of the equation by T_2.

$$V_1 \times \frac{T_2}{T_1} = V_2 \times \frac{\cancel{T_2}}{\cancel{T_2}}$$

$$V_2 = V_1 \times \frac{T_2}{T_1}$$

Now, substitute known values to find the value of V_2

$$V_2 = 0.54 \text{ L} \times \frac{443 \cancel{\text{K}}}{383 \cancel{\text{K}}}$$

$$V_2 = 0.62 \text{ L}$$

kelvins appear in the top and bottom of the equation, and cancel, leaving the desired units, liters. This answer makes sense; because the temperature increases, we expect the volume to increase, too.

13.4 VOLUME AND MOLES: AVOGADRO'S LAW

What Is Avogadro's Law and How Can It Be Used?

Experiments to determine the relationship between the number of moles of a gas and the volume showed that there was a linear relationship between volume and moles. In fact, the volume is directly proportional to the number of moles. This means that if the number of moles is doubled, the volume is doubled, and if the volume is tripled, then the number of moles must be tripled. Avogadro's law holds true only if the temperature and pressure are held constant.

We can write Avogadro's law mathematically as $V = a \times n$, where a is a proportionality constant and n is the number of moles. We can rearrange $V = a \times n$ to give $\frac{V}{n} = a$. Because this relationship holds true for any volume of gas, as long as the temperature and pressure are constant, then

$$\frac{V_1}{n_1} = \text{constant} = \frac{V_2}{n_2}$$

or, we can leave out the constant and the equation is true for a gas undergoing a change in volume or moles

$$\frac{V_1}{n_1} = \frac{V_2}{n_2}$$

We can use this equation to solve problems involving changes in volume or moles, as long as the temperature and pressure do not change.

Example:

At a constant temperature of 235 K and a pressure of 2.0 atm, the number of moles of oxygen gas is increased from 1.54 to 2.06. If the initial volume is 14.8 L, what is the final volume? In this problem, temperature and pressure are constant, and we are provided information about the volume of oxygen, and the number of moles of oxygen. We can use Avogadro's law to solve the problem. We know the initial conditions, V_1 and n_1, and we know the final number of moles, n_2, is 2.06 mol. We need to solve for V_2. Avogadro's law can be written

$$\frac{V_1}{n_1} = \frac{V_2}{n_2}$$

Rearrange to isolate V_2 on one side

$$V_1 \times \frac{n_2}{n_1} = V_2 \times \frac{\cancel{n_2}}{\cancel{n_2}}$$

$$V_2 = V_1 \times \frac{n_2}{n_1}$$

We have algebraically rearranged the equation; now, substitute known values and find V_2.

$$V_2 = 14.8 \text{ L} \times \frac{2.06 \text{ mol}}{1.54 \text{ mol}}$$

$$V_2 = 19.8 \text{ L}$$

Notice that the units mol cancel, leaving liters, which is the desired unit. This answer makes sense. The number of moles of gas has increased, so the volume should increase too.

13.5 THE IDEAL GAS LAW

What Is the Ideal Gas Law and How Can It Be Used?

All of the laws you have studied so far, Boyle's, Charles's and Avogadro's laws, all have a constant as part of their mathematical equations. All three laws can be combined into one, the **ideal gas law**, which can be written

$$PV = nRT$$

In the ideal gas law, all three of the constants from the individual laws are combined into one constant, abbreviated *R*. *R* is called the **universal gas constant** and has the value 0.08206 L atm/K mol. Because the universal gas constant includes units of liters, atmospheres, kelvins, and moles, calculations using this equation must use numbers with the same units or you will be left with uncanceled units.

The ideal gas law contains four variables; pressure (*P*), volume (*V*), temperature (*T*) and number of moles (*n*). If you know the value of any three variables, you can calculate the fourth. The gas constant, *R*, remains the same in all calculations.

Example:

If a sealed 5.50 L canister contains 4.21 mol of the nerve gas phosgene at 25.0 °C, what is the pressure of the phosgene gas in the canister? Analyze the information given before you begin to solve the problem. Volume = 5.50 L, temperature = 25.0 °C, quantity of gas = 4.21 mol. The pressure is unknown. It is possible to use the ideal gas law to solve this problem. Rearrange the equation to isolate *P* on one side.

$$P \times \frac{V}{V} = \frac{nRT}{V}$$

$$P = \frac{nRT}{V}$$

The temperature is given in degrees Celsius and the ideal gas law requires that temperature be in kelvins. You must convert before solving the problem.

$$^{t}K = {}^{t\circ}C + 273$$
$$^{t}K = 25.0 + 273$$
$$^{t}K = 298\ K$$

$$P = \frac{(4.21\ \text{mol})(0.08206\ \text{L atm/K mol})(298\ \text{K})}{5.50\ \text{L}} = 18.7\ \text{atm}$$

Note that the ideal gas law when written as $PV = nRT$ does not work when conditions are changing, that is, when a problem gives initial conditions and final conditions. To use the ideal gas law when conditions are changing, you must rearrange the equation.

Example:

When 0.52 mol of a gas is added to a container with a moveable piston which contains 2.12 mol and 38.3 L of a gas, what is the new volume? The temperature of the gas is 0 °C and the pressure is 1 atm. When you analyze the information in this problem, you can see that the temperature and pressure are not changing, but the number of moles and the volume are. Rearrange the ideal gas law equation so that all of the variables which change are on one side of the equation.

$$\frac{V}{n} = \frac{RT}{P}$$

This relationship is true for any volume of the ideal gas, so we can also write

$$\frac{V_1}{n_1} = \frac{RT}{P} = \frac{V_2}{n_2}$$

or,

$$\frac{V_1}{n_1} = \frac{V_2}{n_2}$$

This form of the ideal gas equation is the same as Avogadro's law. If you can remember the ideal gas law, you can derive all of the other gas laws from it. To solve the problem, rearrange the equation so that V_2 is isolated on one side.

$$V_2 = V_1 \times \frac{n_2}{n_1}$$

$$V_2 = 38.3\ \text{L} \times \frac{2.64\ \text{mol}}{2.12\ \text{mol}} = 47.7\ \text{L}$$

Notice that the unwanted units of mol cancel, leaving the desired unit of volume, the liter.

13.6 DALTON'S LAW OF PARTIAL PRESSURES

What Is Dalton's Law of Partial Pressures and How Can It Be Used?

Dalton's law of partial pressures says that the total pressure of a mixture of gases in a container is equal to the sum of the pressures of the individual gases. The individual pressure of a gas in a mixture is called a **partial pressure**. The partial pressure of an individual gas is the amount of pressure the gas would exert if it were the only gas in the container. Dalton's law of partial pressures can be expressed mathematically as

$$P_{\text{TOTAL}} = P_1 + P_2 + P_3$$

Where P_1, P_2 and P_3 represent the partial pressures for three different gases. The number of partial pressures which make up P_{TOTAL} depends upon the number of gases in the mixture. Each partial pressure in a mixture can be calculated from the ideal gas law.

$$P_1 = \frac{n_1 RT}{V} \text{ and } P_2 = \frac{n_2 RT}{V} \text{ and } P_3 = \frac{n_3 RT}{V}$$

If the number of moles of an individual gas is known or can be calculated, then the partial pressure of that particular gas can be calculated, too. Let's see how Dalton's law of partial pressures can be used.

Example:

A sample of oxygen gas at 26 °C was bubbled through water. In the process, the oxygen picked up water vapor. The volume of the oxygen plus the water vapor was 4.6 L and the partial pressure of the water vapor was 0.03 atm. The total pressure of oxygen plus water vapor was 0.98 atm. How many moles of oxygen are in the 4.6 L of gas? This problem asks for the number of moles of oxygen gas. We can calculate moles of a gas using the ideal gas law in this form.

$$n = \frac{PV}{RT}$$

To calculate moles of oxygen, we need to know the pressure of the oxygen in the container, the volume of gas, and the temperature. All of these values are given to us in the problem, except the partial pressure of oxygen in the oxygen-water vapor mixture. How can the partial pressure of oxygen be calculated? We are given the total pressure of the gas mixture, and the partial pressure of the water vapor. By using Dalton's law of partial pressures, we can calculate the partial pressure of oxygen.

$$P_{\text{TOTAL}} = P_{\text{WATER VAPOR}} + P_{O_2}$$

Rearrange the equation to isolate P_{O_2} on one side.

$$P_{TOTAL} - P_{WATER\ VAPOR} = P_{O_2}$$

Now substitute values for P_{TOTAL} and P_{O_2}.

$$P_{O_2} = 0.98 \text{ atm} - 0.03 \text{ atm}$$
$$P_{O_2} = 0.95 \text{ atm}$$

We now know that the partial pressure of oxygen gas is 0.95 atm. Now use the ideal gas law to calculate the moles of oxygen gas.

$$n = \frac{PV}{RT}$$

$$n_{O_2} = \frac{(0.95 \not{\text{atm}})(4.6 \not{\text{L}})}{(0.08206 \not{\text{L}} \not{\text{atm}}/\not{\text{K}} \text{ mol})(299 \not{\text{K}})}$$

$$n_{O_2} = 0.18 \text{ mol}$$

13.7 LAWS AND MODELS--A REVIEW

What Is the Difference Between a Law and a Theory?

The ideal gas law predicts how gases will act under certain conditions, but does not tell us why gases behave in this manner. Theories or models try to explain the behavior which is predicted by laws. Section 13.8 will introduce a theory which tries to explain the observed behavior of gases.

13.8 THE KINETIC MOLECULAR THEORY OF GASES

What Is the Kinetic Molecular Theory of Gases?

The **kinetic molecular theory** tries to explain why gases behave the way that they do. The theory consists of assumptions about the nature of the gas particles themselves. The main points of the theory are: 1. Gases are composed of tiny particles, either atoms or molecules. 2. The particles in a gas are very small, and very far apart from one another, so that the volume of each particle compared with the overall volume can be considered zero. 3. Gas particles are always moving and colliding with the sides of the container. The force of these collisions produces the pressure in the container. 4. It is assumed that particles are not repelled or attracted to each other. 5. The average kinetic energy (the energy of motion) of the gas particles is directly proportional to the Kelvin temperature of the gas.

13.9 THE IMPLICATIONS OF THE KINETIC MOLECULAR THEORY

What Does Temperature Mean?

Temperature is a measure of the motion of gas particles. As the temperature rises, the kinetic energy of the gas particles increases, and the gas particles collide with the sides of the container more often.

What Is the Relationship Between Temperature and Pressure?

As the temperature of a gas increases, the particles collide with the sides of the container more often. Because the pressure of a gas depends upon the force with which the gas hits the sides of the container, it makes sense that as the temperature of a gas rises, the pressure also rises.

What Is the Relationship Between Volume and Temperature?

If a gas is placed in a container with a moveable piston which is exactly balanced by the pressure inside the container and the pressure outside the container, and the temperature of the gas is raised, what will happen? The gas particles begin to move faster and collide with the sides and top of the container more frequently, causing the pressure inside the container to increase. Because the pressure inside increases, and the pressure outside stays constant, the effect is to cause the piston to move outward until the pressure outside is once more exactly balanced by the pressure inside. This has the effect of causing the volume of the container to increase.

The postulates about the nature of gas particles can explain many of the observations about the actual behavior of gases.

13.10 GAS STOICHIOMETRY

How Can You Use the Ideal Gas Law to Solve Stoichiometric Problems?

The strategies you learned for solving stoichiometry problems in Chapter 10 will be essential for solving stoichiometry problems involving gases. The difference is that you will be asked to find the number of moles of a gas, or the volume of gas produced.

Example:

If a mixed drink containing 13.2 g of alcohol is consumed, what volume of carbon dioxide could be produced by the body during the metabolism of the alcohol? The temperature of the carbon dioxide gas is 37 °C, body temperature, and the pressure is 1.00 atm. The balanced equation for the reaction is

$$C_2H_5OH + 3O_2 \rightarrow 2CO_2 + 3H_2O$$

We are given grams of ethyl alcohol, and want to know the volume of product, CO_2. We will need to know the stoichiometry of the equation, that is, how many moles of CO_2 are produced from 13.2 g of C_2H_5OH. First, use the molar mass of C_2H_5OH to convert between grams of C_2H_5OH and moles. The mole ratio which relates moles of C_2H_5OH to moles of CO_2 is $\frac{2 \text{ mol } CO_2}{1 \text{ mol } C_2H_5OH}$. This mole ratio can be used to calculate the number of moles of CO_2.

$$13.2 \text{ g } C_2H_5OH \times \frac{1 \text{ mol } C_2H_5OH}{46.1 \text{ g } C_2H_5OH} \times \frac{2 \text{ mol } CO_2}{1 \text{ mol } C_2H_5OH} = 0.573 \text{ mol } CO_2$$

Now, we can use the ideal gas law to solve for the volume of carbon dioxide. Remember that temperatures expressed in degrees Celsius must be converted to kelvins. We know that n equals 0.573 mol CO_2, T equals 310. K and P equals 1.00 atm. We can use the ideal gas law to find V.

$$V = \frac{nRT}{P}$$

$$V = \frac{(0.573 \text{ mol } CO_2)(0.08206 \text{ L atm/K mol})(310. \text{ K})}{1.00 \text{ atm}} = 14.6 \text{ L } CO_2$$

LEARNING REVIEW

1. Gas in compressed gas cylinders is usually under a great deal of pressure. If the gas in a particular cylinder has a pressure of 2500 psi, how many torr is this?

2. Convert each of the units of pressure below.

   a. 898.5 mm Hg to atm
   b. 0.408 atm to torr
   c. 68,471 Pa to mm Hg
   d. 50.9 psi to atm

3. The relationship observed by Boyle between volume and pressure is

   a. linear
   b. proportional
   c. inversely proportional
   d. no relationship

4. Examine the cylinder with a moveable piston. If the piston moves downward, causing the volume of the gas to decrease, will the pressure of the gas become larger or smaller?

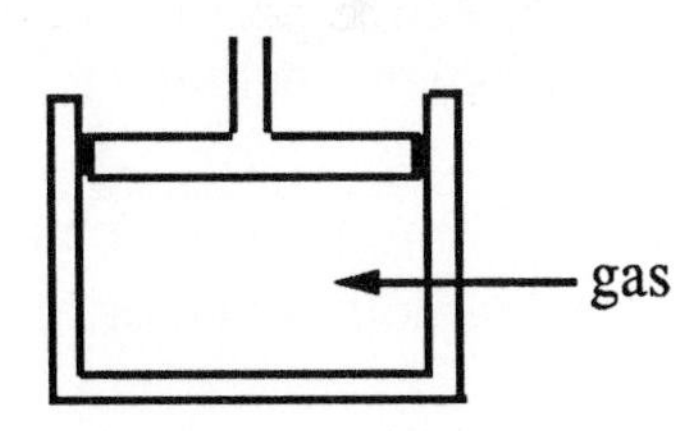

5. A sample of nitrogen gas at 2.4 atm has a volume of 50.3 L. If the pressure is decreased to 1.9 atm, will the new volume be greater or smaller? What is the new volume?

6. Acetylene is a gas used as a fuel for some welding torches. If 0.52 L of acetylene has a pressure of 1824 torr, what is the pressure if the volume is decreased to 0.39 L?

7. Absolute zero is the temperature at which gases have zero volume. But zero volume has never been measured in the laboratory. So how do we know at what temperature the volume of a gas equals zero?

8. Which of the following equations cannot be derived from Charles's law, $V = bT$?

   a. $\frac{V}{T} = b$

   b. $\frac{V_1}{T_1} = \frac{V_2}{T_2}$

   c. $\frac{V_1}{T_1} + \frac{V_2}{T_2} = b$

   d. $\frac{3V}{3T} = b$

9. A container with a moveable piston contains 0.89 L methane gas at 100.5 °C. If the temperature of the gas rises by 11.3 °C, what is the new volume of the gas?

10. A sample of carbon dioxide gas at 0.5 atm and 10.8 °C has a volume of 19.4 L. What is the volume of the same sample of carbon dioxide at 0.5 atm and a temperature of 14.6 °C?

11. Avogadro's law describes the relationship between the amount (number of moles) of a gas and the volume of the gas. Under what conditions of temperature and pressure is Avogadro's law true?

12. A balloon containing helium gas contains 0.32 mol He at 2432 mm Hg and 25 °C. The volume of the helium is 2.45 L. If an additional 0.14 mol of He is injected into the balloon while holding the temperature and pressure constant, what is the volume of the helium?

13. Where does the universal gas constant, *R*, come from?

14. Use the ideal gas law to solve the problems below.

   a. A sample of chlorine gas at 543 torr has a volume of 21.6 L. If the temperature of the chlorine is 0 °C, how many moles of chlorine gas are present?

   b. Poisonous carbon monoxide gas is a product of the internal combustion engine. If 1.2 mol of CO at 11 °C and 102 mm Hg are present in a container, what will be the volume of the CO gas?

   c. 0.020 mol of a gas has a pressure of 0.62 atm at 26.4 °C, and a volume of 0.79 L. If the temperature is lowered by 5.2 °C and the pressure is increased to 0.96 atm, what is the new volume?

   d. 0.45 mol of a gas has a pressure of 299 torr at 300. °C and a volume of 53.8 L. At the same temperature and pressure, the volume of the gas is decreased to 39.7 L. How many moles of gas are present after the volume has changed?

15. Which statements about Dalton's law of partial pressures are true and which are false?

   a. The total pressure (P_{TOTAL}) of a mixture of gases is independent of the sizes of the gas particles.
   b. Attractive forces between gas particles are important in determining P_{TOTAL}.
   c. For ideal gases, P_{TOTAL} depends solely on the total number of moles of gas, for any temperature and volume.

16. Assume that a sample of humid air contains only nitrogen gas, oxygen gas and water vapor. If the atmospheric pressure is 745 mm Hg and the partial pressure of N_2 is 566 mm Hg and of oxygen is 140. mm Hg, what is the partial pressure of water vapor in the air?

17. A 7.5 L mixture of gases is produced by mixing 4.0 L of N_2 at 450 torr, 3.5 L of O_2 at 252 torr, and 0.21 L of CO_2 at 150 torr. If the temperature is held constant at 65 °C, what is the total pressure of the mixture?

18. Mercuric oxide can be decomposed into elemental mercury and molecular oxygen.

$$2HgO \rightarrow 2Hg + O_2$$

A sample of HgO is decomposed and the oxygen is collected by displacing water. The volume of gas collected is 255 mL at 25 °C and at 745 torr. The partial pressure of H_2O at 25 °C is 24 torr.

a. What is the partial pressure of O_2 collected?
b. How many moles of water were present in the collected gas?

19. Explain the difference between a law and a model.

20. Which statements about the Kinetic Molecular Theory are true, and which are false?

a. The postulates of the Kinetic Molecular Theory are true for all gases, at all temperatures and pressures.
b. Gas particles are assumed to either attract or repel each other.
c. The distance between individual gas particles is much greater than the volume of an individual gas particle.
d. As the Kelvin temperature of a gas increases, the average kinetic energy increases.

21. According to the Kinetic Molecular Theory, what are we measuring when we measure the temperature of a gas?

22. According to the Kinetic Molecular Theory, what happens to the pressure when the temperature of a gas in a closed container is increased?

23. Carbon dioxide is produced during the combustion of liquid propane fuel.

$$C_3H_8 + 5O_2 \rightarrow 3CO_2 + 4H_2O$$

If 5.0 kg of propane are burned at 1.000 atm pressure and 400. °C, what volume of CO_2 gas is produced?

24. A sample of fluorine gas has a volume of 19.9 L at STP. How many moles of fluorine are in the sample?

25. Hydrogen gas and chlorine gas react to produce gaseous hydrogen chloride. If the volume of HCl produced is 28.3 L at 630. mm Hg and 200. °C, how many grams of HCl were produced?

ANSWERS TO LEARNING REVIEW

1. There is no conversion factor directly between psi and torr. However, we can convert psi to atm and atm to torr.

$$2500 \text{ psi} \times \frac{1.000 \text{ atm}}{14.69 \text{ psi}} \times \frac{760.0 \text{ torr}}{1.000 \text{ atm}} = 1.3 \times 10^5 \text{ torr}$$

2. a. $898.5 \text{ mm Hg} \times \frac{1.000 \text{ atm}}{760.0 \text{ mm Hg}} = 1.182 \text{ atm}$

   b. $0.408 \text{ atm} \times \frac{760.0 \text{ torr}}{1.000 \text{ atm}} = 310. \text{ torr}$

   c. $68471 \text{ Pa} \times \frac{1.000 \text{ atm}}{101{,}325 \text{ Pa}} \times \frac{760.0 \text{ mm Hg}}{1.000 \text{ atm}} = 513.6 \text{ mm Hg}$

   d. $50.9 \text{ psi} \times \frac{1.000 \text{ atm}}{14.69 \text{ psi}} = 3.46 \text{ atm}$

3. Boyle observed that as the volume increased, the pressure decreased. Pressure and volume are inversely proportional. The correct answer is c.

4. If the volume of gas inside the cylinder becomes smaller, then the pressure of the gas will become larger.

5. This problem provides pressure and volume data. Boyle's law relates pressure to volume. We can use $P_1V_1 = P_2V_2$, which is one way of writing Boyle's law, to solve this problem. In this equation, P_1 and V_1 represent initial, or starting, conditions. P_2 and V_2 represent final or changed conditions. 2.4 atm is the initial pressure, P_1, and 50.3 L is the initial volume, V_1. Pressure has changed so P_2 is 1.9 atm, and we are asked for the new volume, V_2. First, will the new volume be larger or smaller? The pressure decreases from 2.4 atm to 1.9 atm. There is an inverse relationship between temperature and pressure, so if the pressure decreases, we would expect the volume to increase.

   Rearrange the equation to isolate V_2 on one side.

$$P_1V_1 = P_2V_2$$

$$V_1 \times \frac{P_1}{P_2} = V_2 \times \frac{P_2}{P_2}$$

$$V_2 = V_1 \times \frac{P_1}{P_2}$$

Now, substitute values into the equation.

$$V_2 = 50.3 \text{ L N}_2 \times \frac{2.4 \text{ atm}}{1.9 \text{ atm}}$$

$$V_2 = 64 \text{ L N}_2$$

This answer makes sense. The pressure has decreased from 2.4 atm to 1.9 atm so we would expect the volume to increase. The volume has in fact increased from 50.3 L to 64 L. There is an inverse relationship between pressure and volume.

6. This problem presents only pressure and volume data, so we can use Boyle's law. A sample of acetylene gas has a volume of 0.52 L and a pressure of 1824 torr. A second sample of acetylene has a volume of 0.39 L and we are asked to calculate the pressure, P_2. Use Boyle's law in the form

$$P_1 \times V_1 = P_2 \times V_2$$

Rearrange the equation to isolate P_2 on one side of the equation.

$$P_1 \times \frac{V_1}{V_2} = P_2 \times \frac{V_2}{V_2}$$

$$P_2 = P_1 \times \frac{V_1}{V_2}$$

Now substitute values into the equation.

$$P_2 = 1824 \text{ torr} \times \frac{0.52 \text{ L acetylene}}{0.39 \text{ L acetylene}}$$

$$P_2 = 2400 \text{ torr}$$

This answer makes sense. The volume of acetylene has decreased, so the pressure must increase.

7. We can measure the volume of a gas at various temperatures, some of which are very cold, but not quite absolute zero. We can then plot on a graph each temperature and volume pair.

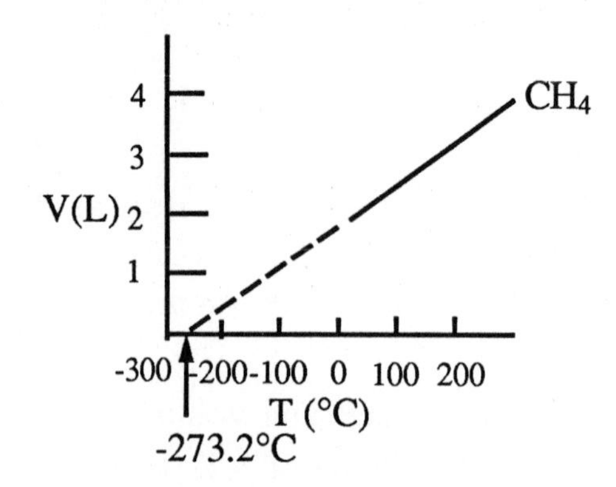

Although we cannot measure the volume of a gas at -273.2 °C, we can determine at what temperature the volume would be zero by drawing a continuation of the straight line of the graph and noting where the temperature axis hits the line. This has been done for many different gases and the result is always the same; the temperature at which the volume of a gas would be zero is -273.2 °C.

8. a. Charles's law can be written as $V = bT$, where V is volume, T is temperature and b is a constant. We can rearrange the equation by dividing both sides by T.

$$\frac{V}{T} = b \times \frac{\cancel{T}}{\cancel{T}}$$

$$b = \frac{V}{T}$$

So, $\frac{V}{T} = b$ can be derived from Charles's law.

b. If $\frac{V_1}{T_1} = b$ and $\frac{V_2}{T_2} = b$, then it is also true that $\frac{V_1}{T_1} = \frac{V_2}{T_2}$, so this relationship can also be derived from $V = bT$.

c. We know that $\frac{V_1}{T_1} = b$ and that $\frac{V_2}{T_2} = b$. We also know that $\frac{V_1}{T_1} = \frac{V_2}{T_2}$, but we cannot add $\frac{V_1}{T_1}$ and $\frac{V_2}{T_2}$ together to equal the constant b. So this relationship cannot be derived from $V = bT$.

d. By multiplying temperature and volume by the same value, we are not changing the equation, because $\frac{3V}{3T}$ and $\frac{V}{T}$ are both equal to the constant b.

$$\frac{3V}{3T} = b = \frac{V}{T}$$

So this relationship can also be derived from $V = \frac{b}{T}$.

9. This problem provides temperature and volume data and asks about the volume of methane gas when the temperature changes. We can use Charles's law to solve the problem. One form of Charles's law is

$$\frac{V_1}{T_1} = \frac{V_2}{T_2}$$

The initial volume of methane is 0.89 L and the initial temperature is 100.5 °C. The rise in temperature is 11.3 °C, so the final temperature is the sum of 100.5 °C plus 11.3 °C, or 111.8 °C. The unknown quantity is the final volume, V_2. Rearrange the equation to isolate V_2 on one side.

$$\frac{V_1}{T_1} = \frac{V_2}{T_2}$$

$$V_1 \times \frac{T_2}{T_1} = V_2 \times \frac{\cancel{T_2}}{\cancel{T_2}}$$

$$V_2 = V_1 \times \frac{T_2}{T_1}$$

Because both the initial and final temperatures are given in °C, we must convert to kelvins.

initial temperature (T_1)
$^{t}K = {}^{t}°C + 273$
$^{t}K = 100.5 + 273$
$^{t}K = 374$

final temperature (T_2)
$^{t}K = 111.8 + 273$
$^{t}K = 385$

Now, substitute values into the equation.

$$V_2 = 0.89 \text{ L methane} \times \frac{385 \text{ K}}{374 \text{ K}}$$

$$V_2 = 0.92 \text{ L methane}$$

This answer makes sense. The temperature has increased, so the volume should also increase. There is a direct relationship between the temperature in kelvins of a gas and its volume.

10. This problem provides a sample of CO_2 gas in which the temperature is changed, but the number of moles of gas and the pressure remains constant. We want to calculate the new volume of the gas. We can use Charles's law to solve this problem. The initial conditions are V_1, 19.4 L, and T_1, 10.8 °C. When the conditions are changed T_2 is 14.6 °C and the volume, V_2, is unknown. We can use Charles's law written as $\frac{V_1}{T_1} = \frac{V_2}{T_2}$ to solve this problem.

Rearrange the equation to isolate V_2 on one side.

$$V_1 \times \frac{T_2}{T_1} = V_2 \times \frac{T_2}{T_2}$$

$$V_2 = V_1 \times \frac{T_2}{T_1}$$

Because the temperature is given in °C, we must convert to kelvins before solving the problem.

$$^{t}\text{K} = {}^{t\circ}\text{C} + 273$$

The initial temperature is

$$^{t}\text{K} = 10.8 + 273$$
$$^{t}\text{K} = 284$$

The final temperature is

$$^{t}\text{K} = 14.6 + 273$$
$$^{t}\text{K} = 288$$

Now, substitute values into the equation.

$$V_2 = 19.4 \text{ L carbon dioxide} \times \frac{288 \text{ K}}{284 \text{ K}}$$

$$V_2 = 19.7 \text{ L carbon dioxide}$$

The volume of the second sample of carbon dioxide is 19.7 L, slightly more than the volume of the first sample. This answer makes sense. The temperature of the second sample is slightly higher than the temperature of the first sample, so the volume should also be slightly more.

11. Avogadro's law, which expresses a relationship between the volume of a gas and the number of moles of that gas, is true only when the gas temperature and pressure are constant.

12. 2.45 L of helium gas in a balloon contains 0.32 mol He. When the quantity of gas inside the balloon is increased by 0.14 mol, we are asked to calculate the new volume of helium. This problem involves both volume and moles, and we are told that the temperature and pressure are held constant, so we can use Avogadro's law to solve for the new volume. The initial volume, V_1, is 2.45 L, and the initial number of moles, n_1, is 0.32 mol. The final volume, V_2, is unknown, and the final number of moles, n_2, is equal to the initial number of moles plus the number of additional moles added to the balloon. So, n_2 is equal to 0.32 mol plus 0.14 mol, for a total of 0.46 mol. Rearrange Avogadro's law to isolate V_2 on one side.

$$\frac{V_1}{n_1} = \frac{V_2}{n_2}$$

$$V_1 \times \frac{n_2}{n_1} = V_2 \times \frac{\not{n_2}}{\not{n_2}}$$

$$V_2 = V_1 \times \frac{n_2}{n_1}$$

Now, substitute values into the equation.

$$V_2 = 2.45 \text{ L} \times \frac{0.46 \text{ m}\not{\text{o}}\text{l He}}{0.32 \text{ m}\not{\text{o}}\text{l He}}$$

$$V_2 = 3.5 \text{ L He}$$

13. When the three laws which relate volume to pressure, to temperature and to the number of moles are combined, the constants which are present in each individual law are combined into one constant, called R, the universal gas constant. When volume is expressed in liters, pressure in atmospheres and temperature in kelvins, R has units of L atm/K mol and a value of 0.08206 L atm/K mol.

14. The ideal gas law can be expressed as $PV = nRT$. Remember that temperature must be

expressed in kelvins, pressure in atmospheres, and volume in liters. If any of the quantities given in the problem are expressed in other units, we must convert before we can use the ideal gas law.

a. This problem provides the pressure of chlorine gas in torr and the temperature in °C. Before we proceed let's convert torr to atmosphere and °C to K.

$$543 \text{ torr} \times \frac{1.000 \text{ atm}}{760.0 \text{ torr}} = 0.714 \text{ atm}$$

$$^{t}K = {}^{t}\,°C + 273$$

$$^{t}K = 0 + 273$$

$$^{t}K = 273$$

We are provided with pressure, volume, and temperature data and we are asked for the number of moles. Rearrange the ideal gas law equation to isolate *n* on one side.

$$PV = nRT$$

$$\frac{PV}{RT} = n \times \frac{RT}{RT}$$

$$n = \frac{PV}{RT}$$

Now, substitute values into the equation.

$$n = \frac{(0.714 \text{ atm})(21.6 \text{ L } Cl_2)}{(0.08206 \text{ L atm/ K mol})(273 \text{ K})}$$

$$n = 0.688 \text{ mol } Cl_2$$

b. This problem asks us to calculate the volume of carbon monoxide gas, CO, given temperature, pressure and the number of moles of gas. The pressure of CO is given in mm Hg and temperature in °C. Before we proceed, we need to convert mm Hg to atmospheres and °C to K.

$$102 \text{ mm Hg} \times \frac{1.000 \text{ atm}}{760.0 \text{ mm Hg}} = 0.134 \text{ atm}$$

$$^{t}K = {}^{t\circ}C + 273$$

$$^{t}K = 11 + 273$$

$$^{t}K = 284$$

We can use the ideal gas law to solve this problem because we are provided with temperature, pressure and the number of moles of gas. Rearrange the ideal gas law to isolate V on one side.

$$PV = nRT$$

$$V \times \frac{\cancel{P}}{\cancel{P}} = \frac{nRT}{P}$$

$$V = \frac{nRT}{P}$$

Now, substitute values into the equation.

$$V = \frac{(1.2 \text{ m}\cancel{o}\text{l CO})(0.008206 \text{ L a}\cancel{tm}/\cancel{K} \text{ m}\cancel{o}\text{l})(284 \cancel{K})}{0.134 \text{ a}\cancel{tm}}$$

$$V = 210 \text{ L CO}$$

c. This problem involves a sample of gas initially under one set of conditions. The conditions are changed and we are asked to calculate the new volume. Notice that the number of moles of gas stays the same because no additional gas is added, nor is any gas removed. We can use the ideal gas law to solve this problem because we know the number of moles, the pressure, and we can calculate the new temperature.

The temperature of the original sample was 26.4 °C, but this temperature was lowered by 5.2 °C. So the final temperature of the gas is 26.4 °C minus 5.2 °C, which is 21.2 °C. Because this temperature is in °C, we must first convert to K before solving the problem.

$$^{t}K = {}^{t\circ}C + 273$$

$$^{t}K = 21.2 + 273$$

$$^{t}K = 294$$

Rearrange the ideal gas law to isolate V on one side.

$$PV = nRT$$

$$V \times \frac{\not{P}}{\not{P}} = \frac{nRT}{P}$$

$$V = \frac{nRT}{P}$$

Now, substitute values into the equation.

$$V = \frac{(0.020\ \text{m}\not{\text{o}}\text{l})(0.08206\ \text{L at}\not{\text{m}}/\not{\text{K}}\ \text{m}\not{\text{o}}\text{l})(294\ \not{\text{K}})}{0.96\ \text{a}\not{\text{t}}\text{m}}$$

$$V = 0.50\ \text{L}$$

d. In this problem, we are provided with information about two different gases. The information about the first gas is complete. That is, we know the number of moles, the temperature, the volume and the pressure of the gas. We know the temperature, the volume and pressure of the second gas and are asked to calculate the number of moles, *n*. We could use Avogadro's law, which relates volume and moles, to solve this problem, but we are asked to use the ideal gas law. Because we know the temperature and the volume and the pressure of the second gas, we can use the ideal gas law to calculate the number of moles. First, convert pressure, which is given in torr, to atm, and temperature, which is given in °C, to K.

$$299\ \text{t}\not{\text{o}}\text{rr} \times \frac{1.000\ \text{atm}}{760.0\ \text{t}\not{\text{o}}\text{rr}} = 0.393\ \text{atm}$$

$$^{t}\text{K} = {}^{t\circ}\text{C} + 273$$

$$^{t}\text{K} = 300. + 273$$

$$^{t}\text{K} = 573$$

Rearrange the ideal gas law to isolate *n* on one side.

$$PV = nRT$$

$$\frac{PV}{RT} = n \times \frac{\not{R}\not{T}}{\not{R}\not{T}}$$

$$n = \frac{PV}{RT}$$

Now, substitute values into the equation.

$$n = \frac{(0.393 \text{ atm})(39.7 \text{ L})}{(0.08206 \text{ L atm/K mol})(573 \text{ K})}$$

$$n = 0.332 \text{ mol}$$

15. a. The total pressure (P_{TOTAL}) depends only on the total quantity of gas, not on the kind and size of the particles. So this statement is true.

    b. Because P_{TOTAL} does not depend on the identity of the gas particles but only on the quantity of particles, forces between gas particles do not exist. This statement is false.

    c. The pressure of a single gas or of a mixture of gases depends on the number of moles, and also the temperature and volume. A mixture of gases follows the ideal gas law, just as a single gas does. So this statement is false.

16. P_{TOTAL} represents the pressure exerted by all the gases in the atmosphere, which in this case is equal to 745 mm Hg. P_{TOTAL} equals $P_{OXYGEN} + P_{NITROGEN} + P_{WATER\ VAPOR}$. The two major components of air are nitrogen and oxygen. If $P_{NITROGEN}$ equals 566 mm Hg and P_{OXYGEN} equals 140. mm Hg, then we can use the following equation to calculate $P_{WATER\ VAPOR}$.

$$P_{TOTAL} = P_{OXYGEN} + P_{NITROGEN} + P_{WATER\ VAPOR}$$

Rearrange the equation to isolate $P_{WATER\ VAPOR}$ on one side.

$$P_{WATER\ VAPOR} = P_{TOTAL} - (P_{OXYGEN} + P_{NITROGEN})$$

Now, substitute values into the equation.

$$P_{WATER\ VAPOR} = 745 \text{ mm Hg} - (140. \text{ mm Hg} + 566 \text{ mm Hg})$$

$$P_{WATER\ VAPOR} = 39 \text{ mm Hg}$$

17. We are asked to calculate P_{TOTAL} for a mixture of three gases. To determine P_{TOTAL}, we need to know the partial pressures for each individual gas. Calculating the partial pressure of any gas by using the ideal gas law requires us to know the number of moles, n, of that gas. So we need to find a way to calculate n. We are given the initial volume of each gas, the initial pressure, and the initial temperature, so we can use the ideal gas law in the form $n = \frac{PV}{RT}$ to find the number of moles. Because the number of moles of each gas does not change when the gases are mixed together, we can use the value of n we calculated for each gas before they were mixed together to calculate the partial pressure of each gas under the conditions present when the gases are mixed. Because pressure is given in torr, and temperature in °C, convert to atmospheres and kelvins before proceeding.

The temperature is held constant at 65 °C, so we only need to convert to kelvins once.

$$^{t}\text{K} = {}^{t\circ}\text{C} + 273$$

$$^{t}\text{K} = 65 + 273$$

$$^{t}\text{K} = 338$$

For nitrogen

$$450 \text{ torr} \times \frac{1.000 \text{ atm}}{760.0 \text{ torr}} = 0.59 \text{ atm}$$

$$n_{N_2} = \frac{(0.59 \text{ atm})(4.0 \text{ L } N_2)}{(0.08206 \text{ L atm/K mol})(338 \text{ K})}$$

$$n_{N_2} = 0.085 \text{ mol}$$

For oxygen

$$252 \text{ torr} \times \frac{1.000 \text{ atm}}{760.0 \text{ torr}} = 0.332 \text{ atm}$$

$$n_{O_2} = \frac{(0.332 \text{ atm})(3.5 \text{ L } O_2)}{(0.08206 \text{ L atm/K mol})(338 \text{ K})}$$

$$n_{O_2} = 0.042 \text{ mol}$$

For carbon dioxide

$$150 \text{ torr} \times \frac{1.000 \text{ atm}}{760.0 \text{ torr}} = 0.20 \text{ atm}$$

$$n_{CO_2} = \frac{(0.197 \text{ atm})(0.21 \text{ L } CO_2)}{(0.08206 \text{ L atm/K mol})(338 \text{ K})}$$

$$n_{CO_2} = 1.5 \times 10^{-3} \text{ mol}$$

So now we know the number of moles of each gas. We can use the ideal gas law to calculate the partial pressure of each gas. We will need to use the volume of the gas mixture, not the volume of each individual gas. Use the ideal gas law with P isolated on one side.

$$P = \frac{nRT}{V}$$

$$P_{N_2} = \frac{(0.085 \text{ mol } N_2)(0.08206 \text{ L atm/K mol})(338 \text{ K})}{7.5 \text{ L}}$$

$$P_{N_2} = 0.31 \text{ atm}$$

$$P_{O_2} = \frac{(0.042 \text{ mol } O_2)(0.08206 \text{ L atm/K mol})(338 \text{ K})}{7.5 \text{ L}}$$

$$P_{O_2} = 0.16 \text{ atm}$$

$$P_{Co_2} = \frac{(1.5 \times 10^{-3} \text{ mol } CO_2)(0.08206 \text{ L atm/K mol})(338 \text{ K})}{7.5 \text{ L}}$$

$$P_{CO_2} = 5.5 \times 10^{-3} \text{ atm}$$

The total pressure, P_{TOTAL}, can be calculated from

$$P_{TOTAL} = P_{N_2} + P_{O_2} + P_{CO_2}$$

$$P_{TOTAL} = 0.31 \text{ atm} + 0.16 \text{ atm} + 5.5 \times 10^{-3} \text{ atm}$$

$$P_{TOTAL} = 0.48 \text{ atm}$$

18. a. We want to know the partial pressure of oxygen produced from the decomposition of HgO and collected by displacing water. We know that P_{O_2} equals P_{TOTAL} minus P_{H_2O}. We are given both P_{H_2O} and P_{TOTAL} in the problem and we can calculate P_{O_2}.

$$P_{O_2} = P_{TOTAL} - P_{H_2O}$$

$$P_{O_2} = 745 \text{ torr} - 24 \text{ torr}$$

$$P_{O_2} = 721 \text{ torr}$$

b. We want to know the number of moles of water, n_{H_2O}, which are present in the mixture of water vapor and oxygen gas. If we have enough information, we can use the ideal gas law to calculate the number of moles of water. We are given volume, pressure (P_{H_2O}), and the temperature. Rearrange the ideal gas law so that n is isolated on one side.

$$n = \frac{PV}{RT}$$

Pressure is given in units of torr, volume is in milliliters, and temperature in °C, so we must first convert units.

pressure

$$24 \text{ torr} \times \frac{1.000 \text{ atm}}{760.0 \text{ torr}} = 0.032 \text{ atm}$$

volume

$$255 \text{ mL} \times \frac{1 \text{L}}{1000 \text{ mL}} = 0.255 \text{ L}$$

temperature

$$t\text{K} = t\text{°C} + 273$$
$$t\text{K} = 25 + 273$$
$$t\text{K} = 298$$

Now, substitute into the ideal gas law.

$$n_{H_2O} = \frac{(0.032 \text{ atm})(0.255 \text{ L})}{(0.08206 \text{ L atm/K mol})(298 \text{ K})}$$

$$n_{H_2O} = 0.00033 \text{ mol} = 3.3 \times 10^{-4} \text{ mol}$$

19. A law is a generalization about behavior which has been drawn from repeated observation about nature. A model is a theory which attempts to explain why a certain behavior is observed.

20. a. False. Gases obey the postulates of the Kinetic Molecular Theory best at high temperatures and/or high pressures. When these conditions are not met, gases can exhibit other kinds of behavior.
    b. False. In postulate four of the Kinetic Molecular Theory it is assumed that gas particles neither attract nor repel each other.
    c. True. In postulate two of the Kinetic Molecular Theory, it is assumed that individual gas particles are very small compared to the distances between the particles.
    d. True. The kinetic energy of a gas is proportional to the Kelvin temperature.

21. When we measure the temperature of a gas, we are actually measuring how fast the gas particles are moving. Postulate five of the Kinetic Molecular Theory results from the observation that as the temperature of a gas increases, the movement of gas particles increases.

22. As the temperature of a gas in a closed container increases, the particles move faster and hit the walls of the container more often. The collisions of the particles with the walls of the container are harder. So an increase in gas pressure when the temperature increases is due to the gas particles hitting the sides of the container with greater frequency and greater force.

23. This problem provides the pressure, the temperature and the quantity in kg of a sample of propane gas. We are asked to find the volume of another gas, carbon dioxide, produced when propane burns. We know some information about carbon dioxide: we know that the pressure is 1 atm and the temperature is 400 °C. If we knew the number of moles of carbon dioxide, we could use the ideal gas law to calculate the volume of carbon dioxide.

$$PV = nRT$$
$$? \quad ?$$

We know the number of kg of propane which react to form CO_2, and we have a balanced equation, so, we can calculate the moles of carbon dioxide using the mole ratio from the balanced equation. But first, let's calculate the number of moles of propane which are equivalent to 5.0 kg of propane.

$$5.0\ \cancel{kg}\ C_3H_8 \times \frac{1000\ \cancel{g}\ C_3H_8}{1\ \cancel{kg}\ C_3H_8} \times \frac{1\ mol\ C_3H_8}{44.09\ \cancel{g}\ C_3H_8} = 110\ mol\ C_3H_8$$

Now that we know the number of moles of propane, we can use the mole ratio to calculate the moles of CO_2 which would be produced. The balanced equation tells us that every mole of C_3H_8 produces 3 moles of CO_2 so the correct mole ratio is $\frac{3\ mol\ CO_2}{1\ mol\ C_3H_8}$.

$$110\ \cancel{mol}\ C_3H_8 \times \frac{3\ mol\ CO_2}{1\ \cancel{mol}\ C_3H_8} = 330\ mol\ CO_2$$

Now we know the number of moles of CO_2, the pressure and the temperature, so we can use the ideal gas law to calculate the volume. First, isolate V on one side.

$$PV = nRT$$

$$V \times \frac{\cancel{P}}{\cancel{P}} = \frac{nRT}{P}$$

$$V = \frac{nRT}{P}$$

Because the temperature is given in °C, we must convert to kelvins.

$$^{t}K = {}^{t\circ}C + 273$$

$$^{t}K = 400 + 273$$

$$^{t}K = 700$$

Now, substitute values into the equation.

$$V = \frac{(330\ \cancel{mol}\ C_3H_8)(0.08206\ L\ \cancel{atm}/\cancel{K}\ \cancel{mol})(700\ \cancel{K})}{1.0\ \cancel{atm}}$$

$$V = 19000\ L\ CO_2$$

24. We can solve this problem two ways. In the first method, we can use the ideal gas law because we know the volume, the temperature and the pressure. Rearrange the equation to isolate moles on one side.

$$PV = nRT$$

$$\frac{PV}{RT} = n \times \frac{\cancel{RT}}{\cancel{RT}}$$

$$n = \frac{PV}{RT}$$

The fluorine gas is at STP, which means that the temperature is 0 °C and the pressure is 1 atm. We need to convert °C to kelvins before we proceed.

$$^{t}K = 0 + 273$$

$$^{t}K = 273$$

Now, substitute values into the equation.

$$n_{F_2} = \frac{(1.00 \text{ a}\cancel{\text{t}}\text{m})\ (19.9 \text{ L})}{(0.08206\ \cancel{\text{L}}\ \text{at}\cancel{\text{m}}/\cancel{\text{K}}\ \text{mol})(273\ \cancel{\text{K}})}$$

$$n_{F_2} = 0.888 \text{ mol}$$

In the second method we can use the conversion 1 mol of an ideal gas = 22.4 liters of gas. This conversion works only if the gas is at STP. In this problem we are told that the temperature and pressure are at STP.

$$19.9 \text{ L } \cancel{F_2} \times \frac{1.00 \text{ mol } F_2}{22.4 \text{ L } \cancel{F_2}} = 0.888 \text{ mol } F_2$$

Both methods gave the same answer, but we can only use the second method when the temperature is 0 °C (273 K) and the pressure is 1.00 atmosphere.

25. We are given the volume, the pressure and the temperature of hydrogen chloride gas, and we are asked for the number of grams of hydrogen chloride. We can use the ideal gas law to calculate the moles of HCl. Once we know the number of moles, we can use the molar mass to calculate grams.

Use the ideal gas law in the form

$$n = \frac{PV}{RT}$$

Because we are given the temperature in °C, and pressure in mm Hg, first determine the equivalent temperature in kelvins, and equivalent atmospheres.

$$^{t}\text{K} = {}^{t\circ}\text{C} + 273$$

$$^{t}\text{K} = 500$$

$$630 \text{ mm Hg} \times \frac{1.000 \text{ atm}}{760.0 \text{ mm Hg}} = 0.83 \text{ atm}$$

Now, substitute values into the equation.

$$n_{HCl} = \frac{(0.83 \text{ atm})(28.3 \text{ L})}{(0.08206 \text{ L atm/K mol})(500 \text{ K})}$$

$$n_{HCl} = 0.6 \text{ mol}$$

Now we can calculate the grams of HCl.

$$0.6 \text{ mol HCl} \times \frac{36.46 \text{ g HCl}}{1 \text{ mol HCl}} = 20 \text{ g HCl}$$

This volume of HCl contains 20 g HCl.

PRACTICE EXAM

1. Which statement about pressure is not true?

   a. 1.000 atm is equal to 760.0 torr.
   b. Mercury stays in a glass barometer tube because of the effects of gravity on the mercury.
   c. Atmospheric pressure is the weight of air pulled toward the earth's center.
   d. The SI unit of pressure is the Pascal.
   e. 14.69 psi is equal to 1.000 atm.

2. A 4.5 L sample of hydrogen gas has a pressure of 500. torr. If the volume of the hydrogen is changed to 3.8 L, what is the new pressure?

a. 590 torr
b. 420 torr
c. 855 torr
d. 2.4×10^{-3} torr
e. 1.7×10^{-3} torr

3. A steam engine cylinder has an initial volume of 16.0 L, and steam occupying the cylinder is at a temperature of 250. °C. If the volume of the cylinder is suddenly increased to 30.0 L by the movement of the piston, what is the new temperature of the steam in the cylinder?

a. 133 K
b. 469 K
c. 0.918 K
d. 981 K
e. 279 K

4. One balloon contains 3.6 L and 1.6 mol of helium gas. If a second balloon contains 3.1 mol of helium, what is the volume of the gas in the second balloon if the temperature and pressure of both balloons is the same?

a. 7.0 L
b. 18 L
c. 0.54 L
d. 0.14 L
e. 1.9 L

5. What is the temperature of a 0.65 L sample of fluorine gas at 620 torr which contains 1.3 mol fluorine?

a. 0.0792 K
b. 0.201 K
c. 60.2 K
d. 3780 K
e. 5.0 K

6. 1.75 mol argon gas is placed inside a container with a volume of 10.0 L at a temperature of 15 °C. The temperature is raised to 65.0 °C. What is the pressure of argon at the new temperature?

a. 0.933 atm
b. 0.718 atm
c. 4.64 atm
d. 4.85 atm
e. 4.14 atm

7. A mixture of methane and carbon monoxide gas at 0 °C has a total volume of 12.9 L and the quantity of gas is 0.90 mol. What is the P_{TOTAL} of the mixture?

a. 2.4 atm
b. 0.99 atm
c. 3.4 atm
d. 1.6 atm
e. 1.4 atm

8. Which of the following statements is not true?

a. As the temperature of a gas increases, the gas particles move faster.
b. For ideal gases, the gas particles neither attract nor repel each other.
c. The temperature of a gas is a measure of the motion of that gas.
d. The pressure of a gas is a result of the small volume of the gas particles.
e. Theories are built to explain observations.

9. Which relationship is not expressed by Boyle's, Charles's, or Avogadro's laws?

a. The volume and pressure of a gas are directly related.
b. Any volume of gas divided by its temperature is equal to a constant.
c. When temperature and pressure are constant, the product of the volume of a gas and its pressure are always equal to a constant.
d. When temperature and pressure are held constant, the volume of a gas is directly proportional to the number of moles of that gas.
e. The volume of a gas is directly proportional to the temperature in kelvins.

10. Magnesium metal reacts with hydrochloric acid to produce magnesium chloride and hydrogen gas.

$$Mg + 2HCl \rightarrow MgCl_2 + H_2$$

What mass of magnesium is needed to react with excess hydrochloric acid to produce 0.25 L hydrogen gas if the temperature is 40 °C and the pressure is 2.5 atm?

a. 0.19 g
b. 0.58 g
c. 41 g
d. 0.024 g
e. 5.3 g

PRACTICE EXAM ANSWERS

1. b (13.1)
2. a (13.2)
3. d (13.3)
4. a (13.4)
5. e (13.5)
6. d (13.5)
7. d (13.6)
8. d (13.7, 13.8, 13.9)
9. a (13.2, 13.3, 13.4)
10. b (13.10)

10. Magnesium metal reacts with hydrochloric acid to produce magnesium chloride and hydrogen gas.

$$Mg + 2HCl \rightarrow MgCl_2 + H_2$$

What mass of magnesium is necessary to react with excess hydrochloric acid to produce [illegible] L of hydrogen gas if the temperature is [illegible] and the pressure is [illegible]

[illegible]

CHAPTER 14: LIQUIDS AND SOLIDS

INTRODUCTION

This chapter discusses the properties of liquids and solids. You will learn what makes the particles in solids stay together and why some liquids boil at higher temperatures than others. One liquid of particular importance is water. Liquid water is one of the most important parts of our environment, and chemists routinely use water to dissolve many kinds of substances. As we will see, water has some unusual properties.

AIMS FOR THIS CHAPTER

1. Be able to explain and interpret a heating/cooling curve for water. (Section 14.1)
2. Know the differences between intramolecular and intermolecular forces. (Section 14.2)
3. Be able to use the molar heat of fusion and the molar heat of vaporization to calculate the energy required to melt or form ice, to boil water and to condense steam. (Section 14.2)
4. Be able to describe dipole-dipole attraction, hydrogen bonding, and London dispersion forces between molecules. (Section 14.3)
5. Know how the intermolecular forces play a role in changes of physical state, that is, from liquid to solid, or liquid to gas. (Section 14.3)
6. Understand the concept of vapor pressure, and be able to predict whether a molecule will have a large or small vapor pressure. (Section 14.4)
7. Know the properties of crystalline solids. (Section 14.5)
8. Know what types of bonding occur in different crystalline solids, and how bonding in metals accounts for the different properties of different metals. (Section 14.6)

QUICK DEFINITIONS

Normal boiling point — The temperature at which a liquid is converted to a gas at 1 atm pressure. (Section 14.1)

Heating/cooling curve — A curve which shows how the temperature of a substance changes as more energy is added. (Section 14.1)

Normal freezing point — The temperature at which a liquid is converted to a solid, or a solid to a liquid, at 1 atm pressure. (Section 14.1)

Intramolecular forces — Forces within a molecule. (Section 14.2)

Intermolecular forces — Forces between two or more molecules. (Section 14.2)

Molar heat of fusion — The amount of energy required to melt one mole of a substance. (Section 14.2)

Molar heat of vaporization — The amount of energy required to convert one mole of liquid to its vapor. (Section 14.2)

Dipole-dipole attractions — Polar molecules have an end which bears a partial positive charge, and an end which bears a partial negative charge. The molecules line up with the negative end of one molecule near the positive end of another molecule, because opposite charges attract each other. These attractions are called dipole-dipole attractions. (Section 14.3)

Hydrogen bonding — A special kind of strong dipole-dipole attraction which occurs among molecules when a hydrogen atom is bonded to an electronegative atom such as nitrogen, oxygen or fluorine. The hydrogen on one molecule is attracted to the N, O, or F on another molecule. (Section 14.3)

London dispersion forces — Weak forces which occur among noble gas atoms or nonpolar molecules. Electrons temporarily shift to one side of the molecule, causing a temporary negative charge on one side and a temporary positive charge on the other. The electrons of another molecule can be attracted to the temporary positive charge, causing attractions between molecules. (Section 14.3)

Vaporization — The evaporation of a liquid. (Section 14.4)

Condensation — The formation of a liquid from a vapor. (Section 14.4)

Vapor pressure — Some of the liquid placed in a closed container vaporizes. If the pressure of the vapor is measured when the rates of vaporization and condensation are equal, the pressure measured is called the vapor pressure. (Section 14.4)

Crystalline solids — Solids whose microscopic arrangement of atoms is regular. Often, the orderly arrangement is reflected in the shape of the crystal visible to the naked eye. (Section 14.5)

Ionic solids — A solid composed of individual ions and held together by the charges on the ions. (Section 14.6)

Molecular solid — A solid composed of individual molecules. The solid is held together by either dipole-dipole forces or London dispersion forces. (Section 14.6)

Atomic solid — A solid composed of individual atoms of only one element. Some atomic solids are held together by London dispersion forces, but some, such as diamond, are held together by strong covalent bonds. (Section 14.6)

Electron sea model — Solid metals are often thought of as having cations packed tightly together, floating in a sea of valence electrons. (Section 14.6)

Alloy — A mixture of elements which is metallic in nature. (Section 14.6)

Substitution alloy — Some of the metal atoms which are packed together in the solid have been replaced with atoms of another metallic element. (Section 14.6)

Interstitial alloy — An element with smaller atoms is added to atoms of the host metal and is able to fill in the small holes between the larger host metal atoms. (Section 14.6)

CONTENT REVIEW

14.1 WATER AND ITS PHASE CHANGES

What Are Some of the Properties of Water?

The **normal freezing point** of water is 0 °C, and the **normal boiling point** of water is 100 °C, both at 1 atm pressure. Water has a wide temperature span, 100 °C, over which it exists as a liquid. The **heating/cooling curve** for water shows what happens when heat energy is applied to solid water, then to liquid water, and finally to water vapor. As heat is applied to solid ice, the temperature increases to 0 °C. Now the ice begins to melt. The temperature ceases to increase until all of the ice is melted. When only liquid water is present, the temperature begins to climb again, until it reaches 100 °C. The temperature remains at 100 °C until all of the liquid water is converted to water vapor. Once the liquid has been converted to steam, the temperature begins to rise again. Note that the amount of heat energy required to convert ice to liquid water is less than the amount required to convert liquid water to steam.

14.2 ENERGY REQUIREMENTS FOR THE CHANGES OF STATE

How Can We Calculate the Energy Requirements For Changing the State of Water Molecules?

The amount of energy needed to melt a mole of any substance is called the **molar heat of fusion.** The molar heat of fusion for ice is 6.02 kJ/mol. The amount of energy needed to convert a mole of a liquid to a vapor is called the **molar heat of vaporization.** The molar heat of vaporization for water is 40.6 kJ/mol. We can use the molar heat of fusion and the molar heat of vaporization to calculate the energy required to melt or vaporize any quantity of water.

Example:

How much energy in joules is required to vaporize 35.5 g of water at 100 °C? We know that to vaporize 1 mol of water, 40.6 kJ are required. In this problem we want to know how many kJ are required for 35.5 g water. We can use the molar mass of water, 18.02 g, to calculate the moles of water in 35.5 g water, and use the molar heat of vaporization for water to calculate the number of kJ required for the vaporization of 35.5 g water.

$$35.5\ \not{g}\ \not{H_2O} \times \frac{1\ \text{mol}\ H_2O}{18.02\ \not{g}\ \not{H_2O}} = 1.97\ \text{mol}\ H_2O$$

Now, use the molar heat of vaporization for water to calculate the number of kJ required for the vaporization of 1.97 moles of water.

$$1.97\ \text{m}\not{o}\text{l}\ \not{H_2O} \times \frac{40.6\ \text{kJ}}{1\ \text{m}\not{o}\text{l}\ \not{H_2O}} = 80.0\ \text{kJ}$$

Example:

How much energy in joules is required to convert 221 g liquid water at 53 °C to steam at 100 °C? Solving this problem requires two steps because liquid water must be heated to the boiling point, and then liquid water at 100 °C must be converted to steam at 100 °C. The first part of the problem requires calculating the energy needed to heat liquid water at 53 °C to liquid water at 100 °C. To calculate the energy needed to raise the temperature of liquid water, a conversion factor involving the specific heat capacity of water is needed. We are given the specific heat capacity of liquid water as 4.18 J/g °C. Specific heat capacity was discussed in detail in Section 3.5. To calculate the energy needed to raise the temperature of liquid water, use the formula

$$Q = c \times m \times \Delta T,$$
$$\text{energy} = \text{specific heat capacity} \times \text{mass of water} \times \text{change in temperature.}$$

We can calculate ΔT, the change in temperature, because we know both the initial and the final temperatures of liquid water. ΔT equals 100 °C minus 53 °C, which is equal to 47 °C. Substitute values into the equation and find Q.

$$Q = 4.18 J/g\,°C \times 221\ g \times 47\ °C = 43{,}000\ J$$

The second part requires calculating the energy needed to convert liquid water at 100 °C to steam at 100 °C. Calculating the amount of energy needed to vaporize 221 g of water requires that we know how many moles of water are in 221 g of water. We can use the molar mass of water, 18.02 g, to calculate the moles of water from grams of water.

$$221\ \cancel{g}\ \cancel{H_2O} \times \frac{1\ mol\ H_2O}{18.02\ \cancel{g}\ \cancel{H_2O}} = 12.3\ mol\ H_2O$$

We can now use the molar heat of vaporization for water to calculate the energy needed for this part of the problem.

$$12.3\ \cancel{mol}\ \cancel{H_2O} \times \frac{40.6\ kJ}{\cancel{mol}\ \cancel{H_2O}} = 499\ kJ$$

We are left with two energy amounts, the first from heating liquid water, and the second from converting liquid water to steam. To determine the total energy required for the process, it is necessary to add the energies determined separately. We wish to add 43,000 J plus 499 kJ. The units are different, however. One is given in joules, and the other is given in kilojoules. We must convert one of the units to the other. Let's convert J to kJ.

$$43{,}000\ \cancel{J} \times \frac{1\ kJ}{1000\ \cancel{J}} = 43\ kJ$$

We can now add the two energy amounts together.

$$43\ kJ + 499\ kJ = 542\ kJ$$

The total amount of energy needed is 542 kJ.

14.3 INTERMOLECULAR FORCES

What Kinds of Intermolecular Forces Exist?

Intermolecular forces are the attractive forces between molecules which hold them together. One kind of intermolecular force is called a **dipole-dipole attraction**. These forces occur when molecules are polar; that is, when one part of the molecule has a partial positive charge, and another part of the molecule has a partial negative charge. The differences in charge are often caused by differences in electronegativity among the atoms in a molecule. Electronegativity is discussed in Section 12.2 of your textbook. Polar molecules line up so that the end with the positive charge is next to the end of another molecule with a negative charge, because opposite charges attract each other. These attractive forces are called dipole-dipole attractions.

Another kind of intermolecular attraction is the **hydrogen bond**. Hydrogen bonds occur when molecules are polar, and in fact, are a special type of dipole-dipole attraction. When a hydrogen atom is bonded to a very electronegative atom such as nitrogen, oxygen or fluorine, the dipole-dipole attractions between the hydrogen of one molecule and the electronegative atom on another molecule are very strong. For example, hydrogen bonds form between molecules of ammonia because ammonia is a very polar molecule. The hydrogen atoms on one ammonia molecule are attracted to the nitrogen atom of another ammonia molecule.

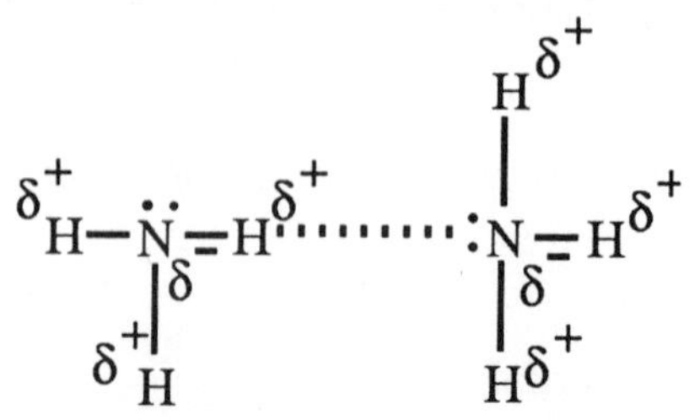

A third type of intermolecular force, called London dispersion forces, occurs between nonpolar molecules. Dipole-dipole attractions and hydrogen bonding are caused when electrons are permanently shifted to one side of the molecule. Nonpolar molecules do not have permanent shifts in electron distribution. Sometimes, however, as the electrons move around, the electron distribution will become temporarily unbalanced, causing a temporary negative charge on one side of the molecule. As the molecule with shifted electrons comes into contact with another molecule, a temporary shift in electrons is induced in the other molecule so that the negative area on one molecule is attracted to the positive area on the other molecule. These forces only last a short period of time, until the electrons are redistributed, and they are very weak, much weaker than dipole-dipole attractions or hydrogen bonds.

What Effect Do Intermolecular Forces Have on the Properties of Substances?

Each of the Group 6 elements will form a compound by combining with two hydrogen atoms. Of the four smallest atoms in this group, oxygen, sulfur, selenium and tellurium, oxygen is the only one which is highly electronegative, and the only one which forms hydrogen bonds. When we look at a graph of boiling point *vs.* the period of the compound formed, we see that water has a much higher boiling point than would be predicted from the molecular weight. The oxygen atom is very electronegative and tends to pull the electrons it shares with hydrogen towards itself. So oxygen has a partial negative charge. The hydrogen atoms are electron deficient and have a partial positive charge. Water has a high boiling point because hydrogen bonds form between water molecules. It takes more heat energy to overcome the forces of the hydrogen bonds, hence the boiling point is higher. As a general rule, molecules which can hydrogen bond with each other will have higher boiling points than molecules which do not hydrogen bond.

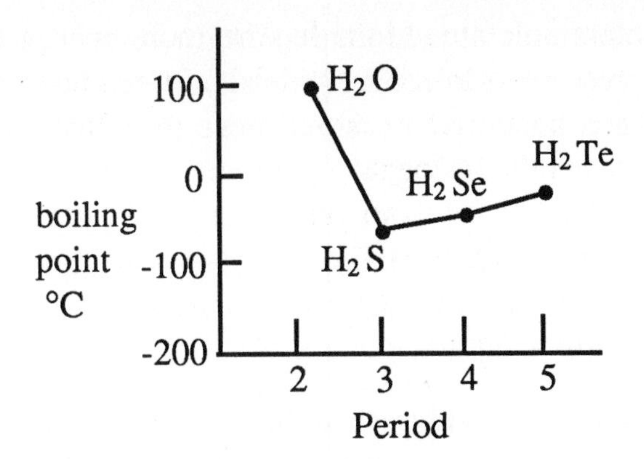

14.4 EVAPORATION AND VAPOR PRESSURE

What Are Vaporization, Condensation and Vapor Pressure?

Vaporization is equivalent to evaporation, and happens when molecules of a liquid substance escape from the surface and become a vapor. For a liquid to evaporate, the molecules must overcome whatever intermolecular forces are in existence. Energy is required to break the intermolecular forces. **Condensation** is the formation of vapor molecules from a liquid.

When the rate of vaporization is equal to the rate of condensation, the number of molecules escaping the liquid is the same as the number of molecules returning to the liquid. When this happens, vaporization and condensation balance each other, and the system is said to be at **equilibrium**. A liquid in a container open to the atmosphere continues to evaporate until all of the liquid has been converted to vapor. This balancing of processes occurs only when a liquid is placed in a closed container.

When liquid in a closed container has reached equilibrium, the rates of vaporization and condensation are equal. The pressure of the vapor above the liquid is called the equilibrium vapor pressure, or just **vapor pressure**.

What Determines Vapor Pressure?

The two major factors which determine vapor pressure are the molecular weight of a molecule and the types of intermolecular forces which are present. If molecules have a high molecular weight, at any given temperature they move slower than smaller molecules. The molecules escape less easily from the surface of the liquid. They have a low vapor pressure. Molecules with low molecular weight move faster at any given temperature, and they escape easily from the surface of the liquid. They have higher vapor pressures. Molecules which are nonpolar and whose only intermolecular forces are London dispersion forces do not require much energy to break the molecules apart and so have relatively high vapor pressures.

Molecules which have dipole-dipole attractions require more energy to overcome these forces, and so have a lower vapor pressure. Molecules which hydrogen bond require the most energy to overcome the strong hydrogen bonding forces and have the lowest vapor pressures, assuming that the molecular weights are relatively equal.

14.5 THE SOLID STATE: TYPES OF SOLIDS

What Are the Classes of Crystalline Solids?

One class of solids is the **crystalline solids**, which have an orderly arrangement of atoms or ions. It is often possible to see the orderly arrangement of the particles in the shape of the crystal. There are three important types of crystalline solids: **ionic solids**, **molecular solids**, and **atomic solids**. Ionic solids are composed of ions packed together in an orderly manner, with anions packed next to cations. Sodium chloride is an example of a crystalline ionic solid. Molecular solids are composed of individual molecules arranged in an orderly manner. Table sugar is a crystalline molecular solid made up of individual sucrose molecules. Atomic solids are composed of individual atoms of the same element. An example of a crystalline atomic solid is diamond, made of carbon atoms covalently bonded to each other.

14.6 BONDING IN SOLIDS

What Forces Hold Together Particles in Crystalline Solids?

In ionic solids, ions are held together in a solid by the strong attractive forces between oppositely charged ions. These forces are strong, and ionic solids typically have very high melting points.

Molecular solids are held together by either dipole-dipole attractions or by London dispersion forces. Both of these forces are weaker than the attractions in ionic solids. Because the forces which hold the molecules together are relatively weak, molecular solids generally have lower melting points than ionic solids.

Atomic solids are held together by a variety of forces. Some atomic solids, such as the noble gases in the solid state, are held together only by London dispersion forces. These solids have low melting points. Other atomic solids, such as diamonds, are held together by strong covalent bonds. These solids have a high melting point.

What Forces Hold Together Metals?

Metals are another example of crystalline atomic solids. Metals are often described as a series of metal cations in a sea of valence electrons. This is called the **electron sea model**. This model explains some of the properties of metals. For example, metals conduct electricity because the valence electrons are free to move and can conduct electricity. The cations within the sea of valence electrons can move easily, and their easy movement accounts for metals being easy to beat into sheets and pull into wires.

It is easy to mix atoms of other elements with the atoms of a pure metal. When the resulting mixture has metallic properties, it is called an **alloy**. In a **substitutional alloy**, some of the atoms of the pure metal are replaced with atoms of another metal. Usually, the replacement metal has atoms close in size to the original metal so that the metallic crystal retains its regular arrangement. In an **interstitial alloy**, atoms of another element are added to the pure metal. Instead of substituting for atoms of the original metal in the crystal, these atoms are smaller and fit into the holes between metal atoms. The size of the new atoms added to the metal must be small enough to fit in the holes without disrupting the crystalline structure of the metal. Carbon is one element used to make interstitial alloys.

LEARNING REVIEW

1. Why does ice float on the surface of liquid water?

2. Which of the following statements about water are true, and which are false?

    a. At 0 °C water can be either a solid or a liquid.
    b. The normal boiling point of water is 97.5 °C at 1 atm pressure.
    c. When liquid water is converted to steam, covalent bonds in the water molecules are broken.
    d. Forces between water molecules are called intramolecular forces.

3. Why does it take more energy to convert liquid water to steam than it does to convert ice to liquid water?

4. How much energy is required when 15.0 g of liquid water at 75 °C is heated to 100 °C, and then converted to steam at 100 °C? The molar heat of vaporization for liquid water is 40.6 kJ/mol and the specific heat capacity of liquid water is 4.18 J/g °C.

5. How much energy is required to convert a block of ice weighing 2.45 kg to liquid water at 50 °C? The molar heat of fusion for ice is 6.02 kJ/mol, and the specific heat capacity of liquid water is 4.18 J/g °C.

6. How much energy is required to convert 6.0 g liquid water to steam at 100 °C?

7. Which bond is stronger, a covalent bond or a hydrogen bond?

8. Draw a diagram showing how three NH_3 molecules can form hydrogen bonds.

9. Why does water, H_2O, boil at a higher temperature than H_2S?

10. Explain how London dispersion forces form between two molecules of argon.

11. Match the following terms with the correct definition.

a.	condensation	the pressure exerted by a liquid at equilibrium with its vapor
b.	vapor pressure	a balance between two opposite processes
c.	vaporization	vapor molecules form a liquid
d.	equilibrium	a liquid becomes a gas

12. Which of the molecules in each pair would have the greater vapor pressure? To help you determine which of these molecules is polar, review Sections 12.2 and 12.3 of your textbook.

    a. CH_4, CH_3OH b. H_2O, H_2S

13. Name the three types of crystalline solids and give an example of each.

14. What intermolecular force is responsible for holding each of the following compounds together in the solid state?

    a. KCl
    b. HF
    c. $SiCl_4$

15. Mixtures of metallic elements form alloys. What are the two kinds of alloys and how do they differ from each other?

ANSWERS TO LEARNING REVIEW

1. Ice is less dense than the water, so it floats on the surface. The density of ice (solid water) is less than the density of liquid water because water expands as it freezes. This means that ice has the same mass as water, but occupies a greater volume. Using the formula density = mass/volume, as the volume becomes greater, the density is less.

2. a. 0 °C is the normal freezing point of water. Both liquids and solids can exist at the same time at the normal freezing point of a liquid. This statement is true.
   b. The normal boiling point of water is 100 °C at 1 atm pressure, not 97.5 °C. This statement is false.
   c. A physical change occurs when liquid water boils to become steam. No chemical bonds are broken. Liquid water contains H_2O molecules, and so does steam. This statement is false.
   d. Forces between water molecules are called **inter**molecular forces. Forces between atoms of a single water molecule are called **intra**molecular forces. This statement is false.

3. Converting ice to liquid water involves overcoming relatively few intermolecular forces. This is true because the molecules in both solid and liquid water are close together, so there is relatively little disruption when the change from solid to liquid occurs. However, going from liquid water to steam requires overcoming the intermolecular forces between water molecules, so a great amount of energy is required. The water molecules in steam are very far apart compared to water molecules in the liquid state.

4. In this problem, there are two changes occurring. Each change must be considered separately when calculating the overall amount of energy needed. In the first change, we heat the liquid water from 75 °C to 100 °C. This conversion requires the use of the specific heat capacity for water along with the equation below.

$$\text{Energy required} = \text{specific heat capacity} \times \text{mass} \times \text{temperature change}$$
$$Q = c \times m \times \Delta T$$

We are given the specific heat capacity for liquid water, the mass of water, and we can calculate the change in temperature. If the final temperature is 100 °C and the beginning temperature is 75 °C, then the change in temperature, ΔT, must be 25 °C.

Now substitute into the equation.

$$Q = c \times m \times \Delta T$$

$$Q = 4.18 \text{ J/g °C} \times 15.0 \text{ g water} \times 25 \text{ °C}$$

$$Q = 1600 \text{ J}$$

In the second change, liquid water at 100 °C is converted to steam at 100 °C. We are going from the liquid state to a gas, but the temperature remains 100 °C. To calculate the amount of energy needed during this change, we can use the molar heat of vaporization for water as a conversion factor.

$$\text{molar heat of vaporization} = 40.6 \text{ kJ/mol } H_2O$$

This conversion factor requires us to know the number of moles of steam. The problem tells us we have 15.0 g steam. We can calculate the number of moles from the molar mass of water.

$$15.0 \text{ } \cancel{H_2O} \text{ } \cancel{g} \times \frac{1 \text{ mol } H_2O}{18.02 \text{ } \cancel{g} \text{ } \cancel{H_2O}} = 0.832 \text{ mol } H_2O$$

Now, use the molar heat of vaporization as a conversion.

$$0.832 \text{ } \cancel{mol} \text{ } \cancel{H_2O} \times \frac{40.6 \text{ kJ}}{\cancel{mol} \text{ } \cancel{H_2O}} = 33.8 \text{ kJ}$$

To convert 15.0 g liquid water from 75 °C to 100 °C requires 1600 J and to convert 15.0 g liquid water at 100 °C to a gas requires 33.8 kJ. To determine the total amount of energy required, we can add together the two quantities of energy. However, the units do not match. One quantity is given in Joules and the other in kiloJoules. Let's convert J to kJ.

$$1600\ \cancel{J} \times \frac{1\ kJ}{1000\ \cancel{J}} = 1.6\ kJ$$

The total amount of energy required is 1.6 kJ plus 33.8 kJ, which is equal to 35.4 kJ.

5. This problem has two parts. The first part is the calculation of the energy required to convert a block of ice to liquid water. The second part is the calculation of the energy required to heat the liquid water from 0 °C to 50 °C. In the first change, we want to melt the block of ice. We are going from the solid state to the liquid state, but the temperature stays the same. To calculate the amount of heat needed during this change of state, we can use the molar heat of fusion for ice as a conversion factor.

molar heat of fusion = 6.02 kJ/mol water

This conversion factor requires us to know the number of moles of water and we are given kilograms of water. We need to convert from kilograms to grams, and from grams to moles using the molar mass.

$$2.45\ \cancel{kg\ H_2O} \times \frac{1000\ \cancel{g\ H_2O}}{1\ \cancel{kg\ H_2O}} \times \frac{1\ mol\ H_2O}{18.02\ \cancel{g\ H_2O}} = 136\ mol\ H_2O$$

Now, use the molar heat of fusion for water as a conversion factor.

$$136\ \cancel{mol\ H_2O} \times \frac{6.02\ kJ}{\cancel{mol\ H_2O}} = 819\ kJ$$

In the second change we want to heat liquid water from 0 °C to 50 °C. This conversion requires us to use the specific heat capacity for water along with the equation below.

$$\text{Energy required} = \text{specific heat capacity} \times \text{mass} \times \text{temperature change}$$
$$Q = c \times m \times \Delta T$$

We are given the specific heat capacity for water (c), the mass in grams of water to be heated (m), and we can calculate the change in temperature. If the final temperature is 50 °C and the initial temperature is 0 °C, then the change in temperature, ΔT, must be 50 °C.

$$Q = 4.18\ J/\cancel{g}\ \cancel{°C} \times 2.45\ \cancel{kg} \times \frac{1000\ \cancel{g}}{1\ \cancel{kg}} \times 50\ \cancel{°C}$$

$$Q = 5.1 \times 10^5 \text{ J}$$

To convert 2.45 kg ice to liquid water at 0 °C requires 819 kJ, and to convert 2.45 kg water at 0 °C to water at 50 °C requires 5.1×10^5 J. To determine the total amount of energy required, we can add together the two quantities of energy. However, the units do not match. One quantity is given in Joules, the other in kiloJoules. Let's convert Joules to kiloJoules. Remember that each kJ equals 1000 (10^3) J.

$$5.1 \times 10^5 \text{ J} \times \frac{1 \text{ kJ}}{1000 \text{ J}} = 510 \text{ kJ}$$

The total amount of energy required is 819 kJ plus 510 kJ, which is equal to 1300 kJ.

6. This problem requires only one step. We want to know how much energy is needed to convert water at 100 °C to steam at 100 °C. We can use the molar heat of vaporization of water as a conversion factor after we convert grams of water to moles of water.

$$6.0 \text{ g water} \times \frac{1 \text{ mol water}}{18.02 \text{ g water}} = 0.33 \text{ mol water}$$

Now, use the molar heat of vaporization to convert mol water to kiloJoules.

$$0.33 \text{ mol water} \times \frac{40.6 \text{ kJ}}{\text{mol water}} = 13 \text{ kJ}$$

The amount of energy required to convert 6.0 g water at 100 °C to steam at 100 °C is 13 kJ.

7. Hydrogen bonds are very strong dipole-dipole forces which occur between molecules which contain a hydrogen atom bonded to either a nitrogen, an oxygen or a fluorine atom. Even relatively strong hydrogen bonds are only one percent as strong as covalent bonds. Covalent bonds are much stronger than hydrogen bonds.

8. Ammonia molecules can form hydrogen bonds because there is a hydrogen atom bonded to a strongly electronegative nitrogen atom, producing a strong dipole.

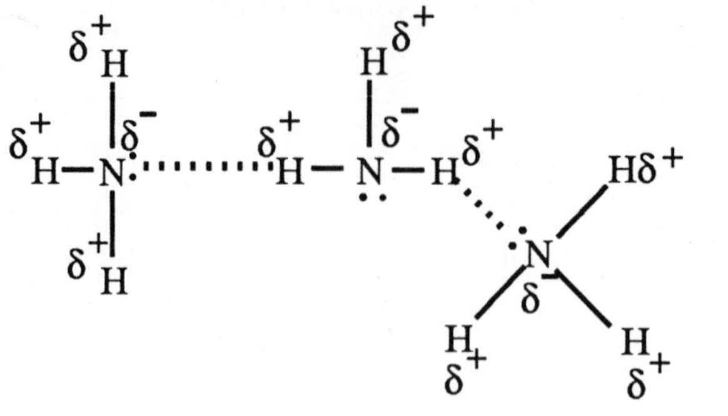

9. Both H_2O and H_2S contain two hydrogen atoms. The difference between the two molecules is the central atom, oxygen or sulfur. Oxygen is an electronegative atom, while sulfur is not very electronegative. Water molecules can form strong hydrogen bonds because the O-H bond is very polar. The difference in electronegativity between oxygen and hydrogen is large. A great deal of heat energy is required to break apart the hydrogen bonds between water molecules so that the molecules can enter the vapor phase. The S-H bond in H_2S is not very polar because the difference in electronegativity between hydrogen and sulfur is small. No hydrogen bonds form between H_2S molecules, so less heat energy is required to boil liquid H_2S.

10. The weak intermolecular forces are called London dispersion forces. They are weaker than hydrogen bonds or dipole-dipole interactions. We normally assume that the electrons in atoms such as argon are distributed somewhat evenly around the atom. However, we cannot predict the path an electron will take as it moves around the atom. Sometimes, more electrons are momentarily found on one side of an argon atom. This causes a small and temporary dipole. This temporary dipole can induce a dipole on an atom which may be nearby. The negative side of one argon atom is attracted to the positive side of another argon atom. These attractive forces between atoms are weak, and they do not last long. As the electrons move, the force dissipates.

11.

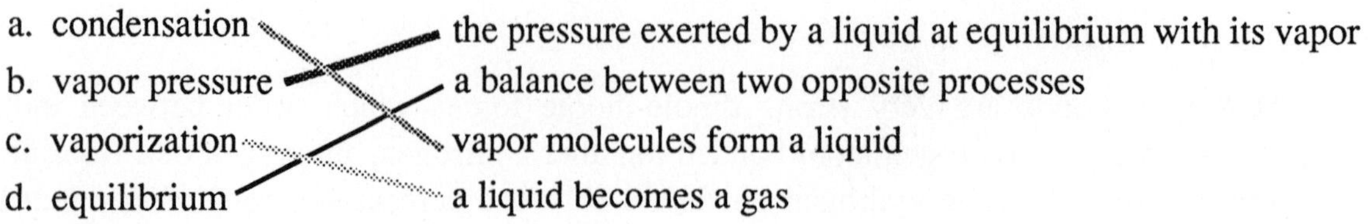

a. condensation	the pressure exerted by a liquid at equilibrium with its vapor
b. vapor pressure	a balance between two opposite processes
c. vaporization	vapor molecules form a liquid
d. equilibrium	a liquid becomes a gas

12. Two factors determine the relative vapor pressure of any two liquids, the molecular weights of the two molecules, and the intermolecular forces.

    a. CH_3OH has a higher molecular weight, 32.0 g/mol, than CH_4, which has a molecular weight of 16.0 g/mol. In a molecule of CH_3OH, a hydrogen atom is bonded to an electronegative oxygen atom. When hydrogen atoms are bonded to electronegative atoms such as oxygen, hydrogen bonds can form among molecules. CH_3OH molecules are attracted to each other by hydrogen bonds. CH_4 molecules contain only carbon-hydrogen bonds. Carbon is not as electronegative as atoms such as oxygen, so molecules of CH_4 are not very polar. There are no hydrogen bonds between CH_4 molecules. The intermolecular forces will be weak London dispersion forces. So the vapor pressure of CH_4 will be higher than CH_3OH.
    b. H_2S has a higher molecular weight, 34.0 g/mol, than H_2O, 18.0 g/mol. Purely on the basis of molecular weight, H_2S would have a lower vapor pressure than water. However, intermolecular forces are also important in determining vapor pressure. Water molecules form relatively strong hydrogen bonds, while H_2S does not. Because of the lack of strong intermolecular forces, H_2S has a higher vapor pressure than H_2O.

13. The three types of crystalline solids are ionic solids, molecular solids, and atomic solids. An ionic solid exists as a collection of cations and anions, held together by the attractive forces between the ions. An example is the salt, potassium chloride. Molecular solids consist of molecules. There are no ions present. An example is table sugar, or sucrose. An atomic solid is made from individual atoms, all the same. Pure copper metal is an example of an atomic solid.

14. a. KCl consists of potassium ions and chloride ions. The ions are held together in solid KCl by the forces which exist between the oppositely charged ions.
    b. When hydrogen is bonded to electronegative fluorine atoms to produce molecules of hydrogen fluoride, hydrogen bonds can form between molecules. The fluorine atom bears a partial negative charge, while the hydrogen atom bears a partial positive charge.
    c. $SiCl_4$ contains only silicon-chlorine bonds. This molecule will not form intermolecular hydrogen bonds because hydrogen bonding only occurs when hydrogen bonds to electronegative atoms such as oxygen, nitrogen or fluorine. Also, this molecule is not polar because it is a balanced molecule. There are no partial positive or negative ends. So there are no hydrogen bonds and no dipole-dipole attractions among $SiCl_4$ molecules, but London dispersion forces cause weak attractive forces between the molecules.

15. An alloy is a substance which contains a mixture of elements and has metallic properties. There are two basic kinds of alloys. Substitutional alloys have some of the metal atoms replaced with other metal atoms which are approximately the same size. Interstitial alloys have some of the small spaces between metal atoms filled with atoms smaller than the metal atoms.

PRACTICE EXAM

1. Which of the following statements about water is **not** true?

   a. When the temperature of steam decreases, energy is removed.
   b. In the winter water pipes break because water expands when it freezes.
   c. The gaseous temperature range of water is between 0 °C and 100 °C.
   d. At the normal freezing point of water, ice and liquid water can coexist.
   e. Water at the boiling point does not increase in temperature until all the liquid water has been converted to steam.

2. Which change shown on the heating curve for water requires the greatest amount of energy?

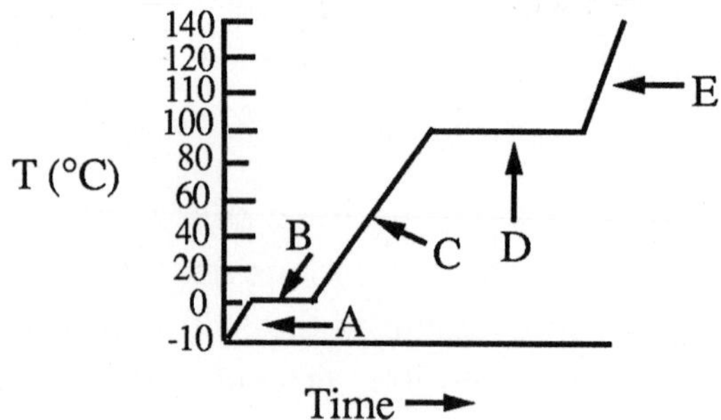

3. How much energy is required to raise the temperature of 38.3 g of liquid sulfuric acid from 25 °C to 75 °C? The specific heat capacity of liquid sulfuric acid is 1.67 J/g °C.

   a. 4.8 kJ
   b. 3.2 kJ
   c. 63 kJ
   d. 1.6 kJ
   e. 84 kJ

4. How much energy is required to convert 20.0 g water at 30 °C to steam at 100 °C? The specific heat capacity of liquid water is 4.18 J/g °C and the molar heat of vaporization is 40.6 kJ/mol.

   a. 818 kJ
   b. 51 kJ
   c. 62 kJ
   d. 5.9 kJ
   e. 54 kJ

5. Which of the molecules below would form intermolecular hydrogen bonds?

a. HCl
b. H_2S
c. CH_4
d. H_2Te
e. NH_3

6. Which of the molecules below would have intermolecular dipole-dipole attractions? To help you determine which of these molecules is polar, you can review Sections 12.2 and 12.3 in your textbook.

a. CH_3-O-CH_3
b. H_2
c. CO_2
d. CCl_4
e. $CaCl_2$

7. Which of the solids below would have the lowest melting point?

a. Cu
b. H_2O
c. C
d. NaCl
e. Cl_2

8. Which statement about vapor pressure is **not** correct?

a. A liquid in a sealed container at equilibrium with its vapor phase undergoes simultaneous evaporation and condensation.
b. Volatile liquids are ones with a high vapor pressure.
c. HCl would have a lower vapor pressure than H_2.
d. Molecules which can form hydrogen bonds tend to have higher vapor pressures than molecules which do not form hydrogen bonds.
e. When water is placed in a sealed container, some of the molecules escape from the liquid surface and enter the vapor phase.

9. Which statement about solid KCl is true?

   a. In solid KCl the ions are packed together much as spheres are packed in a box.
   b. Covalent bonds are responsible for the crystalline structure of solid KCl.
   c. Solid KCl is an atomic solid.
   d. London dispersion forces hold solid KCl together.
   e. The melting point of solid KCl would be about the same as the melting point of solid argon.

10. Which of the substances below forms a molecular solid?

   a. Fe
   b. H_2S
   c. $Ca(NO_3)_2$
   d. Kr
   e. $CuSO_4$

ANSWERS TO PRACTICE EXAM

1. c (14.1)
2. d (14.1)
3. b (14.2)
4. b (14.2)
5. e (14.3)
6. a (14.3)
7. e (14.3 and 14.5)
8. d (14.4)
9. a (14.5 and 14.6)
10. b (14.6)

CHAPTER 15: SOLUTIONS

INTRODUCTION

It is important to know how much solid is dissolved in a liquid. Telling someone that you added a little sugar to their iced tea does not give them much of a clue as to how sweet the tea will be. A "little" sugar to you might mean a "lot" of sugar to someone else. In this chapter you will learn ways to express concentration so that you know just how much dissolved solid is present in a given volume of liquid.

AIMS FOR THIS CHAPTER

1. Know how intermolecular forces affect the solubility of molecules. (Section 15.1)
2. Know how to define terms associated with concentration of a solution. (Section 15.2)
3. Know how to calculate the mass percent of a solution. (Section 15.3)
4. Be able to calculate the molarity, M, of a solution. (Section 15.4)
5. Be able to calculate the moles of solute in a solution. (Section 15.4)
6. Be able to calculate the amount of concentrated stock solution needed to prepare a diluted solution. (Section 15.5)
7. Be able to perform solution stoichiometric calculations using a balanced equation to determine the amount of product formed. (Section 15.6)
8. Be able to perform solution stoichiometric calculations using the balanced equation for a neutralization (acid-base) reaction to determine the amount of reactant needed to neutralize (react with) another reactant. (Section 15.7)
9. Be able to calculate the normality, equivalents, or volume of a solution and use the calculations to determine how much acid or base is needed to react with a given amount of acid or base solution. (Section 15.8)

QUICK DEFINITIONS

Solution — A homogeneous mixture, the components of which are evenly mixed, so that any part of the mixture contains the same components as any other part. (Introduction)

Solvent — The part of a solution present in the largest amount. (Introduction)

Solute — The part of a solution present in the smallest amount, the material dissolved in the solvent. (Introduction)

Aqueous solutions — Solutions in which the solvent is water. (Introduction)

Like dissolves like — Solutes are usually soluble in solvents which have properties similar to the solute. Polar solvents dissolve polar solutes, and nonpolar solvents dissolve nonpolar solutes. (Section 15.1)

Saturated — A solution which contains as much solute as it can hold. No more will dissolve. (Section 15.2)

Unsaturated — A solution which does not contain the maximum amount of solute that it can dissolve. (Section 15.2)

Concentrated — A solution which contains a lot of solute. (Section 15.2)

Dilute — A solution which does not contain much solute. (Section 15.2)

Mass percent — A way of stating the concentration of a solution. Mass percent is the mass of solute in a given mass of solution, times 100 %. (Section 15.3)

Molarity — A way of stating the concentration of a solution. Molarity equals moles solute/liter solution. (Section 15.4)

Standard solution — A solution which has been carefully prepared so that its concentration is accurately known. (Section 15.4)

Dilution — The process of preparing a more dilute solution from a concentrated stock solution. (Section 15.5)

Neutralization reaction — An acid-base reaction. (Section 15.7)

One equivalent of an acid — The amount of acid needed to produce 1 mol of H^+ ions. (Section 15.8)

One equivalent of a base — The amount of base needed to produce 1 mol of OH^- ions. (Section15.8)

Equivalent weight — The mass in grams of an acid or base needed to produce 1 equivalent of H^+ or OH^- ions. (Section 15.8)

Normality — A way of stating the concentration of a solution. Normality equals equivalents solute/liter solution. (Section 15.8)

CONTENT REVIEW

15.1 SOLUBILITY

What Happens When a Substance Dissolves?

Many common ionic and nonionic substances dissolve in water. When an ionic substance dissolves, it breaks apart into ions. For example, when potassium sulfate, K_2SO_4, dissolves, it forms K^+ ions and SO_4^{2-} ions. To understand why a crystal of potassium sulfate breaks apart into ions, we need to consider the nature of the water molecule. Water molecules are polar; that is, one end of the molecule has a partial positive charge, and the other end has a partial negative charge. The positive charge on the potassium cation is attracted to the partial negative charge on the water molecule, and the negative charge on the sulfate anion is attracted to the partial positive charge on the water molecule. The K_2SO_4 crystal is pulled apart by the polarity of the water molecule. Ions in solution are surrounded by oppositely charged ends of water molecules.

Nonionic compounds such as ethyl alcohol, C_2H_5OH, also dissolve in water. The O-H of ethyl alcohol is polar, just as the O-H on the water molecule is. This means that alcohol molecules have a negative end and a positive end just as water molecules do. Alcohol molecules are attracted by water molecules and are dissolved in them.

Molecules which are not water soluble do not have positive or negative ends to be attracted to water.

15.2 SOLUTION COMPOSITION: INTRODUCTION

What Are Some Of The Common Terms Used When Describing Solutions?

Solutions are often described by the amount of solute they contain. A solution which has dissolved as much solute as can dissolve is called a **saturated** solution. A solution which can dissolve more solute is **unsaturated.** A solution which contains a lot of solute is said to be **concentrated,** while a solution which contains a small amount of solute is **dilute.**

15.3 SOLUTION COMPOSITION: MASS PERCENT

How Can the Mass Percent of a Solution Be Calculated?

The mass percent of a solution is the mass of solute present in the **total** mass of solute plus solvent, multiplied by 100 %.

$$\text{mass percent} = \frac{\text{mass of solute}}{\text{mass of solution}} \times 100\ \%$$

Example:

If 15.2 g of K_2SO_4 is dissolved in 255 g of water, what is the mass percent of K_2SO_4 in the solution? We are given the mass of the solute, 15.2 g, and the mass of the solvent, 255 g. Remember that to solve for mass percent, we must add the mass of the solute to that of the solvent so that we use the total mass of the solution to calculate the mass percent.

$$\text{mass percent} = \frac{15.2 \text{ g } K_2SO_4}{15.2 \text{ g } K_2SO_4 + 255 \text{ g } H_2O} \times 100\% = 5.63\% \ K_2SO_4$$

15.4 SOLUTION COMPOSITION: MOLARITY

What is Molarity?

The mass percent of a solution is inconvenient to use when the solvent is a liquid. It is much more convenient to measure the volume of a liquid solvent than it is to measure the mass. The most often used indication of the amount of solute in a given volume of solution is molarity. **Molarity** is equal to the number of moles of solute per volume of solution.

$$\text{Molarity} = \frac{\text{moles of solute}}{\text{liters of solution}}$$

This is often abbreviated as $M = \frac{\text{mol}}{\text{L}}$.

How Can We Calculate the Molarity of a Solution?

Example:

What is the molarity of a solution which contains 5.6 g of NaCl in a total solution volume of 212 mL? Before we begin to solve the problem, examine the information provided. The amount of solute, NaCl, is given in grams. The definition of molarity requires that the solute be expressed in moles. We can convert between grams and moles by using the molar mass. The volume of the solution is given in milliliters. The definition of molarity requires the use of liters. Thus, we will need to perform some conversions before we can correctly calculate the molarity of the solution.

$$5.6 \text{ g NaCl} \times \frac{1 \text{ mol NaCl}}{58.4 \text{ g NaCl}} = 0.096 \text{ mol NaCl}$$

We now know the number of moles of sodium chloride, which we calculated from the number of grams given in the problem.

$$212 \text{ mL} \times \frac{1\text{L}}{1000 \text{ mL}} = 0.212 \text{ L}$$

The volume of the solution is now correctly expressed in liters. We can now solve for molarity, M.

$$M = \frac{\text{mol}}{\text{L}}$$

$$M = \frac{0.096 \text{ mol NaCl}}{0.212 \text{ L solution}}$$

$$M = 0.45 \frac{\text{mol}}{\text{L}} \text{ or } 0.45 \text{ M NaCl}$$

Is the Molarity of Ions In Solution the Same As the Molarity Of the Solution?

When the molarity of a solution is calculated, it is assumed that the solute is in the form it would be in before it dissolved. One mole of Na_2SO_4 when added to water produces two moles of Na^+ ions and one mole of SO_4^{-2} ions.

$$1.0 \text{ mol } Na_2SO_4 \longrightarrow 2.0 \text{ mol } Na^+ + 1.0 \text{ mol } SO_4^{2-}$$

When ionic compounds dissolve in water, they form individual ions. When we say we have a 1.0 M Na_2SO_4 solution, we mean that we consider the condition of the solute before it dissociates into ions. When we say that a solution contains 2.0 M Na^+ ions, we mean that we are considering the concentration of ions in solution, after the sodium sulfate has dissolved.

However, the situation with sucrose is different. A 1 M solution of sucrose contains 1 mol of sucrose. Sucrose molecules do not dissociate into ions.

15.5 DILUTION

What is Dilution?

Dilution is the process of adding more solvent to a solution. If a laboratory has a strong solution, also called a stock solution, on hand, any number of diluted solutions can be prepared from the strong one. Preparing diluted solutions is a common laboratory experience.

How Can We Calculate How Much Stock Solution Is Needed To Prepare a More Dilute Solution?

When we prepare a dilute solution, we are measuring a quantity of stock solution and adding it to water. Once we have added stock solution to water, the quantity of solute in the more dilute solution does not change. All of the solute in the measured portion of stock solution goes into the more dilute solution. In other words, the amount of solute in the measured portion of stock solution is the same as the amount of solute in the more dilute solution. The only thing which has changed is the volume of the solution. We have decreased the concentration of the solution by increasing the volume, but the amount of solute stays the same.

Example:

How can we prepare 5.0 L of 1.5 M HCl from a stock solution which is 12 M? In this problem, we want to take some of the 12 M HCl solution and add it to water, so that the final solution is 1.5 M. We have to take some solute from the stock solution and add it to water. The question is, how much stock solution will it take to make 5.0 L of a 1.5 M HCl solution? Because we know how many liters of dilute solution to prepare and the desired molarity of the dilute solution, we can calculate how many moles of HCl will be present in the final solution. Rearrange the equation below to isolate moles on one side.

$$M = \frac{\text{mol}}{\text{L}}$$

$$M \times V = \frac{\text{mol}}{\not{V}} \times \not{V}$$

$$\text{mol} = M \times V$$

Volume, abbreviated V, is often used in molarity equations to replace the L for liters. Substitute the molarity of the more dilute solution (1.5 M) and volume of the more dilute solution (5.0 L) into the equation.

$$\text{mol HCl needed} = 5.0\ \not{\text{L}}\ \text{sol}\not{\text{u}}\text{tion} \times \frac{1.5\ \text{mol}}{\not{\text{L}}\ \text{sol}\not{\text{u}}\text{tion}} = 7.5\ \text{mol HCl}$$

There will be 7.5 mol HCl in the final dilute solution, so we need to measure out 7.5 mol HCl from the stock solution. The only question we need to answer is, how many liters of stock solution will contain 7.5 mol HCl? We can use the formula for molarity to help with this calculation, too. We want to know the number of liters, the volume, of HCl needed. We already know the molarity of the concentrated solution, 12 M, and the number of moles, 7.5 mol. Use the formula V x M = moles, but rearrange to isolate V on one side.

$$\frac{V \times \not{M}}{\not{M}} = \frac{\text{moles}}{M}$$

$$V = \frac{\text{moles}}{M}$$

Substitute the values into the equation.

$$V = \frac{7.5\ \text{m}\not{\text{o}}\text{l HCl}}{12\ \text{m}\not{\text{o}}\text{l HCl/L}} = 0.63\ \text{L HCl}$$

We can also solve this problem by using the formula $M_1 \times V_1 = M_2 \times V_2$. M_1 represents the molarity of the stock solution, 12 M HCl. V_1 is the volume of the stock solution needed, the volume we want to find. M_2 is the molarity of the dilute solution we wish to make, 1.5 M HCl. V_2 is the volume of the dilute solution, 5.0 L. In this problem we want to know the volume of stock solution needed, so isolate V_1 on one side of the equation by dividing both sides by M_1.

$$M_1 \times V_1 = M_2 \times V_2$$

$$\frac{\cancel{M_1}}{\cancel{M_1}} \times V_1 = V_2 \times \frac{M_2}{M_1}$$

$$V_1 = V_2 \times \frac{M_2}{M_1}$$

Now, substitute values into the equation.

$$V_1 = 5.0 \text{ L} \times \frac{1.5 \cancel{M}}{12 \cancel{M}}$$

$$V_1 = 0.63 \text{ L}$$

The answer to this problem is 0.63 L either way we solve it. So to make 5.0 L of 1.5 M HCl from a stock solution which is 12 M HCl, take 0.63 L of the stock solution and dilute with enough water to bring the final volume to 5.0 L.

15.6 STOICHIOMETRY OF SOLUTION REACTIONS

How Can We Use Stoichiometry to Solve Problems Involving Solutions?

In Chapter 10, you learned to solve stoichiometry problems, problems which required a balanced equation for a reaction. From the balanced equation you could answer questions about the quantity of reactant required, or the quantity of product produced. The same principles are used here, but the reactions which occur are in solution. Use the 5 steps in Section 15.6 of your textbook to solve solution stoichiometry problems.

Example:

How many grams of solid barium sulfate will be formed by the addition of excess sulfuric acid to 85.5 mL of 0.64 M barium chloride?

Step 1: The first step to solve this problem is to write the balanced equation for the reaction. This problem involves ionic compounds, so write the net ionic equation. Writing net ionic equations is covered in detail in Section 7.3 of your textbook.

$$Ba^{2+}(aq) + SO_4^{2-}(aq) \longrightarrow BaSO_4(s)$$

Step 2: The next step is to calculate the number of moles of reactants. We are not told how many moles of sulfuric acid were added, only that it is present in excess, but are told the volume and the concentration of barium chloride. We can calculate the number of moles of barium ion which were added. Because this is a reaction involving ions, it is important to know the concentration of the **ions** present, not just the concentration of barium chloride before the reaction occurs. Each mole of $BaCl_2$ produces one mole of Ba^{2+} and two moles of Cl^-.

$$1 \text{ mol } BaCl_2 \longrightarrow 1 \text{ mol } Ba^{2+} + 2 \text{ mol } Cl^-$$

So in this case, the molarity of $BaCl_2$ equals the molarity of Ba^{2+}. We are given the molarity of $BaCl_2$ and the volume. Using the formula which relates volume and molarity, we can calculate the moles of Ba^{2+}. Do not forget that milliliters must be converted to liters.

$$\text{mol} = V \times M$$

$$\text{mol } Ba^{2+} = \text{volume} \times \frac{\text{mol } Ba^{2+}}{L}$$

$$\text{mol } Ba^{2+} = 85.5 \text{ mL} \times \frac{1 \text{ L}}{1000 \text{ mL}} \times \frac{0.64 \text{ mol } Ba^{2+}}{L} = 0.055 \text{ mol } Ba^{2+}$$

Step 3: We need to determine which of the reactants is the limiting one. We are told in the problem that sulfuric acid is in excess, so Ba^{2+} must be the limiting reactant.

Step 4: We need to determine how many moles of product can be formed from the moles of reactant. Examine the balanced equation and determine the mole ratio. 0.055 mol of Ba^{2+} added as barium chloride will produce 0.055 mol of barium sulfate.

$$0.055 \text{ mol } Ba^{2+} \times \frac{1 \text{ mol } BaSO_4}{1 \text{ mol } Ba^{2+}} = 0.055 \text{ mol } BaSO_4$$

Step 5: Convert moles to the desired units, in this case, the mass of $BaSO_4$. To convert from moles to grams, we can use the molar mass of $BaSO_4$.

$$0.055 \text{ mol } BaSO_4 \times \frac{233.4 \text{ g } BaSO_4}{1 \text{ mol } BaSO_4} = 13 \text{ g } BaSO_4$$

Solving these problems requires knowledge about solutions, as well as about topics covered in previous chapters of your textbook.

15.7 NEUTRALIZATION REACTIONS

How Can We Perform Solution Calculations Involving Acids and Bases?

Remember that strong acids and strong bases form ions when dissolved in water, and that the net ionic equation for the reaction between a strong acid and a strong base is

$$H^+(aq) + OH^-(aq) \longrightarrow H_20(l)$$

A **neutralization** reaction occurs when just enough strong base is added to react completely with all the strong acid which is present in the reaction container.

Example:

What volume of 0.125 M NaOH is needed to react with 26.5 ml of 0.205 M HNO_3? If we follow the five steps given in Section 15.6 of your textbook, we can solve this kind of problem.

Step 1: Write the balanced equation, or the net ionic equation if the reaction involves ionic compounds. The net ionic equation for the reaction between HNO_3 and NaOH is

$$H^+ + OH^- \longrightarrow H_2O$$

Step 2: We need to know the number of moles of reactants. We do not have enough information yet to calculate the moles of sodium hydroxide required, but we have both volume and molarity for the nitric acid and we can calculate the number of moles of nitric acid present from the definition of molarity. Because the definition of molarity is given in moles per liter, we will need to convert milliliters to liters using the equivalence statement 1000 mL equals 1 L.

$$\text{moles } HNO_3 = 26.5 \cancel{\text{mL}} \times \frac{1 \cancel{\text{L}}}{1000 \cancel{\text{mL}}} \times \frac{0.205 \text{ mol}}{\cancel{\text{L}}} = 0.00543 \text{ mol } HNO_3$$

Step 3: We need to determine the limiting reactant. We want to add enough OH^- ion to react exactly with a given amount of H^+ ion, so the nitric acid is limiting.

Step 4: Use the balanced equation to determine the number of moles of OH^- needed. Since there is a 1:1 molar ratio, 0.00543 mol H^+ requires exactly 0.00543 moles of OH^-.

$$0.00543 \text{ mol H}^+ \times \frac{1 \text{ mol OH}^-}{1 \text{ mol H}^+} = 0.00543 \text{ mol OH}^-$$

Step 5: We need to find the volume of NaOH which contains 0.00543 mol of OH^-. We can use the relationship mol = M x V to complete the calculation.

$$0.00543 \text{ mol NaOH} = \text{volume} \times \frac{0.125 \text{ mol}}{\text{L}}$$

Rearrange this to isolate volume on the left side.

$$\text{volume} = \frac{0.00543 \text{ mol NaOH}}{0.125 \text{ mol/L}} = 0.0434 \text{ L NaOH}$$

43.4 milliliters of 0.125 M NaOH will react exactly with 26.5 milliliters of 0.205 M HNO_3.

15.8 SOLUTION COMPOSITION: NORMALITY

What Is Meant By an Equivalent of an Acid or a Base?

An equivalent of an acid is the amount of acid which will provide one mole of hydrogen ions. An equivalent of a base is the amount of base which will provide one mole of hydroxide ions. How much acid or base does it take to provide one mole of hydrogen or hydroxide ions? Acids such as HCl and HNO_3 when dissolved in water provide one mole of H^+ ions for each mole of HCl or HNO_3.

$$1 \text{ mol HCl} \rightarrow 1 \text{ mol H}^+ + 1 \text{ mol Cl}^-$$

For acids which produce one mole of hydrogen ions for each mole of acid, one mole of acid equals one equivalent of acid. Strong bases such as NaOH and KOH when dissolved in water provide one mole of OH^- ions for each mole of NaOH or KOH.

$$1 \text{ mol KOH} \rightarrow 1 \text{ mol K}^+ + 1 \text{ mol OH}^-$$

For bases which produce one mole of hydroxide ions for each mole of strong base, one mole of base equals one equivalent of base.

Equivalent weight is the mass of acid or base in grams which will provide one mole of hydrogen or hydroxide ions. For acids and bases which provide one mole of hydrogen or hydroxide ions per mole of acid or base, the equivalent weight is equal to the molar mass. What is the equivalent weight for acids and bases which provide more than one mole of hydrogen or hydroxide ions per mole of acid or base?

Example:

Phosphoric acid, H_3PO_4, when dissolved in water, provides three moles of H^+ for each mole of H_3PO_4.

$$1 \text{ mol } H_3PO_4 \rightarrow 3 \text{ mol } H^+ + 1 \text{ mol } PO_4^{3-}$$

So, an equivalent of H_3PO_4 would be one third of a mole, because one third of a mole would provide one mole of hydrogen ions. The equivalent weight of phosphoric acid would be molar mass of phosporic acid divided by three, or 97.99 g $H_3PO_4/3$, which is 32.66 g H_3PO_4.

What Is Normality And How Can We Use It In Calculations?

Normality is the number of equivalents per liter of solution, $N = \frac{\text{equiv}}{\text{L}}$. It is a convenient way of expressing the concentration of acids and bases.

Example:

How many grams of H_3PO_4 are required to prepare 250. mL of a 0.600 N solution? We can solve this problem by using the definition for normality, $N = \frac{\text{equiv}}{\text{L}}$. We know the normality and the volume of the solution, and we want to know the number of equivalents. Rearrange this equation to isolate equivalents on one side.

$$\text{equiv} = \text{volume} \times \text{N}$$

Because normality is defined as equivalents per liter, we will need to convert volume given in milliliters to liters.

$$\text{equivalents } H_3PO_4 = 250. \cancel{\text{mL}} \times \frac{1 \cancel{\text{L}}}{1000 \cancel{\text{mL}}} \times \frac{0.600 \text{ equiv}}{\cancel{\text{L}}} = 0.150 \text{ equiv } H_3PO_4$$

We can calculate grams of H_3PO_4 from the equivalent weight, and we can determine the equivalent weight from the molar mass, as long as we know how many equivalents are present in each molar mass. For H_3PO_4, there are three equivalent weights for each molar mass.

$$0.150 \cancel{\text{equiv } H_3PO_4} \times \frac{1 \cancel{\text{mol } H_3PO_4}}{3 \cancel{\text{equiv } H_3PO_4}} \times \frac{97.994 \text{ g } H_3PO_4}{1 \cancel{\text{mol } H_3PO_4}} = 4.90 \text{ g } H_3PO_4$$

LEARNING REVIEW

1. 150 mL of ethyl alcohol is mixed with 1 L of water. Which is the solute, ethyl alcohol or water?

2. Which of the molecules below would you predict to be soluble in water?

   a.

$$\begin{array}{l} CH_2-OH \\ | \\ CH-OH \\ | \\ CH_2-OH \end{array}$$

   b. $CH_3CH_2CH_3$

   c. Na_2SO_4

3. Three solutions are prepared by mixing the quantities of sodium chloride given below in a volume of 500 mL of solution. Which solution is the most concentrated?

   a. 55 g NaCl in 500 mL solution
   b. 127 g NaCl in 500 mL solution
   c. 105 g NaCl in 500 mL solution

4. A student stirred 5.0 g of table sugar into 250. g of hot coffee. What is the mass percent of sugar in the coffee?

5. A certain table wine contains 11.8% ethyl alcohol by mass. How many grams of ethyl alcohol are in 1500 g of wine?

6. Calculate mass percents for the solutions below.

   a. 6.5 g KOH in 250. g water
   b. 0.40 g baking soda in 2000.0 g flour
   c. 150. g acetone in 438 g water

7. A solution of HCl in water is 0.15 M. How many mol/L of HCl are present?

8. 150.5 g of NaOH are dissolved in water. The final volume of the solution is 3.8 L. What is the molarity of the solution?

9. Calculate the molarity of each of the solutions below.

a. 0.62 g $AgNO_3$ in a final volume of 1.5 L solution
b. 10.6 g NaCl in a final volume of 286 mL solution
c. 152 g $CaSO_4$ in a final volume of 0.92 L solution

10. 2.5 L of a solution of KI in water has a concentration of 0.15 M. How many grams of KI are in the solution?

11. What is the concentration of each ion in the following solutions?

a. 2.0 M H_2SO_4
b. 0.6 M Na_3PO_4
c. 1.5 M $AlCl_3$

12. How many moles of KCl are present in 1.5 L of a 0.48 M solution of KCl in water?

13. How many moles of Cl^- are present in 520 mL of a 1.5 M solution of $FeCl_3$ in water?

14. How many grams of NaOH are needed to make 1.50 L of a 0.650 M NaOH solution?

15. How many grams of K_2SO_4 are needed to make 250 mL of a 0.150 M K_2SO_4 solution?

16. Sodium fluoride is added to many water supplies to prevent tooth decay. How many grams of NaF must be added to a water supply so that 2.0×10^6 L of water contain 3.0×10^{-6} M NaF?

17. What volume of 12 M HCl solution is needed to make 2.5 L of 1.0 M HCl?

18. What volume of 18 M H_2SO_4 stock solution is needed to make 1855 mL of 0.65 M H_2SO_4?

19. If 2.5 L of solution contains 0.10 M $CaCl_2$, how many grams of Na_3PO_4 are needed to exactly precipitate all of the calcium as $Ca_3(PO_4)_2$?

20. How many grams of $Fe(OH)_3$ can be produced by the addition of 0.25 moles of $FeCl_3$ to 1.2 L of a 0.85 M NaOH solution?

21. How many grams of $PbCl_2$ can be produced from the addition of 25 mL of a 0.50 M $Pb(NO_3)_2$ solution to 150 mL of 0.45 M NaCl?

22. What volume of 0.15 M NaOH solution will react completely with 150 mL of 0.25 M HCl?

23. What volume of 0.30 M H_2SO_4 solution is needed to react completely with 1.2 L of 0.85 M NaOH?

24. For each of the strong acids and strong bases below, give the number of equivalents present in 1 mole, and the equivalent weight of each.

material	**molar mass**	**equivalents**	**equivalent weight**
1.0 mol HCl	36.46		
1.0 mol H_2SO_4	98.08		
1.0 mol KOH	56.11		

25. If 5.6 grams of phosphoric acid, H_3PO_4, are added to water so that the final volume is 125 mL, what is the normality of the solution?

26. What volume of 0.20 N H_2SO_4 is needed to react exactly with 2.5 L of 0.125 N NaOH?

ANSWERS TO LEARNING REVIEW

1. Ethyl alcohol is the solute. Water is the solvent, because it is present in the largest amount.

2. a. This molecule contains three polar O-H bonds. Each of these O-H bonds can form hydrogen bonds with water. This molecule is soluble in water.
   b. This molecule has no polar bonds. There is no part of the molecule which will interact with polar water molecules. This molecule is not soluble in water.
   c. Sodium sulfate is an ionic compound. Many ionic compounds dissolve in water because the charged ions are pulled from the crystal by the polar water molecules. This molecule is soluble in water.

3. Solution b, 127 g NaCl per 500 mL solution, is the most concentrated because the amount of solute per amount of solution is the greatest.

4. The mass percent of a solution can be calculated by

$$\text{mass percent} = \frac{\text{mass of solute}}{\text{mass of a solution}} \times 100\ \%$$

The mass of solute is 5.0 g sugar; and the mass of the solution is the mass of solute plus the mass of the solvent, 5.0 g sugar plus 250. g coffee equals 255 g solution. The mass percent sugar is

$$\frac{5.0\ \not{g}\ \text{sugar}}{255\ \not{g}\ \text{solution}} \times 100\ \% = 2.0\ \%\ \text{sugar}$$

5. We can use the definition of mass percent to solve this problem. 11.8 % ethyl alcohol by mass means that there are 11.8 grams ethyl alcohol per 100 grams of wine. We are asked for the grams of ethyl alcohol in 1500 grams wine.

$$\text{mass}\ \% = \frac{\text{mass of solute}}{\text{mass of solution}} \times 100\ \%$$

Rearrange this equation to isolate mass of solute on one side.

$$\frac{\text{mass}\ \not{\%} \times \text{mass of solution}}{100\ \not{\%}} = \text{mass of solute}$$

$$\text{mass of ethyl alcohol} = \frac{11.8\ \% \times 1500\ \text{g}}{100\ \%} = 180\ \text{g}$$

6. a. The mass of solute is 6.5 g KOH and the mass of solution is 6.5 g KOH plus 250. g water, which is 257 g. The mass percent is

$$\frac{\text{mass of solute}}{\text{mass of solution}} = \frac{6.5\ \not{g}\ \text{KOH}}{257\ \not{g}\ \text{solution}} \times 100\ \% = 2.5\ \%\ \text{KOH}$$

b. The mass of solute is 0.40 g and the mass of solution is 0.40 g baking soda plus 2000.0 g flour, which is equal to 2000.4 g. The mass percent is

$$\frac{0.40\ \not{g}\ \text{baking soda}}{2000.4\ \not{g}\ \text{solution}} \times 100\ \% = 0.02\ \%\ \text{baking soda}$$

c. The mass of solute is 150. g acetone and the mass of solution is 150. g of acetone plus 438 g water, which is 588 g. The mass percent is

$$\frac{150.\ \not{g}\ \text{acetone}}{588\ \not{g}\ \text{solution}} \times 100\ \% = 25.5\ \%\ \text{acetone}$$

7. The definition of molarity, M, is moles solute/liter solution. An HCl solution which is 0.15 M would contain 0.15 mol HCl/liter solution.

$$0.15\ \text{M HCl} = \frac{0.15\ \text{mol HCl}}{\text{L}}$$

8. The molarity of a solution is equal to the moles solute/liter solution. In this problem we have 150.5 g solute, NaOH. We do not know the number of moles. By using the molar mass for NaOH, we can calculate the number of moles.

$$150.5\ \cancel{g}\ Na\cancel{O}H \times \frac{1\ mol\ NaOH}{40.00\ \cancel{g}\ Na\cancel{O}H} = 3.763\ mol\ NaOH$$

Now, we can find the molarity.

$$M = \frac{moles\ solute}{liter\ solution}$$

$$M = \frac{3.763\ mol\ NaOH}{3.8\ L}$$

$$M = 0.99\ M\ NaOH$$

9. a. First, calculate the moles of $AgNO_3$.

$$0.62\ \cancel{g}\ Ag\cancel{N}O_3 \times \frac{1\ mol\ AgNO_3}{169.91\ \cancel{g}\ Ag\cancel{N}O_3} = 0.0036\ mol\ AgNO_3$$

Now, calculate the molarity.

$$M = \frac{0.0036\ mol\ AgNO_3}{1.5\ L}$$

$$M = 2.4 \times 10^{-3}\ M\ AgNO_3$$

b. First, calculate the moles of NaCl.

$$10.6\ \cancel{g}\ N\cancel{a}Cl \times \frac{1\ mol\ NaCl}{58.44\ \cancel{g}\ N\cancel{a}Cl} = 0.181\ mol\ NaCl$$

The volume of the solution is given in milliliters. We need to know the number of liters.

$$286\ \cancel{m}L\ solution \times \frac{1\ L\ solution}{1000\ \cancel{m}L\ solution} = 0.286\ L\ solution$$

Now, calculate the molarity.

$$M = \frac{0.181\ mol\ NaCl}{0.286\ L\ solution}$$

$$M = 0.633\ M\ NaCl$$

c. First, calculate the moles of $CaSO_4$.

$$152\ \cancel{g}\ Ca\cancel{S}O_4 \times \frac{1\ mol\ CaSO_4}{136.14\ \cancel{g}\ Ca\cancel{S}O_4} = 1.12\ mol\ CaSO_4$$

Now, calculate the molarity.

$$M = \frac{1.12 \text{ mol } CaSO_4}{0.92 \text{ L solution}}$$

$$M = 1.2 \text{ M } CaSO_4$$

10. This problem gives us the number of liters of solution and the concentration of the solution in molarity. From the definition of molarity, moles solute/liter solution, we can calculate the number of moles of solute, KI.

$$2.5 \not{L} \text{ solution} \times \frac{0.15 \text{ mol KI}}{\not{L} \text{ solution}} = 0.38 \text{ mol KI}$$

Now use the molar mass of KI to calculate the grams of KI.

$$0.38 \text{ m}\not{o}\text{l } \not{K}\text{I} \times \frac{166.0 \text{ g KI}}{\text{m}\not{o}\text{l } \not{K}\text{I}} = 63 \text{ g KI}$$

11. a. Sulfuric acid produces 2 moles of hydrogen ions for each mole of sulfuric acid, and 1 mole of sulfate ions for each mole of sulfuric acid.

$$1 \text{ mol } H_2SO_4(aq) \longrightarrow 2 \text{ mol } H^+(aq) + 1 \text{ mol } SO_4^{2-}(aq)$$

A 2.0 M solution of sulfuric acid would contain 2(2.0 mol H^+) per liter, or 4.0 M H^+ total, and 2(1.0 mol SO_4^{2-}) per liter, or a total of 2.0 M SO_4^{2-}.

b. Sodium phosphate produces 3 moles of sodium ions for each mole of sodium phosphate, and 1 mole of phosphate ions for each mole of sodium phosphate.

$$1 \text{ mol } Na_3PO_4(aq) \longrightarrow 3 \text{ mol } Na^+(aq) + 1 \text{ mol } PO_4^{3-}(aq)$$

A 0.6 M solution of sodium phosphate would contain 3(0.6 mol Na^+) per liter, or 1.8 M Na^+ total, and 1(0.6 mol PO_4^{3-}) per liter, or 0.6 M PO_4^{3-} total.

c. Aluminum chloride produces 1 mole of aluminum ions for each mole of aluminum chloride, and 3 moles of chloride ions for each mole of aluminum chloride.

$$1 \text{ mol } AlCl_3(aq) \longrightarrow 1 \text{ mol } Al^{3+}(aq) + 3 \text{ mol } Cl^-(aq)$$

A 1.5 M solution of aluminum chloride would contain 1(1.5 mol Al^{3+}) per liter, or 1.5 M Al^{3+} total, and 3(1.5 mol Cl^-) per liter, or 4.5 M Cl^- total.

12. We are given the concentration and the volume of a solution containing KCl and water, and are asked for the number of moles of solute, KCl. The number of moles of KCl can be calculated from the definition of molarity, moles solute/liter solution.

$$1.5 \text{ L solution} \times \frac{0.48 \text{ mol KCl}}{\text{L solution}} = 0.72 \text{ mol KCl}$$

13. We are given the concentration of a solution containing $FeCl_3$ and water, and the volume of the solution in milliliters. We are asked for the moles of chloride ion. The number of moles of $FeCl_3$ can be calculated from the definition of molarity, which is moles solute/liter solution. Because the volume is given in milliliters, convert milliliters of solution to liters.

$$520 \text{ mL solution} \times \frac{1 \text{ L solution}}{1000 \text{ mL}} \times \frac{1.5 \text{ mol } FeCl_3}{\text{L solution}} = 0.78 \text{ mol } FeCl_3$$

There are three moles of chloride ions and one mole of iron(III) ions for each mole of $FeCl_3$.

$$0.78 \text{ mol } FeCl_3 \times \frac{3 \text{ mol } Cl^-}{\text{mol } FeCl_3} = 2.3 \text{ mol } Cl^-$$

14. We are given the concentration of a solution containing NaOH and water and we are asked for the number of grams of NaOH needed to make 1.50 L of solution. The number of moles of NaOH can be calculated from the definition of molarity, moles solute per liter of solution. The grams of NaOH can be calculated using the molar mass of NaOH.

$$1.50 \text{ L solution} \times \frac{0.650 \text{ mol NaOH}}{\text{L solution}} \times \frac{40.00 \text{ g NaOH}}{\text{mol NaOH}} = 39.0 \text{ g NaOH}$$

15. We are given the concentration of a solution containing K_2SO_4 and water, and we are asked for the number of grams of K_2SO_4 needed to make 250 mL of solution. The number of moles of K_2SO_4 can be calculated from the definition of molarity. The grams of K_2SO_4 can be calculated using the molar mass of K_2SO_4. We will need to convert the given units of volume, milliliters, to liters.

$$250 \text{ mL solution} \times \frac{1 \text{ L solution}}{1000 \text{ mL solution}} \times \frac{0.150 \text{ mol } K_2SO_4}{\text{L solution}} \times \frac{174.26 \text{ g } K_2SO_4}{\text{mol } K_2SO_4} = 6.5 \text{ g } K_2SO_4$$

16. We are given the concentration of a solution containing NaF and water, and we are asked for the number of grams of NaF needed to make 2.0×10^6 L of solution. The number of moles of NaF can be calculated from the definition of molarity. The grams of NaF can be calculated from the molar mass of NaF.

$$2.0 \times 10^6 \text{ L solution} \times \frac{3.0 \times 10^{-6} \text{ mol NaF}}{\text{L solution}} \times \frac{41.99 \text{ g NaF}}{\text{mol NaF}} = 2.5 \times 10^2 \text{ g NaF}$$

17. In this problem we are asked to calculate how much of a concentrated stock solution, which is 12 M HCl, is needed to prepare a dilute HCl solution. We will need to know how many moles of HCl are present in 2.5 L of 1.0 M HCl, that is, in the dilute solution. Then we need find a volume of the concentrated solution which contains this same number of moles. We can use this procedure because the number of moles of solute in the dilute solution is the same as the number of moles of solute in the volume of stock solution. Only the volume of water changes. First, find the number of moles of HCl which will be present in the dilute solution by multiplying the volume by the molarity.

$$2.5 \text{ L solution} \times \frac{1.0 \text{ mol HCl}}{\text{L solution}} = 2.5 \text{ mol HCl}$$

So the dilute solution will contain 2.5 mol HCl, and the volume of stock solution we need will also contain 2.5 mol HCl. The volume of stock solution multiplied by the molarity of the stock solution equals the number of moles of HCl which will be in the dilute solution.

$$\text{volume of stock solution} \times \frac{\text{mol HCl in stock solution}}{\text{L stock solution}} = \text{mol HCl in dilute solution}$$

Now, substitute values into the equation.

$$V \times \frac{12 \text{ mol HCl}}{\text{L solution}} = 2.5 \text{ mol HCl}$$

Rearrange the equation to isolate V on one side.

$$V = \frac{2.5 \text{ mol HCl}}{12 \text{ mol HCl/L solution}}$$

$$V = 0.21 \text{ L HCl}$$

So to make 2.5 L of 1.0 M HCl, use 0.21 L of 12 M HCl, and add enough water to bring the total volume to 2.5 L.

A different way to approach this problem is to use the formula

$$M_1 \times V_1 = M_2 \times V_2$$

M_1 represents the molarity of the stock solution; V_1, the volume of the stock solution needed; M_2, the molarity of the dilute solution we wish to make; and V_2, the volume of the dilute solution. In this case we want to know the volume of stock solution needed, so isolate V_1 on one side of the equation by dividing both sides by M_1.

$$M_1 \times V_1 = M_2 \times V_2$$

$$\frac{\not{M}_1}{\not{M}_1} \times V_1 = V_2 \times \frac{M_2}{M_1}$$

$$V_1 = V_2 \times \frac{M_2}{M_1}$$

Now, substitute values into the equation.

$$V_1 = 2.5\ \text{L} \times \frac{1.0\ \not{M}}{12\ \not{M}}$$

$$V_1 = 0.21\ \text{L}$$

The answer to this problem is the same, either way we solve it.

18. We want to know how much concentrated stock solution is needed to make 1855 mL of 0.65 M H_2SO_4. We can use the formula

$$M_1 \times V_1 = M_2 \times V_2$$

M_1 is the molarity of the stock solution, V_1 is the volume of the stock solution, M_2 is the molarity of the dilute solution and V_2 is the volume of the dilute solution. We want to know how much stock solution is needed, so isolate V_1 on one side of the equation.

$$V_1 = V_2 \times \frac{M_2}{M_1}$$

The volume of the dilute solution, V_2, is given in milliliters. We will need to convert milliliters to liters.

$$1855\ \not{m}\text{L} \times \frac{1\ \text{L}}{1000\ \not{m}\text{L}} = 1.855\ \text{L}$$

Now, substitute values into the equation.

$$V_1 = 1.855 \text{ L} \times \frac{0.65 \text{ M } H_2SO_4}{18 \text{ M } H_2SO_4}$$

$$V_1 = 0.067 \text{ L}$$

So, 0.067 L of 18 M H_2SO_4 diluted to 1.855 L would produce a 0.65 M H_2SO_4 solution.

19. In this problem a solution of $CaCl_2$ is mixed with a solution of Na_3PO_4. A reaction occurs. We are asked for the number of grams of Na_3PO_4 which will react with the $CaCl_2$. Section 15.6 of your textbook gives 5 steps for solving problems like this one, so let's follow those same steps here.

Step 1: First, write the balanced molecular equation for this reaction. Because this is a reaction between ionic compounds, we also should write the net ionic equation. The balanced molecular equation is

$$3CaCl_2(aq) + 2\,Na_3PO_4(aq) \rightarrow Ca_3(PO_4)_2(s) + 6NaCl(aq)$$

The net ionic equation is the solid product, $Ca_3(PO_4)_2$, and the ions which react to form the solid product.,

$$3Ca^{2+}(aq) + 2PO_4^{3-}(aq) \rightarrow Ca_3(PO_4)_2(s)$$

Step 2: We need to add just enough PO_4^{3-} to react with all the Ca^{2+}. We need to know how many moles of Ca^{2+} there are in the $CaCl_2$ solution. From the volume and the molarity of the $CaCl_2$ solution, we can calculate the number of moles of Ca^{2+}.

$$V \times M = \text{mol } CaCl_2$$

$$2.5 \text{ L } CaCl_2 \times \frac{0.1 \text{ mol } CaCl_2}{\text{L } CaCl_2} = 0.25 \text{ mol } CaCl_2$$

Each mole of $CaCl_2$ produces 1 mole of Ca^{2+}.

$$0.25 \text{ mol } CaCl_2 \times \frac{1 \text{ mol } Ca^{2+}}{1 \text{ mol } CaCl_2} = 0.25 \text{ mol } Ca^{2+}$$

Step 3: In this problem Ca^{2+} is limiting. We want to add just enough PO_4^{3-} to react with all the Ca^{2+}.

Step 4: We need to know how many moles of PO_4^{3-} will react with 0.25 mol Ca^{2+}. We can use the mole ratio from the balanced equation to calculate the moles of PO_4^{3-} which are needed.

$$0.25 \text{ mol } Ca^{2+} \times \frac{2 \text{ mol } PO_4^{3-}}{3 \text{ mol } Ca^{2+}} = 0.17 \text{ mol } PO_4^{3-}$$

So 0.17 mol PO_4^{3-} will react with 0.25 mol Ca^{2+}.

Step 5. We are asked for grams of Na_3PO_4, not moles of PO_4^{3-}, so convert moles of PO_4^{3-} to grams of Na_3PO_4. Each mole of Na_3PO_4 contains 1 mole of PO_4^{3-} ions. We can use the molar mass of Na_3PO_4 to convert from moles Na_3PO_4 to grams Na_3PO_4.

$$0.17 \text{ mol } PO_4^{3-} \times \frac{1 \text{ mol } Na_3PO_4}{1 \text{ mol } PO_4^{3-}} \times \frac{163.94 \text{ g } Na_3PO_4}{1 \text{ mol } Na_3PO_4} = 28 \text{ g } Na_3PO_4$$

20. We want to know how many grams of product, $Fe(OH)_3$, can be produced when two aqueous solutions are mixed together.

Step 1: Write and balance the equation for this reaction.

$$FeCl_3(aq) + 3NaOH(aq) \rightarrow Fe(OH)_3(s) + 3NaCl(aq)$$

From the balanced equation, write the net ionic equation.

$$Fe^{3+}(aq) + 3OH^-(aq) \rightarrow Fe(OH)_3(s)$$

Step 2: We need to know the number of moles of reactant present in each solution. The solution of $FeCl_3$ contains 0.25 mol $FeCl_3$, and each mole of $FeCl_3$ contains 1 mole of Fe^{3+}.

$$0.25 \text{ mol } FeCl_3 \times \frac{1 \text{ mol } Fe^{3+}}{1 \text{ mol } FeCl_3} = 0.25 \text{ mol } Fe^{3+}$$

We need to know the number of moles of OH^- which are present.

$$V \times M = \text{moles NaOH}$$

$$1.2 \text{ L NaOH} \times \frac{0.85 \text{ mol NaOH}}{\text{L NaOH}} = 1.0 \text{ mol NaOH}$$

Each mole of NaOH contains 1 mole of OH^-.

$$1.0 \text{ mol NaOH} \times \frac{1 \text{ mol } OH^-}{1 \text{ mol NaOH}} = 1.0 \text{ mol } OH^-$$

Step 3: 0.25 mol Fe^{3+} is mixed with 1.0 mol OH^-. Because each mole of Fe^{3+} requires 3 moles of OH^-, 0.25 mol Fe^{3+} requires 3(0.25 mol OH^-) or 0.75 mol OH^-. Because we have 1.0 mol OH^-, the amount of product which forms is limited by the amount of Fe^{3+}.

Step 4: From the mole ratio, each mole of Fe^{3+} produces 1 mole of $Fe(OH)_3$.

$$0.25 \text{ mol } Fe^{3+} \times \frac{1 \text{ mol } Fe(OH)_3}{1 \text{ mol } Fe^{3+}} = 0.25 \text{ mol } Fe(OH)_3$$

Step 5: We want to know the number of grams of $Fe(OH)_3$, so use the molar mass of $Fe(OH)_3$ to convert from moles to grams.

$$0.25 \text{ mol } Fe(OH)_3 \times \frac{106.87 \text{ g } Fe(OH)_3}{1 \text{ mol } Fe(OH)_3} = 27 \text{ g } Fe(OH)_3$$

21. We want to know how many grams of $PbCl_2$ can be produced from the addition of a solution containing $Pb(NO_3)_2$ to a solution containing NaCl. We are given the volume and molarity for both solutions, so calculate the number of moles of each solute which are being mixed.

Step 1: Write the balanced equation for this reaction.

$$Pb(NO_3)_2(aq) + 2NaCl(aq) \longrightarrow PbCl_2(s) + 2NaNO_3(aq)$$

Now, write the net ionic equation.

$$Pb^{2+}(aq) + 2Cl^-(aq) \longrightarrow PbCl_2(s)$$

Step 2: Calculate the moles of reactant present in each solution.

for $Pb(NO_3)_2$

$$V \times M = \text{moles}$$

$$25 \text{ mL} \times \frac{1 \text{ L } Pb(NO_3)}{1000 \text{ mL}} \times \frac{0.50 \text{ mol } Pb^{2+}}{\text{L } Pb(NO_3)_2} = 0.013 \text{ mol } Pb^{2+}$$

for NaCl

$$150 \text{ mL} \times \frac{1 \text{ L NaCl}}{1000 \text{ mL}} \times \frac{0.45 \text{ mol Cl}^-}{\text{L NaCl}} = 0.068 \text{ mol Cl}^-$$

Step 3: Because moles of one reactant are being mixed with moles of another reactant, we need to determine which reactant is limiting. 0.013 mol Pb^{2+} is mixed with 0.068 mol Cl^-. Because each mole of Pb^{2+} requires 2 moles of Cl^-, 0.013 mol Pb^{2+} requires 2(0.013 mol Cl^-) or 0.026 mol Cl^-. Because we have 0.068 mol Cl^-, the amount of $PbCl_2$ which forms is limited by the amount of Pb^{2+}.

Step 4: From the net ionic equation, each mole of Pb^{2+} produces 1 mol $PbCl_2$.

$$0.013 \text{ mol Pb}^{2+} \times \frac{1 \text{ mol PbCl}_2}{1 \text{ mol Pb}^{2+}} = 0.013 \text{ mol PbCl}_2$$

Step 5: We want to know the number of grams of $PbCl_2$, so we can use the molar mass of $PbCl_2$ to convert from moles to grams.

$$0.013 \text{ mol PbCl}_2 \times \frac{278.10 \text{ g PbCl}_2}{1 \text{ mol PbCl}_2} = 3.6 \text{ g PbCl}_2$$

22. In this problem we are mixing a solution of HCl of known volume and molarity with a solution of NaOH of known molarity and an unknown volume. We are asked to determine the volume of NaOH which will react with the HCl. We can follow the same steps we have used previously.

Step 1. Write the balanced equation for this reaction.

$$NaOH(aq) + HCl(aq) \rightarrow NaCl(aq) + H_2O(l)$$

Now, write the net ionic equation.

$$H^+(aq) + OH^-(aq) \rightarrow H_2O(l)$$

Step 2. Calculate the moles of HCl using the formula V x M = moles.

$$150 \text{ mL} \times \frac{1 \text{ L HCl}}{1000 \text{ mL}} \times \frac{0.25 \text{ mol H}^+}{\text{L HCl}} = 0.038 \text{ mol H}^+$$

Step 3. This problem requires mixing just enough OH^- to react with all the H^+ which is present. The moles of H^+ determine how much OH^- is to be added. The H^+ ions are limiting.

Step 4: From the net ionic equation we can determine how many moles of OH^- are needed to react with all the H^+.

$$0.038 \text{ mol } H^+ \times \frac{1 \text{ mol } OH^-}{1 \text{ mol } H^+} = 0.038 \text{ mol } OH^-$$

Step 5: We now know the moles of OH^- and the molarity. We can use the formula V x M equals moles to calculate the volume of NaOH. Rearrange the equation to isolate V on one side.

$$V \times \frac{M}{M} = \frac{\text{moles}}{M}$$

$$V = \frac{\text{moles}}{M}$$

Now, substitute values into the equation.

$$V = \frac{0.038 \text{ mol } OH^-}{0.15 \text{ mol NaOH/L NaOH}}$$

$$V = 0.25 \text{ L NaOH}$$

So, 0.25 L of 0.15 M NaOH will completely react with 150 mL of 0.25 M HCl.

23. In this problem we are mixing a solution of NaOH of known volume and molarity with a solution of H_2SO_4 of known molarity and unknown volume. We are asked to determine the volume of H_2SO_4 which will react with the NaOH. Follow the five steps we have previously used.

Step 1: Write the balanced equation for the reaction.

$$2NaOH(aq) + H_2SO_4(aq) \rightarrow Na_2SO_4(aq) + 2H_2O(l)$$

Now, write the net ionic equation.

$$OH^-(aq) + H^+(aq) \rightarrow H_2O(l)$$

Step 2: Calculate the moles of NaOH using the formula

$$V \times M = \text{moles NaOH}$$

$$1.2 \text{ L NaOH} \times \frac{0.85 \text{ mol } OH^-}{\text{L NaOH}} = 1.0 \text{ mol } OH^-$$

Step 3: This problem requires mixing just enough H^+ to react with all the OH^- which is present. The moles of OH^- determine how much H^+ is added. The OH^- ions are limiting.

Step 4: From the net ionic equation we can determine how many moles of H^+ are needed to react with all the OH^-.

$$1.0 \text{ mol } OH^- \times \frac{1 \text{ mol } H^+}{1 \text{ mol } OH^-} = 1.0 \text{ mol } H^+$$

H_2SO_4 produces 2 moles of H^+ for each mole of H_2SO_4. So let's convert moles of H^+ to moles of H_2SO_4.

$$1.0 \text{ mol } H^+ \times \frac{1 \text{ mol } H_2SO_4}{2 \text{ mol } H^+} = 0.50 \text{ mol } H_2SO_4$$

Step 5: We know the number of moles of H_2SO_4 and the molarity. We can use the formula V x M = moles to calculate the volume of H_2SO_4. Rearrange the equation to isolate V on one side.

$$V \times \frac{M}{M} = \frac{\text{moles}}{M}$$

$$V = \frac{\text{moles}}{M}$$

Now, substitute values into the equation.

$$V = \frac{0.50 \text{ mol } H_2SO_4}{0.30 \text{ mol } H_2SO_4/\text{L } H_2SO_4}$$

$$V = 1.7 \text{ L } H_2SO_4$$

1.7 L of 0.30 M H_2SO_4 would react completely with 1.2 L of 0.85 M NaOH.

24.

material	molar mass	equivalents	equivalent wt
1.0 mol HCl	36.46	1	36.46
1.0 mol H_2SO_4	98.08	2	49.04
1.0 mol KOH	56.11	1	56.11

25. We want to calculate the normality of a solution of phosphoric acid in water. To do so, we need to know the number of equivalents of phosphoric acid present in 5.6 g phosphoric acid. The equivalent weight of phosphoric acid is

$$\text{equivalent weight } H_3PO_4 = \frac{\text{molar mass } H_3PO_4}{3}$$

$$\text{equivalent weight } H_3PO_4 = \frac{97.99 \text{ g}}{3}$$

$$\text{equivalent weight } H_3PO_4 = 32.66 \text{ g}$$

We can now calculate the equivalents of H_3PO_4 present in 5.6 g H_3PO_4.

$$5.6 \cancel{\text{g } H_3PO_4} \times \frac{1 \text{ equiv } H_3PO_4}{32.66 \cancel{\text{g } H_3PO_4}} = 0.17 \text{ equiv } H_3PO_4$$

The definition of normality is $N = \frac{\text{equiv}}{L}$. We can use this equation to calculate the normality of the H_3PO_4 solution.

$$N = \frac{0.17 \text{ equiv } H_3PO_4}{125 \cancel{\text{mL}}} \times \frac{1000 \cancel{\text{mL}}}{L} = 1.36 \text{ equiv } H_3PO_4$$

This solution is 1.36 N H_3PO_4.

26. We are asked to determine what volume of H_2SO_4 of known normality is needed to react exactly with a solution of NaOH of known volume and known normality. When a neutralization reaction occurs, the equivalents of acid which react are equal to the equivalents of base. Because equiv = N x V, we can say that

$$N_{acid} \times V_{acid} = \text{equiv} = N_{base} \times V_{base}$$

or

$$N_{acid} \times V_{acid} = N_{base} \times V_{base}$$

We wish to know the volume of acid which reacts, so rearrange this equation to isolate V_{acid} on one side.

$$V_{acid} \times \frac{\cancel{N_{acid}}}{\cancel{N_{acid}}} = V_{base} \times \frac{N_{base}}{N_{acid}}$$

$$V_{acid} = V_{base} \times \frac{N_{base}}{N_{acid}}$$

Now, substitute values into the equation.

$$V_{acid} = \frac{(2.5\ L)(1.25\ equiv/L)}{(0.20\ equiv/L)}$$

$$V = 1.6\ L\ H_2SO_4$$

PRACTICE EXAM

1. Which molecule would most likely be soluble in the nonpolar solvent hexane, which has the formula $CH_3CH_2CH_2CH_2CH_2CH_3$?

   a. K_2SO_4
   b. H_2O
   c. CCl_4
   d.
   $CH_2—OH$
   |
   $CH—OH$
   |
   $CH_2—OH$
   e. CH_3OH

2. Which statement about the composition of solutions is not true?

   a. The mass percent of a solution is the mass of solute divided by mass of solution times 100 %.
   b. A solution which has 50 g NaCl in 100 g solution has a higher mass percent than a solution which has 225 g NaCl in 750 g solution.
   c. In some solutions the solvent is a gas.
   d. An unsaturated solution has dissolved as much solute as it can.
   e. A concentrated solution has a relatively large amount of solute dissolved in the solvent.

3. What is the molarity of a solution which contains 28.6 g $CuSO_4$ in 500. mL of water?

   a. 0.358 M
   b. 11.2 M
   c. 0.512 M
   d. 2.79 M
   e. 5.72 M

4. How many grams of KNO_3 are needed to prepare 1.4 L of a 0.800 M solution of KNO_3 in water?

a. 110 g
b. 77 g
c. 58 g
d. 180 g
e. 120 g

5. What volume of 18 M H_2SO_4 is needed to make 6.5 L of 0.75 M H_2SO_4 solution?

a. 0.04 L
b. 0.27 L
c. 0.37 L
d. 2.08 L
e. 4.9 L

6. How much Ag_2CrO_4 can be formed when an excess of solid Na_2CrO_4 is added to 0.850 L of 1.5 M $AgNO_3$?

a. 430 g
b. 95 g
c. 300 g
d. 210 g
e. 860 g

7. How much $Ba(OH)_2$ can be formed when 1.5 L of 0.25 M $Ba(NO_3)_2$ is mixed with 2.5 L of 0.35 M NaOH?

a. 150 g
b. 64 g
c. 1200 g
d. 130 g
e. 32 g

8. How much 0.100 M H_2SO_4 is needed to neutralize 0.635 L of 0.855 M KOH?

a. 6.73 L
b. 5.43 L
c. 3.71 L
d. 4.28 L
e. 2.71 L

9. What is the normality of a solution of sulfuric acid which contains 3.5 g H_2SO_4 in 1.5 L solution?

   a. 0.036 N
   b. 0.024 N
   c. 0.012 N
   d. 0.048 N
   e. 0.071 N

10. What volume of 0.25 N H_2SO_4 reacts exactly with 0.65 L of 0.38 N NaOH?

   a. 2.6 L
   b. 0.062 L
   c. 0.43 L
   d. 0.99 L
   e. 0.15 L

PRACTICE EXAM ANSWERS

1. c (15.1)
2. d (15.2 and 15.3)
3. a (15.4)
4. a (15.4)
5. b (15.5)
6. d (15.6)
7. b (15.6)
8. e (15.7)
9. d (15.8)
10. d (15.8)

CHAPTER 16: EQUILIBRIUM

INTRODUCTION

Because atoms and molecules are so tiny, it is hard to imagine what happens when they react and form new products. In this chapter we will learn what is necessary for a reaction to occur, why some reactions stop before all the reactants have been used up, and how to speed up a reaction. Learning how to control chemical reactions has led to many important applications, such as new ways to keep food from spoiling.

AIMS FOR THIS CHAPTER

1. Know how the collision model explains how chemical reactions occur. (Section 16.1)
2. Know what factors can change the rate of a chemical reaction and why these factors cause a change in the rate of reaction. (Section 16.2)
3. Understand what is meant by equilibrium. (Sections 16.3 and 16.4)
4. Know how to write an equilibrium expression for a balanced chemical equation and understand what the equilibrium constant is. (Section 16.5)
5. Know how to write an equilibrium expression for a heterogeneous equilibrium reaction. (Section 16.6)
6. Understand how Le Chatelier's Principle can be used to predict the effect of changes in concentration, volume or temperature on an equilibrium. (Section 16.7)
7. Be able to use the equilibrium constant to predict whether reactants or products predominate, and use the equilibrium expression to calculate concentrations of reactants products or the equilibrium constant itself. (Section 16.8)
8. Know what a solubility product constant is, and how you can use it. (Section 16.9)

QUICK DEFINITIONS

Collision model — A theory which states that molecules react only if they collide with one another. (Section 16.1)

Activation energy — A minimum amount of energy which colliding molecules must have in order to react. Abbreviated E_a. (Section 16.2)

Catalyst — A substance which can speed up a reaction without being used up. A catalyst works by lowering the activation energy. (Section 16.2)

Enzymes — Catalysts found in living tissues which make reactions fast enough to support life. (Section 16.2)

Equilibrium — A state which occurs when the rate of a forward process is exactly equal to the rate of the reverse process. (Section 16.3)

Chemical equilibrium — A dynamic state where the concentrations of chemical components stay the same. (Section 16.3)

Reversible reaction — Reactions can occur in both the forward and the reverse direction and are called reversible. (Section 16.3)

Law of chemical equilibrium — Describes the relationship between the concentration of reactants, products, and the equilibrium constant. (Section 16.5)

Equilibrium expression — A mathematical expression, $K = \frac{[C]^c [D]^d}{[A]^a [B]^b}$ where the letters in square brackets represent molar concentrations of reactants and products at equilibrium, and the small letters represent the coefficients in the balanced equation. (Section 16.5)

Equilibrium constant — The ratio of the concentrations of products to reactants at equilibrium. (Section 16.5)

Equilibrium position — Each set of concentrations for a given set of reactants and products at a specific temperature represent a different equilibrium position. (Section 16.5)

Homogeneous equilibria — Equilibria where the reactants and products are all gases. (Section 16.6)

Heterogeneous equilibria — Equilibria where the reactants and products are mixtures of gases, solids or liquids. (Section 16.6)

Le Chatelier's principle — When a change is imposed on a system at equilibrium, the position of the equilibrium shifts in a direction that tends to reduce the effect of that change. (Section 16.7)

Exothermic reaction — A reaction which produces heat. (Section 16.7)

Endothermic reaction — A reaction which absorbs heat. (Section 16.7)

Solubility product constant When a solid is added to water, the solubility product constant is the product of the molar concentration of the cation raised to the power of its coefficient and the molar concentraton of the anion raised to the power of its coefficient. Can be represented mathematically as $K_{sp} = [A^+]^a[B^-]^b$, where $[A^+]$ represents the molar concentration of cation, and $[B^-]$ represents the molar concentration of anion, both at equilibrium. (Section 16.9)

CONTENT REVIEW

16.1 HOW CHEMICAL REACTIONS OCCUR

What Is the Collision Model?

The collision model says that in order for molecules to react with each other, they must first collide. Increases in the temperature and concentration of reactants brings about more collisions, and the rate of reaction increases. The collision model explains many observations about reactions.

16.2 CONDITIONS THAT AFFECT REACTION RATES

How Do Changes In Temperature Affect the Reaction Rate?

Not all molecules which collide react. Colliding molecules must have a minumum amount of energy in order for a reaction to occur. This minumum amount of energy is called the **activation energy**, abbreviated E_a. As the temperature of a reaction increases, the molecules absorb more and more heat energy. They move around faster, and when they collide, they are more likely to have enough energy to react than they would have at lower temperatures. Therefore, as the temperature increases, the reaction rate increases. Some substances can cause the reaction rate to increase without increasing the temperature. These substances are called **catalysts**. Catalysts are useful because they increase the reaction rate without necessitating an increase in temperature or concentration.

16.3 THE EQUILIBRIUM CONDITION

Why Don't All Reactions Run Until Reactants Run Out?

Many reactions do not continue until all of the reactants have been converted to products. This is because reactions are reversible. A **reversible** reaction means that reactants form products, and products can also form reactants. When a reaction in a closed container appears to stop before the reactants run out, a condition of equilibrium has been reached where the forward reaction and the reverse reaction occur at the same rate. Because both the forward and the reverse reactions occur at the same rate, the concentration of product does not increase, and the concentration of reactant does not decrease. **Equilibrium** is the exact balancing of two processes, one of which is the opposite of the other.

16.4 CHEMICAL EQUILIBRIUM--A DYNAMIC CONDITION

What Is Meant By a Dynamic Equilibrium?

Water vapor reacts with chlorine gas to produce hydrogen chloride gas and oxygen gas according to the balanced equation below.

$$2H_2O(g) + 2Cl_2(g) \leftrightarrows 4HCl(g) + O_2(g)$$

If we mix together 2 moles of H_2O and 2 moles of Cl_2, we have been led to expect that we can produce 4 moles of HCl and 1 mole of O_2. But many reactions appear to stop before all of the reactants, in this case, H_2O and Cl_2, are used up. Why does this happen? When we first mix H_2O and Cl_2 in a closed container, they react and produce HCl and O_2. We now have a mixture of both products and reactants in the container. The collision model suggests that reactants must collide with each other in order to produce products. Eventually though, there are enough molecules of product in the container that they too collide to produce reactants. The reaction is reversible and can occur in both the forward and the reverse direction. At some point, it appears that no more product is being produced and that the reaction has stopped. This is called the equilibrium point and happens when the rate of the forward reaction (reaction of H_2O with Cl_2) is equal to the rate of the reverse reaction (reaction of HCl with O_2). At equilibrium, the reaction has not really stopped, but for every collision which produces products, there is one which produces reactants. Because reactions continue to occur, this is called a **dynamic equilibrium.**

16.5 THE EQUILIBRIUM CONSTANT--AN INTRODUCTION

How Can We Write an Equilibrium Expression For a Reaction?

The results of measuring the concentrations of reactants and products for many reversible reactions led scientists to formulate the **law of chemical equilibrium**. The law of chemical equilibrium can be stated mathematically as $K = \frac{[C]^c [D]^d}{[A]^a [B]^b}$. The letters C and D inside the square brackets represent the molar concentration of products, and the letters A and B inside the square brackets represent the molar concentration of reactants, all at equilibrium. The small letters are powers, taken from the coefficients in the balanced equation of each reactant or product. The letter K is a constant called the **equilibrium constant**, which remains the same as long as the temperature is held constant, even though the concentrations inside the square brackets may change. We can write the equilibrium expression for any equation, as long as the equation is balanced.

Example:

What is the equilibrium expression for the reaction of two carbon dioxide molecules to produce carbon monoxide and oxygen?

$$2CO_2(g) \leftrightharpoons 2CO(g) + O_2(g)$$

First, make sure that the equation is balanced. Then, identify the parts of the equation which will be used in the equilibrium expression--products, reactants, and coefficients. Products and their coefficients appear in the numerator of the expression, and reactants and their coefficients appear in the denominator of the expression. The coefficient of each product or reactant becomes the power to which the concentration is raised. CO_2 has a coefficient of two, so in the equilibrium expression $[CO_2]$ is raised to the power of two, or squared. When reactants or products have a coefficient of one, do not use a superscript one to indicate the power; it is understood. So the equilibrium expression for this reaction is

$$K = \frac{[CO]^2 [O_2]}{[CO_2]^2}$$

What Does the Equilibrium Constant Mean?

It is possible to set up five reaction vessels, each containing different concentrations of CO_2, and after a period of time, measure the concentration of CO_2, CO and O_2.

$$2CO_2(g) \leftrightharpoons 2CO(g) + O_2(g)$$

If you measure the concentrations and calculate the equlibrium constant, K, for each of the five reactions, you will find that K is the same for each of the five, even though the concentrations of CO_2, CO and O_2 in each of the five vessels is different. K depends on the ratios of the concentrations at equilibrium, and is independent of the amount of starting material in the reaction, as long as the temperature is held constant.

16.6 HETEROGENEOUS EQUILIBRIA

How Does a Heterogeneous Equilibrium Affect the Equilibrium Expression?

A **homogeneous** equilibrium has only gases as reactants and products. A **heterogeneous** equilibrium can be composed of gases, liquids, or solids and can exist in aqueous solutions. When we write an equilibrium expression for an equilibrium involving solids or pure liquids, we do not include the solid or the pure liquid in the expression. The concentration of solid or pure liquid is constant so their appearance does not affect the equilibrium constant.

Example:

$$Fe_2O_3(s) + 3H_2(g) \leftrightharpoons 3H_2O(g) + 2Fe(s)$$

In this reaction, there are two solids, Fe_2O_3 and Fe. When we write the equilibrium expression for this reaction, include only gases.

$$K = \frac{[H_2O]^3}{[H_2]^3}$$

16.7 LE CHATELIER'S PRINCIPLE

What Is Le Chatelier's Principle?

Le Chatelier's principle helps to predict what happens to systems at equilibrium, when conditions are changed. **Le Chatelier's** principle says that when a system at equilibrium is changed, the system will shift its equilibrium position in order to reduce the change. The most common changes are changes in concentration, volume, and temperature.

What Happens To an Equilibrium When the Concentration of Reactants or Products Is Changed?

When an equilibrium condition is changed by increasing the concentration of a reactant, the reaction begins removing the extra reactant, in order to re-establish equilibrium. In the following reaction,

$$2CO_2(g) \leftrightharpoons 2CO(g) + O_2(g)$$

increasing the concentration of CO_2 by injecting some CO_2 into the reaction vessel would start the reaction proceeding to the right, to use up the extra CO_2, and to re-establish equilibrium. If additional O_2 were added to the reaction vessel at equilibrium, the reaction would proceed to the left, to use up the O_2 added, and to re-establish equilibrium. Note that when we add more CO_2 to the mixture and a new equilibrium is established, K does not change.

stays the same → $K = \frac{[CO]^2[O_2]}{[CO_2]^2}$ ← addition of more O_2 ← produces more CO_2

Reactants and products can also be removed from a reaction vessel, and this also has an effect on the equilibrium. If CO_2 were removed, the reaction would proceed to the left, in order to produce enough CO_2 to re-establish equilibrium. In all cases, the reaction proceeds in the direction which would diminish the change.

What Happens To an Equilibrium When the Volume Is Changed?

Decreasing the volume of a reaction vessel by pushing inward on a moveable piston increases the pressure inside the vessel. The gas molecules are in a smaller volume than they were before, and they hit the walls of the container more often. Le Chatelier's principle predicts that a reaction will occur in the direction which will decrease the change to the system, that is, in the direction which will decrease the pressure. How can the pressure decrease? Look at the balanced equation for the reaction of CO_2 to form CO and O_2.

$$2CO_2(g) \leftrightharpoons 2CO(g) + O_2(g)$$

The reactant contributes two molecules to the gas in the vessel, while the products contribute three molecules to the gas in the vessel. It is possible to decrease the pressure by having fewer molecules in the gas. If the reaction proceeds to the left, the three product molecules will react to form two reactant molecules, thereby decreasing the pressure. When the volume decreases and pressure increases, the system responds by moving in the direction which will form fewer gas molecules. K still stays the same, but the concentrations change.

What Happens To an Equilibrium When the Temperature Is Changed?

Although a change in the temperature of an equilibrium causes the value of K to change, we can still use Le Chatelier's principle to predict the effect of the change in temperature. To do this, we need to know whether the reaction gives off heat (is **exothermic**) or absorbs heat (is **endothermic**). In the reaction between phosphorus trichloride and chlorine gas to form phosphorus pentachloride, 92.5 kJ of energy is produced. In other words, the reaction is exothermic.

$$PCl_3(g) + Cl_2(g) \leftrightharpoons PCl_5(g) + 92.5 \text{ kJ}$$

We can treat heat just like a product or a reactant. For reactions which absorb heat, the heat appears as a reactant. For reactions which produce heat, heat appears as a product, as in the reaction above. Here PCl_5 and 92.5 kJ are both shown as products of the reaction. Therefore, if we add heat to PCl_5 by raising the temperature, the equilibrium will shift to try to use up the added heat. The reaction will shift to the left, toward the production of PCl_3 and Cl_2.

16.8 APPLICATIONS INVOLVING THE EQUILIBRIUM CONSTANT

How Can We Use Equilibrium Expressions in Calculations?

If you are given equilibrium concentrations of all reactants and products, and if you know the balanced equation for the reaction, you can calculate the equilibrium constant from the equilibrium expression. Also, if you know the equilibrium constant, and all equilibrium concentrations except one, you can use the equilibrium expression to calculate the unknown concentration. This ability can be useful when you need to know the concentration of some product or reactant, but the measurement is difficult.

Example:

In the reaction between H_2 gas and I_2 gas to produce HI,

$$H_2(g) + I_2(g) \leftrightharpoons 2HI(g)$$

the equilibrium concentration of H_2 is 0.19 mol/L, of HI is 1.6 mol/L and the equilibrium constant is 71. What was the equilibrium concentration of I_2? We need to write an equilibrium expression for this reaction, then isolate the unknown concentration, I_2, on one side. The equilibrium expression is

$$K = \frac{[HI]^2}{[H_2][I_2]}$$

To rearrange this equation in order to isolate the I_2, multiply both sides of the equation by $[I_2]$.

$$K \times [I_2] = \frac{[HI]^2}{[H_2]\cancel{[I_2]}} \times \cancel{[I_2]}$$

$$K \times [I_2] = \frac{[HI]^2}{[H_2]}$$

Then divide both sides by K so that I_2 is isolated on the left side of the equation.

$$\frac{\not{K} \times [I_2]}{\not{K}} = \frac{[HI]^2}{[H_2]K}$$

$$[I_2] = \frac{[HI]^2}{[H_2]\ K}$$

Now, substitute values into the equation and find $[I_2]$.

$$[I_2] = \frac{(1.6\ mol/L)^2}{(0.19\ mol/L)(71)} = 0.19\ mol/L$$

At equilibrium, the concentration of I_2 is 0.19 mol/L.

16.9 SOLUBILITY EQUILIBRIA

What Is the Solubility Product Of an Ionic Compound, and How Can It Be Used In Calculations?

When soluble ionic compounds are placed in water, they begin dissolving into individual ions. After some of the solid has dissolved, the individual ions reform the solid compound. Finally, equilibrium is reached, and the rate of dissolving is equal to the rate of formation of solid. We can write an equilibrium expression for this reversible reaction.

Example:

Solid $Al(OH)_3$ dissolves in water to produce Al^{3+} ions and OH^- ions.

$$Al(OH)_3(s) \leftrightharpoons Al^{3+}(aq) + 3OH^-(aq)$$

The equilibrium constant for this type of solubility reaction is called the K_{sp} or **solubility product**. The equilibrium expression is

$$K_{sp} = [Al^{3+}][OH^-]^3$$

This expression describes the concentration of ions at equilibrium. Notice that the solid $Al(OH)_3$ does not appear in the equilibrium expression. Solids and pure liquids do not affect the equilibrium position, and so do not appear in the equilibrium expression.

Given the solubility product for a reaction, we can calculate the solubility in mol/L of an ionic compound.

Example:

The K_{sp} of AgCl is 2.11 x 10^{-11}; what is its solubility in g/L? First, write the balanced equation and the equilibrium expression for the reaction.

$$AgCl(s) \leftrightharpoons Ag^{+}(aq) + Cl^{-}(aq)$$

The reaction above produces an expression $K_{sp} = [Ag^{+}][Cl^{-}]$. Remember that solid AgCl does not appear in the equilibrium expression.

Next, substitute into the equation the value of K_{sp} for this reaction.

$$K_{sp} = 2.11 \times 10^{-11} = [Ag^{+}][Cl^{-}]$$

$x\frac{\text{mol}}{\text{L}}$ of AgCl(s) dissociates into $x\frac{\text{mol}}{\text{L}}$ Ag^{+}(aq) and $x\frac{\text{mol}}{\text{L}}$ Cl^{-}(aq).
At equilibrium $[Ag^{+}] = x\frac{\text{mol}}{\text{L}}$ and $[Cl^{-}] = x\frac{\text{mol}}{\text{L}}$.
We can substitute these values into the equilibrium expression.

$$K_{sp} = 2.11 \times 10^{-11} = [Ag^{+}][Cl^{-}] = (x)(x) = x^2$$

We can say that

$$x^2 = 2.11 \times 10^{-11}$$

$$x = \sqrt{2.11 \times 10^{-11}} = 4.59 \times 10^{-6} \text{ mol/L}$$

The solubility of AgCl is 4.59 x 10^{-6} mol/L. The problem asks for solubility in g/L, not mol/l. We must use the molar mass of AgCl to convert from moles to grams.

$$4.59 \times 10^{-6} \text{ m}\not{\text{o}}\text{l/L AgCl} \times \frac{143.35 \text{ g AgCl}}{\text{m}\not{\text{o}}\text{l AgCl}} = 6.58 \times 10^{-4} \text{ g/L}$$

Thus, the solubility of AgCl is 6.58 x 10^{-4} g/L.

LEARNING REVIEW

1. Why does an increase in the concentration of reactant cause a reaction to speed up?

2. Why does an increase in the temperature cause a reaction to speed up?

3. a. Which letter represents the energy of the products of a reaction?
   b. Which letter represents the energy of the reactants?
   c. Which letter represents the activation energy (E_a) of a reaction?
   d. Which letter represents the catalyzed reaction pathway?

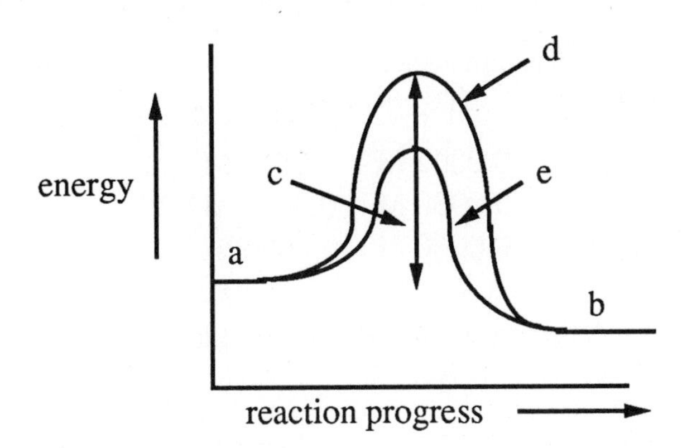

4. What is a catalyst and how does it work?

5. A beaker of liquid water in a sealed container is allowed to reach equilibrium vapor pressure. What is happening to the concentration of water vapor in the beaker?

6. Which of the following statements about equilibrium are true?

   a. After equilibrium is established, the rate of the forward reaction is greater than the rate of the reverse reaction.
   b. Before equilibrium is reached, the concentration of products increases as time passes.
   c. Before equilibrium is reached, the concentration of reactants increases as time passes.

7. Write equilibrium constant expressions for each of the reactions below.

   a. $I_2(g) + Cl_2(g) \rightleftharpoons 2ICl(g)$
   b. $2NO_2(g) + 7H_2(g) \rightleftharpoons 2NH_3(g) + 4H_2O(g)$
   c. $4HCl(g) + O_2(g) \rightleftharpoons 2H_2O(g) + 2Cl_2(g)$
   d. $4NH_3(g) + 5O_2(g) \rightleftharpoons 4NO(g) + 6H_2O(g)$

8. Write the equilibrium expression and calculate the equilibrium constant, K, for the reaction below, at each set of equilibrium concentrations.

$$C_2H_4O_2 \leftrightharpoons C_2H_3O_2^- + H^+$$

Experiment	Equilibrium concentrations			Eq expression	K
	$[C_2H_4O_2]$	$[H^+]$	$[C_2H_3O_2^-]$		
I	1.0 M	0.0042 M	0.0042 M		
II	0.50 M	0.0030 M	0.0030 M		
III	2.0 M	0.0060 M	0.0060 M		

9. Why do we not include the concentration of solids or pure liquids in the equilibrium expression?

10. Write the equilibrium constant expression for each of the reactions below.

a. $Pb(OH)_2(s) \leftrightharpoons Pb^{2+}(aq) + 2OH^-(aq)$
b. $2Sb(s) + 3Cl_2(g) \leftrightharpoons 2SbCl_3(g)$
c. $Fe_2O_3(s) + 3H_2(g) \leftrightharpoons 3H_2O(g) + 2Fe(s)$
d. $Mg(OH)_2(s) \leftrightharpoons MgO(s) + H_2O(g)$

11. What would happen to the position of the equilibrium when the following changes are made to the equilibrium reaction below?

$$2SO_2(g) + O_2(g) \leftrightharpoons 2SO_3(g)$$

a. SO_2 is removed from the reaction vessel.
b. SO_3 is added to the reaction vessel.
c. Oxygen is removed from the reaction vessel.

12. What would happen to the position of the equilibrium when the following changes are made to the reaction below?

$$2HgO(s) \leftrightharpoons Hg(l) + O_2(g)$$

a. Solid HgO(s) is added to the reaction vessel.
b. The pressure in the reaction vessel is increased.

13. When the volume of the following mixture of gases is increased, what will be the effect on the equlibrium position?

$$4HCl(g) + O_2(g) \leftrightharpoons 2H_2O(g) + 2Cl_2(g)$$

14. Predict the effect of decreasing the volume of the container on the position of each equilibrium below.

a. $SiF_4(g) + 2H_2O(g) \rightleftharpoons SiO_2(s) + 4HF(g)$
b. $2H_2(g) + 2NO(g) \rightleftharpoons 2H_2O(g) + N_2(g)$
c. $C(s) + H_2O(g) \rightleftharpoons CO(g) + H_2(g)$

15. Predict the effect of increasing the temperature on the position of each equilibrium below.

a. $H_2(g) + Cl_2(g) \rightleftharpoons 2HCl(g) + \text{heat}$ exothermic
b. $2NH_3(g) + \text{heat} \rightleftharpoons N_2(g) + 3H_2(g)$ endothermic
c. $CO_2(g) + H_2(g) + \text{heat} \rightleftharpoons CO(g) + H_2O(g)$ endothermic

16. If the equilibrium constant for the reaction below is 51.47, the concentration of HI is 0.50 M, and the concentration of H_2 is 0.069 M, what is the concentration of I_2?

$$H_2(g) + I_2(g) \rightleftharpoons 2HI(g)$$

17. Write the balanced equation describing the dissolution of the solids below. Then write the K_{sp} expression.

a. $Ca_3(PO_4)_2$
b. FeS
c. $Sn(ClO_3)_2$
d. $Al(OH)_3$

18. The solubility of $PbSO_4(s)$ is 1.3×10^{-4} mol/L at 25 °C. What is the K_{sp} of $PbSO_4(s)$?

19. Copper(II) sulfide has a K_{sp} of 8.0×10^{-45}. What is the solubility of CuS(s) in water at 25 °C?

ANSWERS TO LEARNING REVIEW

1. The collision model says for reactions to occur, the reactants must first collide with each other. Higher concentrations of reactants cause the reaction rate to speed up because the number of collisions increases as the concentration of reactants increases.

2. Not all collisions possess enough energy to break bonds. A minimum amount of energy, called the activation energy, is needed before a reaction can occur. As the temperature increases, the speed at which the molecules move increases. When the molecules collide, the collisions are more energetic and more of them possess the minimum amount of energy to break bonds, and so more reactions occur.

3. a. Letter b represents the energy of the products. In this part of the graph, the reactants have achieved the activation energy, and products have formed.
   b. Letter a represents the energy of the reactants. The average energy of the reactants is lower than the activation energy.
   c. Letter c represents the activation energy, the minimum amount of energy needed for a reaction to occur.
   d. Letter e represents the catalyzed reaction pathway. A catalyst lowers the activation energy for a reaction.

4. A catalyst is a substance added to the reaction which causes the reaction to speed up. Catalysts are not used up during the reaction. They are not reactants and do not form products. A catalyst works by providing a new path for the reaction, which lowers the activation energy for that reaction.

5. When water is first sealed in the beaker, the level of water in the beaker decreases, as water from the beaker enters the vapor phase. After some period of time, the level of water stops decreasing and stays at the same level. At this point a balance occurs between the processes of evaporation and condensation. The system is at equilibrium and the concentration of water vapor does not change.

6. a. This statement is false. At equilibrium the rate of forward reaction equals the rate of the reverse reaction.
   b. This statement is true. When reactants are mixed, they continue to react to form product. The amount of product increases until equilibrium is reached.
   c. This statement is false. The concentration of reactants decreases until equilibrium is reached, at which point the concentration of reactants remains constant.

7. When writing equilibrium expressions, coefficients become powers and products appear in the numerator, reactants in the denominator.

   a. $$K = \frac{[ICl]^2}{[I_2][Cl_2]}$$

   b. $$K = \frac{[H_2O]^4[NH_3]^2}{[NO_2]^2[H_2]^7}$$

   c. $$K = \frac{[H_2O]^2[Cl_2]^2}{[HCl]^4[O_2]}$$

   d. $$K = \frac{[NO]^4[H_2O]^6}{[NH_3]^4[O_2]^5}$$

8.

Experiment	Equilibrium concentrations $[C_2H_4O_2]$	$[H^+]$	$[C_2H_3O_2^-]$	Eq expression	K
I	1.0 M	0.0042 M	0.0042 M	$K = \frac{[H^+][C_2H_3O_2^-]}{[C_2H_4O_2]}$	1.8×10^{-5}
II	0.50 M	0.0030 M	0.0030 M	$K = \frac{[H^+][C_2H_3O_2^-]}{[C_2H_4O_2]}$	1.8×10^{-5}
III	2.0 M	0.0060 M	0.0060 M	$K = \frac{[H^+][C_2H_3O_2^-]}{[C_2H_4O_2]}$	1.8×10^{-5}

9. The concentrations of pure solids and liquids are constant and do not change. They are not shown in the equilibrium expression.

10. a. $K = [Pb^{2+}][OH^-]^2$

b. $K = \frac{[SbCl_3]^2}{[Cl_2]^3}$

c. $K = \frac{[H_2O]^3}{[H_2]^3}$

d. $K = [H_2O]$

11. a. The equilibrium will shift to the left and begin producing SO_2.
b. The equilibrium will shift to the left and begin consuming SO_3.
c. The equilibrium will shift to the left and begin producing O_2.

12. a. Pure solids and liquids have no affect on the equilibrium, so adding or removing HgO(s) or Hg(l) will not effect the position of the equilibrium.

b. The equilibrium will shift toward the left to reduce the pressure inside the system by consuming oxygen gas.

13. On the left side there are five gaseous molecules and on the right there are four. When the volume is increased, the pressure decreases and the equilibrium will shift toward the side which increases the pressure, to the left.

14. Decreasing the volume causes an increase in pressure. The equilibrium shifts in the direction to relieve the pressure, or toward the side with fewer numbers of gaseous molecules.

    a. The equilibrium would shift toward the left because the left has three gaseous molecules while the right has four gaseous molecules.
    b. The equilibrium would not shift in either direction because the number of gaseous molecules on the right and the left is the same.
    c. The equilibrium would shift toward the left because the left has one gaseous molecule and the right has two gaseous molecules. Solid carbon on the left does not affect the equilibrium.

15. For exothermic reactions, treat heat energy as a product of the reaction. If heat energy is added to the system by raising the temperature, then the equilibrium shifts to the left to consume the heat energy. For endothermic reactions, treat heat energy as a reactant. If heat energy is added to the system by raising the temperature, then the equilibrium shifts to the right to consume the heat energy.

    a. This is an exothermic reaction. Raising the temperature would shift the equilibrium to the left to consume the energy.
    b. This is an endothermic reaction. Raising the temperature would shift the equilibrium to the right to consume excess energy.
    c. This is an endothermic reaction. Raising the temperature would shift the equilibrium to the right to consume the energy.

16. First, write the equilibrium expression for this reaction.

$$K = \frac{[HI]^2}{[H_2][I_2]}$$

Rearrange this equation to isolate I_2 on one side. Divide both sides by $[HI]^2$.

$$\frac{K}{[HI]^2} = \frac{\cancel{[HI]^2}}{\cancel{[HI]^2}} \times \frac{1}{[H_2][I_2]}$$

$$\frac{K}{[HI]^2} = \frac{1}{[H_2][I_2]}$$

Now, multiply both sides by $[H_2]$.

$$\frac{K[H_2]}{[HI]^2} = \frac{1}{[I_2]}$$

Take the inverse of both sides.

$$[I_2] = \frac{[HI]^2}{K[H_2]}$$

Substitute values into the equation and find $[I_2]$.

$$I_2 = \frac{0.25\ (mol/L)^2}{(51.47)\ (0.069\ mol/L)}$$

$$I_2 = 0.070\ mol/L$$

17. The K_{sp} expression does not include the concentration of the solid salt, only the ions in solution.

a. $K_{sp} = [Ca^{2+}]^3[PO_4^{3-}]^2$
b. $K_{sp} = [Fe^{2+}][S^{2-}]$
c. $K_{sp} = [Sn^{2+}][ClO_3^-]^2$
d. $K_{sp} = [Al^{3+}][OH^-]^3$

18. This problem asks us to calculate the K_{sp} for $PbSO_4$, given the solubility of $PbSO_4$. When $PbSO_4$ dissolves in water, lead ions and sulfate ions are released into the aqueous environment.

$$PbSO_4(s) \leftrightharpoons Pb^{2+}(aq) + SO_4^{2-}(aq)$$

The K_{sp} expression for the reaction is

$$K_{sp} = [Pb^{2+}][SO_4^{2-}]$$

We need to know the molar concentrations of Pb^{2+} and SO_4^{2-} to find the K_{sp}. We know that the solubility of $PbSO_4$ is 1.3×10^{-4} mol/L. This means that 1.3×10^{-4} mol $PbSO_4$ dissolves per liter of solution. Each 1.3×10^{-4} mol $PbSO_4$ produces 1.3×10^{-4} mol Pb^{2+} and 1.3×10^{-4} mol SO_4^{2-}. The concentration of Pb^{2+} is 1.3×10^{-4} mol/L and the concentration of SO_4^{2-} is 1.3×10^{-4} mol/L. We can use these concentrations to calculate the K_{sp}.

$$K_{sp} = [Pb^{2+}][SO_4^{2-}]$$

$$K_{sp} = (1.3 \times 10^{-4}\ mol/L)(1.3 \times 10^{-4}\ mol/L)$$

$$K_{sp} = 1.7 \times 10^{-8}\ mol^2/L^2$$

The units for K_{sp} are usually omitted, so the K_{sp} would be reported as 1.7×10^{-8}.

19. This problem asks us to find the solubility of copper(II) sulfide. We are given the K_{sp}, which is 8.0×10^{-45}. When CuS dissolves in water, each mole of CuS which dissolves produces one mole of Cu^{2+} and one mole of S^{2-}.

$$CuS(s) \leftrightharpoons Cu^{2+}(aq) + S^{2-}(aq)$$

The K_{sp} expression for the dissolution of CuS in water is

$$K_{sp} = [Cu^{2+}][S^{2-}]$$

$x\frac{mol}{L}$ of CuS(s) dissociates into $x\frac{mol}{L}$ $Cu^{2+}(aq)$ and $x\frac{mol}{L}$ $S^{2-}(aq)$.
At equilibrium $[Cu^{2+}] = x\frac{mol}{L}$ and $[S^{2+}] = x\frac{mol}{L}$.
We can substitute these values into the equilibrium expression.

$$K_{sp} = 8.0 \times 10^{-45} = [Cu^{2+}][S^{2-}] = (x)(x) = x^2$$

We can say that

$$x^2 = 8.0 \times 10^{-45}$$

$$x = \sqrt{8.0 \times 10^{-45}} = 8.9 \times 10^{-23}\ mol/L$$

The solubility of CuS is 8.9×10^{-23} mol/L.

PRACTICE EXAM

1. When H_2 and Cl_2 are mixed together, which of the following statements is not true?

$$H_2 + Cl_2 \leftrightharpoons 2HCl$$

a. A catalyst for this reaction would lower the activation energy.
b. A catalyst for this reaction would speed up the reaction, yet not be consumed itself.
c. In any sample of H_2 and Cl_2, all the molecules collide with enough energy to react and form HCl.
d. According to the collision model, H_2 molecules and Cl_2 molecules must collide before they can form HCl molecules.
e. Increasing the temperature of a reaction can cause an increase in the speed of a reaction.

2. Which of the statements below does not describe a system at equilibrium?

a. The rate of vapor condensation is greater than the rate of liquid evaporation in a sealed beaker.
b. When NO_2 and N_2O_4 have established equilibrium, they continue to react.
c. The concentration of products and reactants is constant, even after long periods of time.
d. In the reaction $2NO_2 \leftrightharpoons N_2O_4$, the rate of formation of N_2O_4 equals the rate of formation of NO_2.
e. The level of liquid in a sealed beaker neither increases nor decreases.

3. The figure shows the reaction rates over time for the reaction below. Which statement about the figure is true?

$$4H_2O(l) + 3Fe(s) \leftrightharpoons 4H_2(g) + Fe_3O_4(s)$$

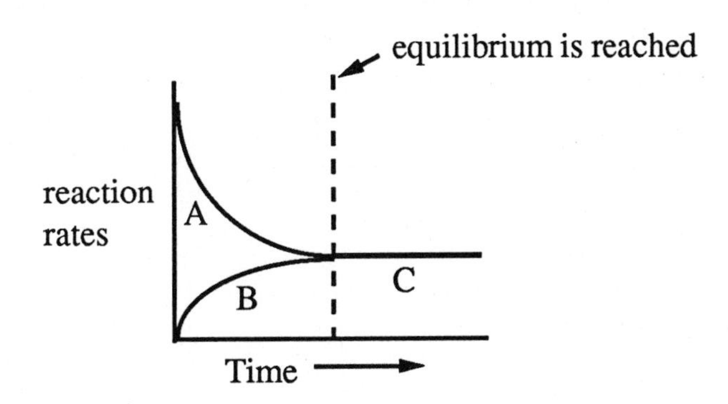

a. A represents the reaction between $H_2(g)$ and $Fe_3O_4(s)$.
b. In C, the concentration of $H_2O(l)$ and $Fe(s)$ equals the concentration of $H_2(g)$ and $Fe_3O_4(s)$.
c. In A, the concentrations of $H_2O(l)$ and $Fe(s)$ are increasing.
d. In B, the concentrations of $H_2(g)$ and $Fe_3O_4(s)$ are increasing.
e. In C, the rate of formation of $H_2(g)$ and $Fe_3O_4(s)$ is greater than the rate of formation of $H_2O(l)$ and $Fe(s)$.

4. What is the correct equilibrium constant expression for the reaction below?

$$4HCl(g) + O_2(g) \leftrightharpoons 2H_2O(g) + 2Cl_2(g)$$

a. $K = \frac{[H_2O]^2[Cl_2]^2}{[HCl]^4[O_2]^2}$

b. $K = \frac{[HCl]^4[O_2]}{[H_2O]^2[Cl_2]^2}$

c. $K = \frac{[H_2O]^2[Cl_2]^2}{[HCl]^4[O_2]}$

d. $K = \frac{[H_2O][Cl_2]}{[HCl][O_2]}$

e. $K = \frac{2[H_2O]\ 2[Cl_2]}{4[HCl]\ [O_2]}$

5. What is the correct equilibrium constant expression for the reaction below?

$$2Ag(s) + Cl_2(g) \leftrightharpoons 2AgCl(s)$$

a. $K = \frac{[AgCl]^2}{[Cl_2]}$

b. $K = \frac{[AgCl]^2}{[Ag]^2[Cl_2]}$

c. $K = \frac{[AgCl]}{[Ag]^2[Cl_2]}$

d. $K = [Ag]^2[Cl_2]$

e. $K = \frac{1}{[Cl_2]}$

6. For the reaction below, which change would not cause the equilibrium to shift to the left?

$$CH_4(g) + 2O_2(g) \leftrightharpoons CO_2(g) + 2H_2O(g)$$

a. CH_4 is added to the system.
b. The concentration of CO_2 is increased.
c. The concentrations of CO_2 and H_2O are increased.
d. O_2 is removed from the system.
e. $H_2O(g)$ is added to the system.

7. For the reaction below, which change would cause the equilibrium to shift to the right? The reaction between CH_4 and H_2S is endothermic.

$$CH_4(g) + 2H_2S(g) + heat \leftrightharpoons CS_2(g) + 4H_2(g)$$

a. Increase the pressure of the system.
b. Decrease the H_2S concentration.
c. Increase the CS_2 concentration.
d. Increase the temperature of the system.
e. Decrease the CH_4 concentration.

8. For the reaction below the equilibrium constant is 5.81, the equilibrium concentration of NH_3 is 1.53 M and of N_2 is 1.10 M. What is the equilibrium concentration of H_2?

$$N_2(g) + 3H_2(g) \leftrightharpoons 2NH_3(g)$$

a. 1.40 M
b. 0.715 M
c. 0.366 M
d. 1.29 M
e. 9.78 M

9. Which is the correct K_{sp} expression for the dissolving of $Pb(OH)_2$?

a. $K_{sp} = \frac{[Pb^{2+}][OH^-]^2}{[Pb(OH)_2]}$

b. $K_{sp} = [Pb^{2+}][OH^-]$

c. $K_{sp} = [Pb^{2+}][OH^-]^2$

d. $K_{sp} = \frac{[Pb(OH)_2]}{[Pb^{2+}][OH^-]^2}$

e. $K_{sp} = \frac{[Pb^{2+}][OH^-]}{[Pb(OH)_2]}$

10. What is the K_{sp} of $PbCO_3$ if the solubility of $PbCO_3$ is 3.0×10^{-7} mol/L at 25 °C?

a. 2.0×10^{-7}
b. 9.0×10^{-14}
c. 7.8×10^{-7}
d. 5.9×10^{-20}
e. 3.9×10^{-7}

PRACTICE EXAM ANSWERS

1. c (16.1 and 16.2)
2. a (16.3 and 16.4)
3. d (16.4)
4. c (16.5)
5. e (16.6)
6. a (16.7)
7. d (16.7)
8. b (16.8)
9. c (16.9)
10. b (16.9)

CHAPTER 17: ACIDS AND BASES

INTRODUCTION

In this chapter you will learn about the properties of acids and bases. You know about some of the properties of acids already. Substances such as lemon juice and vinegar contain acids, and their sour taste is from the acid each contains. You will learn how to determine the properties of acids and how to measure the acidity of a solution.

AIMS FOR THIS CHAPTER

1. Know the Arrhenius and Brönsted-Lowry models of acids and bases, and be able to identify a conjugate acid-base pair. (Section 17.1)
2. Be able to state the properties of weak and strong acids. (Section 17.2)
3. Understand how water can be both an acid and a base, and how the ion-product constant is derived. (Section 17.3)
4. Understand the pH scale and be able to use a calculator to calculate the pH or pOH of a solution, given $[H^+]$ or $[OH^-]$. (Section 17.4)
5. Be able to use a calculator to calculate $[H^+]$ or $[OH^-]$, given the pH or pOH. (Section 17.4)
6. Be able to calculate the pH of a strong acid solution of known molarity. (Section 17.5)
7. Know the characteristics of a buffer, and how a buffer works to resist changes in pH. (Section 17.6)

QUICK DEFINITIONS

Alkalis — Alkalis have a bitter taste and a slippery feel. They are also called bases. (Section 17.1)

Arrhenius concept of acid and base — An acid produces hydrogen ions and a base produces hydroxide ions when in solution. (Section 17.1)

Brönsted-Lowry model — An acid is a proton (H^+) donor, and a base is a proton acceptor. (Section 17.1)

Conjugate acid — When a strong acid is dissolved in water, the species which bonds to the proton is called the conjugate acid. (Section 17.1)

Conjugate base — When a strong acid is dissolved in water, the part of the acid molecule which is left after the proton is gone is called the conjugate base. (Section 17.1)

Conjugate acid-base pair — Two substances related to each other by donating and accepting a proton, H^+. (Section 17.1)

Hydronium ion — The ion H_3O^+, which is a water molecule which has accepted a proton, H^+. (Section 17.1)

Strong acid — An acid that completely dissociates in water to become ions. The equilibrium $HA(aq) + H_2O(l) \leftrightharpoons H_3O^+(aq) + A^-(aq)$ lies far to the right for a strong acid. (Section 17.2)

Weak acid — An acid that does not completely dissociate in water to form ions. The equilibrium $HA(aq) + H_2O(l) \leftrightharpoons H_3O^+(aq) + A^-(aq)$ lies far to the left for a weak acid. (Section 17.2)

Diprotic acid — An acid which can donate two protons for each acid molecule. (Section 17.2)

Oxyacids — Acids which have the acidic proton bonded to an oxygen atom. (Section 17.2)

Organic acids — Acids which contain a carbon skeleton, and often a carboxyl group. (Section 17.2)

Carboxyl group — A carboxyl group is the acidic part of molecules with carbon chains. It consists of a carbon atom doubly bonded to an oxygen and also bonded to an OH.

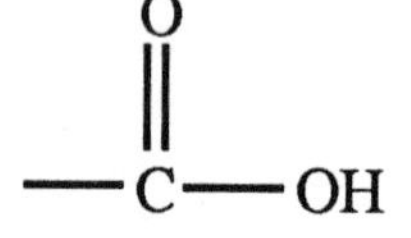

Amphoteric substance — A substance which can behave as an acid or a base. (Section 17.3)

Ionization of water — Dissociation of a water molecule, with the transfer of the proton to a second water molecule to produce $H_3O^+ + OH^-$. (Section 17.3)

Ion-product constant — A constant K_w whose value is 1.0×10^{-14}, and is equal to the molar concentration of hydrogen ion times the molar concentration of hydroxide ion, $[H^+][OH^-]$. (Section 17.3)

Neutral solution — A solution where the hydrogen ion concentration equals the hydroxide ion concentration, $[H^+] = [OH^-]$. The pH of a neutral solution is 7.0. (Section 17.3)

Acidic solution — A solution where the hydrogen ion concentration is greater than the hydroxide ion concentration, $[H^+] > [OH^-]$. The pH is less than 7.0. (Section 17.3)

Basic solution — A solution where the hydrogen ion concentration is less than the hydroxide ion concentration, $[H^+] < [OH^-]$. The pH is greater than 7.0. (Section 17.3)

pH scale — A convenient way to express small hydrogen ion concentrations. $pH = -\log[H^+]$. (Section 17.4)

Buffered solution — A solution which resists pH change even when a strong base or strong acid is added. A buffered solution contains a weak acid, plus the conjugate base of the weak acid. (Section 17.6)

CONTENT REVIEW

17.1 ACIDS AND BASES

What Are Acids and Bases?

Arrhenius proposed that an acid was anything which produced hydrogen ions in aqueous solution, and a base was anything which produced hydroxide ions. Scientists discovered that this definition of acids and bases was too restrictive, that there were other bases besides hydroxide ions. Brönsted and Lowry proposed that an acid was a proton (hydrogen ion) donor, and a base was a proton acceptor. This was a much broader definition of acids and bases.

What Happens When an Acid Is Dissolved in Water?

According to the Brönsted-Lowry model of acids and bases, when an acid dissolves in water, a reaction occurs between the acid (the proton donor) and water (the proton acceptor). Two products are formed as a result of the reaction, a water molecule with an extra proton called a **hydonium ion**, and the remains of the acid after the proton has left.

$$HA(aq) + H_2O(l) \rightarrow H_3O^+(aq) + A^-(aq)$$

The protonated water is called a **conjugate acid**, and the remaining part of the acid (A^-) is called the **conjugate base.** HA and A^- form a pair called the acid-conjugate base pair. H_2O and H_3O^+ form a pair called the base-conjugate acid pair. When an acid or a base react, it is useful to identify the conjugate base or the conjugate acid. What is the conjugate base of H_2S? H_2S can donate a proton to water by the following reaction.

$$H_2S + H_2O \rightarrow H_3O^+ + HS^-$$

The conjugate base of H_2S is HS^-.

17.2 ACID STRENGTH

Why Are Some Acids Weak and Some Strong?

The reaction of an acid with water is a reversible reaction. The proton can be attached to the water molecule, or to the conjugate base.

$$HA + H_2O \rightleftharpoons H_3O^+ + A^-$$

The relative attractions of H_2O and A^- for the proton determines whether HA or H_3O^+ predominates. If A^- attracts the proton more strongly than does H_2O, then the equlibrium lies to the left, and there is relatively little H_3O^+. Acids in which the equlibrium lies to the left are called **weak acids** because they exist mostly in the HA form. If H_2O attracts the proton more strongly than does A^-, then the equilibrium lies to the right. These are called **strong acids** because virtually all of the protons have dissociated from the conjugate base and are attached to a water molecule.

17.3 WATER AS AN ACID AND A BASE

How Can Water Be Both an Acid And a Base?

Water is a substance which can act both as an acid and a base. Such substances are said to be **amphoteric**. The amphoteric nature of water can be seen when two water molecules react. One water molecule donates a proton; it is acting as an acid. The other water molecule accepts the proton; it is acting as a base.

$$H_2O + H_2O \rightarrow H_3O^+ + OH^-$$

We can write an equilibrium expression for the ionization of water by showing just one water molecule.

$$H_2O(l) \rightarrow H^+(aq) + OH^-(aq) \quad K_w = [H^+][OH^-]$$

The equilibrium constant for this reaction is K_w and is called the **ion-product constant**, or **dissociation constant**. Liquid water does not appear in the equilibrium expression because the concentration of water changes very little. It has been shown experimentally that K_w has the value 1.0×10^{-14} mol²/L².

$$K_w = [H^+][OH^-] = (1.0 \times 10^{-7} \text{ mol/L})\ (1.0 \times 10^{-7} \text{ mol/L}) = 1.0 \times 10^{-14} \text{ mol}^2/\text{L}^2$$

For any solution, the product of $[H^+]$ and $[OH^-]$ is always 1.0×10^{-14}. For an aqueous solution of a strong acid where the $[H^+]$ is very high, the $[OH^-]$ must be very low, so that when the two concentrations are multiplied, the result is 1.0×10^{-14}. If we know the concentration of either hydrogen ion or hydroxide ion, we can calculate the other, using the equilibrium expression for the ionization of water.

Example:

If the hydroxide ion concentration of a solution is 2.0×10^{-2} M, what is the hydrogen ion concentration? First, write the equilibrium expression for the ionization of water.

$$K_w = [H^+][OH^-]$$

Then isolate the unknown value, $[H^+]$, on one side of the equation by dividing both sides by $[OH^-]$.

$$\frac{K_w}{[OH]} = \frac{[H^+][\not{O}H^-]}{[\not{O}H^-]}$$

$$[H^+] = \frac{K_w}{[OH^-]}$$

Then, substitute the values into the equation.

$$[H^+] = \frac{1.0 \times 10^{-14}}{2.0 \times 10^{-2}} = 5.0 \times 10^{-13} \text{ M}$$

The concentration of hydrogen ion is 5.0×10^{-13} M. You can check your calculations by multiplying $(2.0 \times 10^{-2})\ (5.0 \times 10^{-13})$.

$$(2.0 \times 10^{-2})(5.0 \times 10^{-13}) = 10 \times 10^{-15} = 1.0 \times 10^{-14}$$

The result is 1.0×10^{-14}.

17.4 THE pH SCALE

How Can We Use Logarithms To Calculate the pH of a Solution?

The hydrogen ion concentration of a solution is represented by very small numbers. Using scientific notation is a good way to represent small numbers, but calculating and using the pH of a solution is another easy way to represent small numbers. Taking the p of any number means we take the logarithm (log) of that number, and multiply the result by -1.

$$pN = (-1) \times \log N$$

When we find the pH of a solution, we take the log of the hydrogen ion concentration in mol/L and multiply by -1.

Example:

What is the pH of a solution with a hydrogen ion concentration of 2.3×10^{-2} mol/L? First, find the log of 2.3×10^{-2} by entering the number on your calculator, and pressing the key for log.

$$\log 2.3 \times 10^{-2} = -1.638$$

Then, press the ± key to multiply -1.638 by -1. Multiplying any number by -1 changes the sign, but not the value of the number.

$$-1.638 \times -1 = 1.638$$

Next, express the number to the correct number of significant figures. When using logarithms, the part of the number to the right of the decimal point should have the same number of digits as the hydrogen ion concentration has significant figures. The number 2.3×10^{-2} has two significant figures, so the pH should have two numbers to the right of the decimal point. 1.638 would become 1.64. The pH of a solution with 2.3×10^{-2} mol/L hydrogen ion is 1.64.

Can a Logarithmic Scale Be Used To Express Hydroxide Ion Concentration?

A log scale can also be used to express the hydroxide ion concentration. pOH is calculated the same way as pH. Start with the hydroxide ion concentration in mol/L. Find the log of that number, then multiply by -1. For example, if the $[OH^-]$ is 6.5×10^{-5} mol/L, the pOH is

$$pOH = -\log (6.5 \times 10^{-5}) = -4.19$$

$$-4.19 \times -1 = 4.19$$

The pOH is 4.19. Because 6.5×10^{-5} has two significant figures, there should be two digits to the right of the decimal point in the pOH value.

What Is the Relationship Between pH and pOH?

$[H^+]$ and $[OH^-]$ are related to each other through the equilibrium expression for the dissociation of water, and the ion product constant, K_w.

$$[H^+]\,[OH^-] = K_w$$

If we take the p of each of the parts of this equation we have

$$-\log[H^+]\ -\log[OH^-] = -\log(1.0 \times 10^{-14})$$

or,

$$pH + pOH = 14.00$$

This means that for any solution, the sum of pH and pOH will always equal 14. If we know the pH, we can easily find the pOH, and vice versa.

Example:

If the pOH of a solution is 8.23, what is the pH? Use the equation pH + pOH = 14 to solve this problem. First, isolate the unknown quantity, pH, on one side of the equation. Then, substitute with the values we know.

$$pH = 14 - pOH$$

$$pH = 14.00 - 8.23 = 5.77$$

The pH of the solution is 5.77.

17.5 CALCULATING THE pH OF STRONG ACID SOLUTIONS

How Can We Calculate the pH of a Strong Acid?

We can calculate the pH of any solution if we know the hydrogen ion concentration. When strong acids dissolve in water, they dissociate completely into H^+ and A^-. If we know the molarity of the acid solution, we know the $[H^+]$.

$$0.5\text{ M HCl} \longrightarrow 0.5\text{ M } H^+ + 0.5\text{ M } Cl^-$$

We know that the $[H^+]$ is 0.5 mol/L because all of the HCl dissociates into H^+ and Cl^-. From the $[H^+]$ we can calculate the pH.

$$pH = -\log(0.5) = 0.30$$

The pH is 0.30. This is a strongly acidic solution.

17.6 BUFFERED SOLUTIONS

What Is a Buffered Solution and How Does It Work?

A buffered solution resists large changes in pH even when strong acid or strong base is added. The addition of even small amounts of strong acid or base can greatly lower or raise the pH of a nonbuffered solution. Living systems contain many differents kinds of buffers to help keep fluids and tissues at the correct pH, even under stressful conditions. Most buffers are made from a weak acid, plus the soluble salt of the weak acid, the conjugate base. Acetic acid and sodium acetate are often used for this purpose. When strong acid is added to a buffered solution containing acetic acid and sodium acetate, the following reaction occurs.

$$C_2H_3O_2^-(aq) + H^+(aq) \rightarrow HC_2H_3O_2(aq)$$

The excess hydrogen ion, which would ordinarily lower the pH of the solution, attaches to the acetate anion to form acetic acid. Remember that weak acids form strong conjugate bases, so the acetate anion has a high affinity for hydrogen ions. The hydrogen ions are removed from the solution, and the pH does not decrease as it would if there were no buffer present. When strong base is added to the solution, the H^+ reacts with the acetate ion in the following way.

$$HC_2H_3O_2(aq) + OH^-(aq) \rightarrow H_2O(l) + C_2H_3O_2^-(aq)$$

The excess hydroxide ion from the strong base reacts with acetic acid to form a water molecule and the conjugate base, the acetate anion. The pH does not rise very much, because the OH^- ions are no longer free in solution. Almost any weak acid and its conjugate base will act as a buffer.

LEARNING REVIEW

1. Explain the differences between the Arrhenius concept of an acid and a base, and the concept of Brönsted and Lowry.

2. For the reaction of perbromic acid with water:

$$HBrO_4(aq) + H_2O(l) \rightarrow H_3O^+(aq) + BrO_4^-(aq)$$

   a. Which two substances are an acid-conjugate base pair?
   b. Which two substances are a base-conjugate acid pair?

3. Write equations to show what happens when each of the acids below reacts with water.

   a. H_2S
   b. HNO_2

4. Show how acetic acid, $HC_2H_3O_2$, reacts with water by drawing Lewis structures for water and its conjugate acid.

5. When formic acid, HCOOH, is mixed with water, the resulting solution weakly conducts an electric current.

   a. Is formic acid a strong or a weak acid?
   b. Toward which side of the reaction does the equilibrium lie?
   c. Which species is the stronger base, H_2O or $HCOO^-$?

6. Which of the acids below are strong acids and which are weak acids?

   a. HF
   b. H_2SO_4
   c. $HC_2H_3O_2$
   d. $HClO_4$

7. The expression for the dissociation of water is $K_w = [H^+][OH^-]$. Why does liquid water not appear in this expression?

8. All the aqueous solutions below are at a temperature of 25 °C.

   a. What is the $[H^+]$ of a solution for which $[OH^-] = 1.5 \times 10^{-6}$ M?
   b. What is the $[H^+]$ of a solution for which $[OH^-] = 6.3 \times 10^{-3}$ M?
   c. What is the $[OH^-]$ of a solution for which $[H^+] = 3.25 \times 10^{-1}$ M?

9. What is the $[H^+]$ of a 0.1 M solution of NaOH in water at 25 °C?

10. What is the pH of each of the solutions below?

   a. A solution in which $[H^+] = 3.0 \times 10^{-3}$ M
   b. A solution in which $[H^+] = 5.2 \times 10^{-6}$ M
   c. A solution in which $[OH^-] = 1.4 \times 10^{-1}$ M

11. What is the pOH of each solution below?

   a. $[OH^-] = 4.89 \times 10^{-10}$ M
   b. $[OH^-] = 3.2 \times 10^{-5}$ M
   c. $[H^+] = 1.6 \times 10^{-8}$ M

12. For any solution, what is the relationship between pH and pOH?

13. The pH of lemon juice is 2.1. What is the pOH?

14. The pOH of black coffee is 9.0. What is the pH?

15. Calculate the $[H^+]$ or $[OH^-]$ for the solutions below.

   a. Milk has a pH of 6.9. What is the $[H^+]$?
   b. Oven cleaner has a pH of 13.4. What is the $[OH^-]$?
   c. Some phosphate-containing detergents have a pOH of 4.7. What is the $[OH^-]$?

16. What is the pH of each of the following solutions?

   a. 0.02 M HCl
   b. 3.5×10^{-3} M HNO_3

17. HCl is added to a solution containing H_2CO_3 and $NaHCO_3$.

   a. Use an equation to show what would happen to the hydrogen ions from the HCl.
   b. Why would the pH of this solution not change drastically when the HCl is added?

ANSWERS TO LEARNING REVIEW

1. Arrhenius's model of acids and bases proposes that acids produce hydrogen ions in aqueous solution, while bases produce hydroxide ions. The Brönsted-Lowry model of acids and bases proposes that acids are proton donors, while bases are proton acceptors.

2. a. For the reaction of perbromic acid with water, perbromic acid/perbromate ($HBrO_4/BrO_4^-$) is the acid-conjugate base pair.

   b. Water/hydronium ion (H_2O/H_3O^+) is the base-conjugate acid pair.

3. One model of an acid postulates that acids donate protons in aqueous solutions. One proton acceptor is the base, water. In aqueous solutions, acids donate protons to the base water to form the hydronium ion and the conjugate base.

a. $H_2S(aq) + H_2O(l) \rightarrow HS^-(aq) + H_3O^+(aq)$

b. $HNO_2(aq) + H_2O(l) \rightarrow NO_2^-(aq) + H_3O^+(aq)$

4.

$$H{-}\ddot{O}{:}(H) + HC_2H_3O_2 \longrightarrow C_2H_3O_2^- \left[H{-}\ddot{O}(H){-}H\right]^+$$

5. a. When formic acid is mixed with water, a weak electric current is generated. An electrical current requires ions. We know that only a few ions are in this solution because the current is weak. So formic acid is a weak acid.

b. Because there are only a few formate and hydronium ions formed, the equilibrium lies toward the left, that is, toward undissociated formic acid.

c. A base can be defined as a proton acceptor. The formate ion has a stronger attraction for hydrogen ions than water does because most of the formic acid is undissociated. The formate ion has pulled the hydrogen ion away from the hydronium ion so $HCOO^-$ is a stronger base than water is.

6. The common strong acids are HCl, HNO_3, H_2SO_4 and $HClO_4$. So, b, H_2SO_4, and d, $HClO_4$, are strong acids. HF and $HC_2H_3O_2$ are weak acids.

7. The concentration of liquid water does not appear in the expression for the dissociation of water because the concentration of water changes very little when dissociation occurs. The concentration of water is considered a constant.

8. a. The product of $[H^+][OH^-]$ is always equal to 1.0×10^{-14}.

$$K_w = 1.0 \times 10^{-14} = [H^+][OH^-]$$

When we are given either $[H^+]$ or $[OH^-]$, we can calculate the other if we remember that K_w is equal to 1.0×10^{-14}. In this problem, we know that $[OH^-] = 1.5 \times 10^{-6}$ M and we want to know $[H^+]$. Rearrange the equation to isolate $[H^+]$ on one side by dividing both sides by $[OH^-]$.

$$K_w = [H^+][OH^-]$$

Divide both sides by $[OH^-]$.

$$\frac{K_w}{[OH^-]} = \frac{[H^+][\not{O}H^-]}{[\not{O}H^-]}$$

$$[H^+] = \frac{K_w}{[OH^-]}$$

Substitute values into the equation.

$$[H^+] = \frac{1.0 \times 10^{-14}}{1.5 \times 10^{-6}}$$

$$[H^+] = 6.7 \times 10^{-9}\ M$$

b. We are given $[OH^-]$ and asked for $[H^+]$. We can use the K_w expression to find $[H^+]$. Rearrange the equation to isolate $[H^+]$ on one side.

$$K_w = [H^+][OH^-]$$

$$\frac{K_w}{[OH^-]} = \frac{[H^+][\not{O}H^-]}{[\not{O}H^-]}$$

$$[H^+] = \frac{K_w}{[OH^-]}$$

Substitute values into the equation.

$$[H^+] = \frac{1.0 \times 10^{-14}}{6.3 \times 10^{-3}}$$

$$[H^+] = 1.6 \times 10^{-12}\ M$$

c. We are given $[H^+]$ are asked for $[OH^-]$. Use the K_w expression to find $[OH^-]$. Rearrange the K_w expression to isolate $[OH^-]$ on one side by dividing both sides by $[H^+]$.

$$K_w = [H^+][OH^-]$$

$$\frac{K_w}{[H^+]} = \frac{[\not{H}^+][OH^-]}{[\not{H}^+]}$$

$$[OH^-] = \frac{K_w}{[H^+]}$$

Now, substitute values into the equation.

$$[OH^-] = \frac{1.0 \times 10^{-14}}{3.25 \times 10^{-1}}$$

$$[OH^-] = 3.08 \times 10^{-14}\ M$$

9. When NaOH dissolves in water, each mole of NaOH forms a mole of Na^+ ions and a mole of OH^- ions. A solution which is 0.1 M NaOH is also 0.1 M OH^-. We can use the K_w expression to find $[H^+]$.

$$K_w = [H^+][OH^-]$$

Rearrange the equation to isolate $[H^+]$ on one side.

$$\frac{K_w}{[OH^-]} = \frac{[H^+][\not{O}H^-]}{[\not{O}H^-]}$$

$$[H^+] = \frac{K_w}{[OH^-]}$$

Substitute values into the equation and find $[H^+]$.

$$[H^+] = \frac{1.0 \times 10^{-14}}{0.1}$$

$$[H^+] = 1 \times 10^{-13}\ M$$

The amount of H^+ is very small compared with the amount of OH^-.

10. The pH scale is a way to write very small numbers in a more convenient form. pH is equal to the $-\log[H^+]$.

    a. To find the pH of a solution in which $[H^+]$ equals 3.0×10^{-3} M, first find the log of 3.0×10^{-3}. Enter 3.0×10^{-3} on your calculator and press the log button.

$$\log(3.0 \times 10^{-3}) = -2.52$$

Now press the ± key to multiply -2.52 times -1. The result is 2.52. So the pH of a solution which contains 3.0×10^{-3} M H^+ is 2.52.

b. Enter 5.2×10^{-6} on your calculator and press the log key to find the log of 5.2×10^{-6}.

$$\log 5.2 \times 10^{-6} = -5.28$$

Now press the ± key to multiply -5.28 times -1. The pH of a solution which contains 5.2×10^{-6} M is 5.28

c. We are given $[OH^-]$ and are asked for pH. We must find $[H^+]$ before we can find the pH. Use the K_w expression to find $[H^+]$.

$$K_w = [H^+][OH^-]$$

$$[H^+] = \frac{K_w}{[OH^-]}$$

$$[H^+] = \frac{1.0 \times 10^{-14}}{1.4 \times 10^{-1}}$$

$$[H^+] = 7.14 \times 10^{-14}$$

Now that we know $[H^+]$, we can find the pH. Enter 7.14×10^{-14} in your calculator and press the log key.

$$\log 7.14 \times 10^{-14} = -13.15$$

Now press the ± key to multiply -13.15 times -1. The result is 13.15. The pH of a solution which contains 1.4×10^{-1} $[OH^-]$ is 13.15.

11. a. The pOH of a solution is equal to -log $[OH^-]$. For a solution with an $[OH^-]$ of 4.89×10^{-10} M, enter 4.89×10^{-10} in your calculator and press the log key.

$$\log 4.89 \times 10^{-10} = -9.311$$

Now press the ± key to multiply -9.311 times -1. The result is 9.311. The pOH of a solution which contains 4.89×10^{-10} M OH^- is 9.311.

b. The pOH of a solution is equal to -log$[OH^-]$. For a solution with an $[OH^-]$ of 3.2×10^{-5} M, enter 3.2×10^{-5} in your calculator and press the log key.

$$\log 3.2 \times 10^{-5} = -4.49$$

Now press the ± key to multiply -4.49 times -1. The result is 4.49. The pOH of a solution which contains 3.2×10^{-5} M OH^- is 4.49.

c. To find the pOH of this solution, we must find the OH^- concentration. Use the K_w expression to find $[OH^-]$.

$$K_w = [H^+][OH^-]$$

$$[OH^-] = \frac{K_w}{[H^+]}$$

$$[OH^-] = \frac{1.0 \times 10^{-14}}{1.6 \times 10^{-8}}$$

$$[OH^-] = 6.3 \times 10^{-7} \text{ M}$$

The concentration of OH^- is 6.3×10^{-7} M. Enter 6.3×10^{-7} in your calculator and press the log key.

$$\log 6.3 \times 10^{-7} = -6.20$$

Now, multiply -6.20 times -1. The result is 6.20. The pOH of a solution which contains 1.6×10^{-8} M H^+ is 6.20.

12. For all solutions the sum of pH and pOH must equal 14.00. This relationship is derived from the K_w expression, $[H^+][OH^-] = 1.0 \times 10^{-14}$. If we know either the pH or the pOH, we can find the other using this relationship. For example, if the pH of a solution is 4.5, then the pOH is 14.00 minus 4.5, which equals 9.5.

13. If we know either the pH of a solution or the pOH, we can calculate the other. pH and pOH are related by the expression

$$pH + pOH = 14.00$$

This equation is derived from

$$[H^+][OH^-] = K_w = 1.0 \times 10^{-14}$$

If the pH of lemon juice is 2.1, then the pOH is

$$pOH = 14.00 - pH$$

$$pOH = 11.9$$

14. pOH and pH are related to each other by the equation

$$pH + pOH = 14$$

If the pOH of black coffee is 9.0, then the pH is

$$pH = 14.00 - pOH$$

$$pH = 5.0$$

15. a. We want to find $[H^+]$ of milk, and the pH of milk is 6.9. To find the pH of milk from $[H^+]$, we take minus the log of $[H^+]$, so to find $[H^+]$ from pH we must go backward and undo the log$[H^+]$. We can do this by finding the inverse log of -pH.

$$[H^+] = \text{inverse log}(-pH)$$

To do this, first enter the pH in your calculator. Then press the ± key to multiply the pH by -1. Then use the keys on your calculator which will generate the inverse log.

$$[H^+] = \text{inverse log}(-pH)$$

$$[H^+] = \text{inverse log}(-6.9)$$

$$[H^+] = 1 \times 10^{-7}\ M$$

b. We want to find the $[OH^-]$ of oven cleaner which has a pH of 13.4. pH and pOH are related to each other by the expression pH + pOH = 14.00.

$$pOH = 14.00 - pH$$

$$pOH = 14.00 - 13.4$$

$$pOH = 0.6$$

Now we need to go backward from pOH to $[OH^-]$. Enter 0.6 in your calculator and press ± to multiply 0.6 times -1. Then use the keys on your calculator to generate the inverse log.

$$[OH^-] = \text{inverse log}(-0.6)$$

$$[OH^-] = 3 \times 10^{-1}\ M$$

c. We want to know the $[OH^-]$ of a phosphate-containing detergent with a pOH of 4.7. We must go backward from pOH to $[OH^-]$. Enter 4.7 in your calculator and press the ± key to multiply the pH by -1. Then use the keys on your calculator to generate the inverse log.

$$[OH^-] = \text{inverse log}(-4.7)$$

$$[OH^-] = 2 \times 10^{-5}\ M$$

16. a. HCl is a strong acid. This means that when HCl is dissolved in water, only H^+ ions and Cl^- ions are present. If a solution is described as 0.02 M HCl, then it actually contains 0.02 M H^+ and 0.02 M Cl^-. Because the pH depends on the hydrogen ion concentration, we can calculate the pH of a solution of HCl if we know the molar concentration of HCl.

$$0.02\ \text{mol/L HCl} = 0.02\ \text{mol/L}\ H^+ \text{ and } 0.02\ \text{mol/L}\ Cl^-$$

$$pH = -\log[H^+]$$

$$pH = -\log(0.02)$$

$$pH = 1.7$$

b. 3.5×10^{-3} mol/L HNO_3 dissolved in water produces 3.5×10^{-3} mol/L H^+ and 3.5×10^{-3} mol/L NO_3^-. Use the hydrogen ion concentration to calculate the pH.

$$pH = -\log[H^+]$$

$$pH = -\log(3.5 \times 10^{-3})$$

$$pH = 2.46$$

17. a. Hydrogen ions produced when HCl is dissolved in water would react with the bicarbonate ion, HCO_3^-, which has a high affinity for hydrogen ions.

$$HCO_3^- + H^+ \longrightarrow H_2CO_3$$

b. The pH of a solution is determined by the number of hydrogen ions in a solution. In a solution which contains a mixture of $NaHCO_3$ and H_2CO_3, the $[H^+]$ remains steady. H_2CO_3 is a weak acid, so most of the acid is in the undissociated form and not much hydrogen ion is produced. But most of the $NaCO_3$ is dissociated as Na^+

and HCO_3^-. When an outside source of hydrogen ion is added to a buffered solution, the hydrogen ions are removed from the solution by reacting with the conjugate base, HCO_3^-, which has a high affinity for hydrogen ions. The hydrogen ions are removed from the solution and the pH of the solution does not change much.

PRACTICE EXAM

1. Which of the following is not a conjugate acid-base pair?

   a. HNO_3, NO_3^-
   b. H_3O^+, H_2O
   c. H_2S, HS^-
   d. $HClO_4$, H_2O
   e. H_2SO_4, HSO_4^-

2. Which statement about strong acids is true?

   a. For strong acids, the dissociation equilibrium lies far to the left.
   b. The amount of undissociated strong acid is the same before and after equilibrium with water.
   c. When a strong acid reacts with water, it forms a strong conjugate base.
   d. Strong acids are completely dissociated in aqueous solution.
   e. Strong acids are weak electrolytes.

3. What is the $[OH^-]$ if $[H^+] = 1.8 \times 10^{-6}$ M? Is this solution acidic, neutral, or basic?

   a. 1.8×10^{-8} M, basic
   b. 8.0×10^{-13} M, basic
   c. 8.0×10^{-3} M, acidic
   d. 5.6×10^{-9} M, acidic
   e. 5.6×10^{-9} M, basic

4. What is the pH of a solution in which $[OH^-] = 6.8 \times 10^{-8}$ M?

   a. 8.83
   b. 6.83
   c. 5.17
   d. 7.17
   e. 5.48

5. What is the $[H^+]$ of a solution in which pOH = 7.9?

a. 1×10^{-8} M
b. 1×10^{-14} M
c. 9×10^{-1} M
d. 8×10^{-1} M
e. 8×10^{-7} M

6. What is the pH of a 0.0253 M solution of the strong acid $HClO_4$?

a. 0.600
b. 1.597
c. 1.145
d. 12.403
e. 13.975

7. Which of the following pairs would make a buffer?

a. HCOOH, $HCOO^-$
b. H_2SO_4, SO_4^{2-}
c. $C_2H_2O_2^-$, H^+
d. NO_3^-, H^+
e. Na^+, $C_2H_2O_2^-$

8. Which is not a characteristic of a buffer?

a. Any added OH^- reacts with the weak acid.
b. Any added H^+ reacts with the conjugate base of the weak acid.
c. Added H^+ or OH^- does not accumulate in solution.
d. A buffer contains both a weak acid and its conjugate base.
e. Added H^+ reacts only with H_2O to form hydronium ions.

9. Which statement about acids is not true?

a. Organic acids contain the carboxyl group.
b. When sulfuric acid dissociates, the anion, HSO_4^-, is a weak acid.
c. Diprotic acids furnish 2 hydrogen ions for every molecule of acid.
d. Hydrofluoric acid, HF, is an oxyacid.
e. Most organic acids are weak acids.

10. Which statement about pH is not true?

a. A solution with a pH of 8 has a higher $[OH^-]$ than a solution with a pH of 3.
b. Each change of 1 pH unit represents a ten fold change in $[H^+]$.
c. pH can be measured in the laboratory with an instrument called a pH meter.
d. As the pH of a solution increases, the hydrogen ion concentration increases.
e. A solution with a pH of 7.0 is neither acidic nor basic.

PRACTICE EXAM ANSWERS

1. d (17.1)
2. d (17.2)
3. d (17.3)
4. b (17.4)
5. e (17.4)
6. b (17.5)
7. a (17.6)
8. e (17.6)
9. d (17.2)
10. d (17.4)

CHAPTER 18: OXIDATION-REDUCTION REACTIONS AND ELECTROCHEMISTRY

INTRODUCTION

There are many important oxidation-reduction reactions. These reactions are characterized by electron transfer. One of the most annoying and costly of the oxidation-reduction reactions is the rusting of automobile bodies. In this chapter you will learn what actually happens during an oxidation-reduction reaction, and what means are available to keep your car from rusting.

AIMS FOR THIS CHAPTER

1. Define oxidation and reduction, and identify oxidation-reduction reactions between a metal and a nonmetal. (Section 18.1)
2. Be able to assign oxidation states to all atoms in molecules and in ions. (Section 18.2)
3. Be able to determine which elements are oxidized and which are reduced, and which species is the oxidizing agent and which is the reducing agent. (Section 18.3)
4. Be able to balance a redox reaction by the half-reaction method. (Section 18.4)
5. Know how a redox reaction can be separated into half reactions for the purpose of generating electrical current. (Section 18.5)
6. Understand how the lead storage battery works, and how common dry cell batteries work. (Section 18.6)
7. Understand how corrosion occurs, and what methods are available to combat corrosion. (Section 18.7)
8. Understand how electrolysis is used to produce a chemical reaction which does not occur naturally. (Section 18.8)

QUICK DEFINITIONS

Oxidation-reduction reactions — Also called redox reactions, they are reactions where one species loses electrons, and another species gains electrons. (Section 18.1)

Oxidation — The loss of electrons. Oxidation is loss (OIL). Also defined as an increase in oxidation state. (Section 18.1)

Reduction — The gain of electrons. Reduction is gain (RIG). Also defined as a decrease in oxidation state. (Section 18.1)

Oxidation states — An imaginary assignment of electrons in a molecule or ion to the most electronegative atom to help determine where electrons in a redox reaction are lost or gained. (Section 18.2)

Oxidizing agent
The species in a redox reaction which is reduced, the electron acceptor. (Section 18.3)

Reducing agent
The species in a redox reaction which is oxidized, the electron donor. (Section 18.3)

Half-reaction
An individual reduction or oxidation taken from a complete equation. Electrons are shown as either reactants or products. (Section 18.4)

Electrochemistry
The study of the methods for converting chemical energy to electrical energy. (Section 18.5)

Galvanic cell
Also called an electrochemical battery. A device which produces electrical current by separating the oxidation and the reduction half-reaction and forcing the electrons to move between the two by a wire. (Section 18.5)

Anode
The electrode in a galvanic cell where oxidation (loss of electrons) occurs. (Section 18.5)

Cathode
The electrode in a galvanic cell where reduction (gain of electrons) occurs. (Section 18.5)

Electrolysis
The process of adding electrical energy to a cell to cause a chemical reaction which by itself would not occur. (Sections 18.5 and 18.8)

Lead storage battery
The automobile battery which depends upon the oxidation of Pb and the reduction of PbO_2 to generate energy. (Section 18.6)

Potential
The "pressure" on electrons to flow from the anode to the cathode. (Section 18.6)

Dry cell batteries
Small efficient batteries which do not use a liquid electrolyte. (Section 18.6)

Corrosion
The oxidation of a metal. This term is usually associated with the process which converts metals from a useful form, such as solid iron, to a non-useful form, such as rust. (Section 18.7)

Cathodic protection A method of protecting metals from corrosion. A metal more easily oxidized than the metal you are protecting is attached with an insulated wire to the protected metal. The more easily oxidized metal is oxidized, leaving the protected metal intact. (Section 18.7)

CONTENT REVIEW

18.1 OXIDATION-REDUCTION REACTIONS

How Does an Oxidation-Reduction Reaction Occur Between a Metal and a Nonmetal?

Reactions between metals and nonmetals involve the transfer of electrons. When the metal Li reacts with Br_2, the result is the ionic compound LiBr.

$$2Li + Br_2 \rightarrow 2LiBr$$

Both Li and Br_2 began as neutral species, but after the reaction, both were ions. Li became the Li^+ cation, and Br_2 became the Br^- anion. Electrons must have been transferred between lithium and bromine for the reaction to have occurred. Reactions of this type are called **oxidation-reduction reactions**, or often **redox reactions**. In this reaction, Li has lost an electron to become the Li^+ cation. This process is called **oxidation**. Each Br atom has gained an electron to form Br^-. This process is called **reduction**. In every oxidation-reduction reaction, one species is oxidized, and another is reduced.

18.2 OXIDATION STATES

How Are Oxidation States Useful, and How Are They Determined?

In some redox reactions it is not easy to tell which species has been oxidized and which has been reduced. The assignment of **oxidation states** or **oxidation numbers** can help you determine which species is oxidized and which reduced, and whether a reaction is really a redox reaction or not. An oxidation state is an imaginary number assigned to each element in a chemical reaction. The number comes from the charge that element would have if it were an ion. How the electrons are assigned to the atoms is governed by a set of rules, but is basically determined by the electronegativity of the atom. Some species are easy to assign oxidation states to. All of the metals which form cations with a 1+ charge have oxidation states of 1+. So the oxidation state of K^+ would be 1+. Many atoms in chemical reactions are not ions, but are covalently bonded to other atoms; that is, the electrons are shared between two atoms. Assign the electrons as though the atoms were ions. The most electronegative atom is assigned both the shared electrons. Because molecules are electrically neutral, the sum of the oxidation states of all the atoms must be zero.

Example:

The oxidation states of N and H in ammonia, NH_3, can be determined by assigning each of the pairs of shared electrons to the most electronegative atom, N. One of the rules says that hydrogen almost always has a 1+ oxidation state, so each hydrogen in the ammonia molecule is 1+. Because the molecule must be electrically neutral, nitrogen has a 3- oxidation state to balance the 3+ from the three hydrogen atoms. All the rules for determining oxidation state are presented in detail in Section 18.2, but are summarized below.

Rules for Assigning Oxidation Numbers

1. Uncombined elements have oxidation states of 0.
2. All monatomic ions have the same oxidation state as their charge.
3. Oxygen has an oxidation state of 2- except when combined as a peroxide.
4. Hydrogen usually has an oxidation state of 1+, except when combined with a metal.
5. In compounds between two elements, assign the shared electrons to the most electronegative atom. The charge on the electronegative atom will be equal to the charge on the ion, when that atom forms ions.
6. The sum of all oxidation states in a molecule must be zero.
7. The sum of all oxidation states in an ion must be equal to the charge on the ion.

18.3 OXIDATION-REDUCTION REACTIONS BETWEEN NONMETALS

How Can We Tell Which Atoms Are Oxidized or Reduced in Covalent Compounds?

In reactions between metals and nonmetals, it is often easy to determine which atoms are oxidized and which are reduced, because the reactants are often pure elements which have oxidation states of 0. The oxidation states of the ions formed in the reaction are the same as the charges on the ions. But in the case of reactions between nonmetals, the oxidation states of the reactants are often not zero, and the compounds are often covalently bonded. It is not always possible to tell by inspection what is oxidized and what is reduced. By assigning oxidation states to each atom in the reaction, and comparing the oxidation states on each side of the equation, it is possible to decide which element is oxidized and which is reduced. When the oxidation state increases, an element has been oxidized, and when the oxidation state decreases, the element has been reduced.

Example:

The net ionic equation for a reaction between $Na_2S_2O_8$ and NaI is

$$S_2O_8^{2-} + 2I^- \rightarrow I_2 + 2SO_4^{2-}$$

We can determine which element is oxidized and which is reduced by assigning oxidation numbers to each of the atoms. On the left side of the equation the I^- ion has an oxidation state of 1-, the same as the charge on the ion. In the $S_2O_8^{-2}$ ion, the persulfate ion, oxygen has an oxidation state of 2-. There are 8 oxygens, so the total number for oxygens is 16-. There are two sulfur atoms in the ion; each one of them must have an oxidation state of 7+ because two times the oxidation state of sulfur plus eight times the oxidation state of oxygen equals the charge on the thiosulfate ion, which is 2-. There is a net 2- charge on the ion, so the sum of all oxidation numbers must equal 2-. On the right side of the equation, I_2 has an oxidation number of 0. Oxygen in the sulfate ion is equal to 2-. There are four oxygens, for a total number of 8-. The oxidation state of sulfur must therefore be 6+ because four times the oxidation state of oxygen plus the oxidation state of sulfur equals the charge on the sulfate ion, which is 2-.

$$S_2O_8^{2-} + 2I^- \rightarrow I_2 + 2SO_4^{2-}$$

7+ each (S), 2- each (O), 1- each (I⁻), 0 each (I₂), 6+ (S), 2- each (O)

In this reaction, iodine increases in oxidation state, from 1- to 0. It is oxidized. Sulfur decreases in oxidation state, from 7+ to 6+. It is reduced. We say that the species which is oxidized is the **reducing agent**, because it causes the reduction of another species. I^- is oxidized so I^- is the reducing agent. The species which is reduced is called the **oxidizing agent** because it causes the oxidation of another species. $S_2O_8^{2-}$ is the oxidizing agent.

18.4 BALANCING OXIDATION-REDUCTION REACTIONS BY THE HALF-REACTION METHOD

How Can Redox Reactions Be Balanced?

In Chapter 6 you learned how to balance simple chemical reactions by inspection. Balancing redox reactions is more difficult, and can rarely be done by inspection. Another method is needed for balancing these reactions. One method for balancing redox reactions is called the half-reaction method. A half-reaction is part of a complete chemical equation. In a redox reaction, there is always a reduction half-reaction, and an oxidation half-reaction. We can write each of them separately, and balance the difference in oxidation states on each side of the equation by adding electrons to either the right or the left side. Your textbook gives on page 599 some general steps to be used when balancing redox reactions, and five specific steps to use when balancing redox reactions which take place in acidic solution. We will use the five specific steps to balance the reaction below.

Example:

Balance the redox reaction below which takes place in acidic solution.

$$Au + Cl^- + NO_3^- \rightarrow AuCl_4^- + NO_2$$

Step 1: Write the equations for the oxidation and reduction half-reactions.

First, decide which element is oxidized and which is reduced by assigning oxidation states to each element. In this reaction, Au is oxidized. It increases in oxidation state, from 0 to 3+. Cl^- participates in the reaction, but does not change in oxidation state. Nitrogen is reduced. It decreases in oxidation state, from 5+ to 4+.

$$\underset{0}{Au} + \underset{1-}{Cl^-} + \underset{5+\ 2-}{NO_3^-} \rightarrow \underset{3+\ 1-}{AuCl_4^-} + \underset{4+\ 2-}{NO_2}$$

We can now write the half-reactions.

$$Au + Cl^- \rightarrow AuCl_4^- \quad \text{oxidation half-reaction}$$

$$NO_3^- \rightarrow NO_2 \quad \text{reduction half-reaction}$$

Step 2a: Balance all the elements except hydrogen and oxygen.

There are four chlorine atoms on the right, so we must add four on the left. The unbalanced equation shows that the chlorine atoms are present as chloride ions.

$$Au + 4Cl^- \rightarrow AuCl_4^-$$

The reduction half-reaction has all atoms except hydrogen and oxygen balanced.

Step 2b: Balance oxygen using H_2O.

The oxidation half-reaction contains no oxygen atoms. Add a water molecule to the right side of the reduction half-reaction to balance the three oxygen atoms on the left.

$$NO_3^- \rightarrow NO_2 + H_2O$$

Step 2c: Balance hydrogen using H^+.

The oxidation half-reaction contains no hydrogen atoms. Add two hydrogen atoms to the left side of the reduction half-reaction to balance the two hydrogens on the right.

$$2H^+ + NO_3^- \rightarrow NO_2 + H_2O$$

Step 2d: Balance the charge using electrons.
In the oxidation half-reaction the total charge on the left is 4- and on the right is 1-, so add three electrons to the right.

$$Au + 4Cl^- \longrightarrow AuCl_4^-$$

charge 4- $\longrightarrow$ 1-

$$Au + 4Cl^- \longrightarrow AuCl_4^- + 3e^-$$

In the reduction half-reaction the total charge on the left is 1+ and on the right is 0, so add an electron to the left side.

$$2H^+ + NO_3^- \longrightarrow NO_2 + H_2O$$

charge 1+ $\longrightarrow$ 0

$$e^- + 2H^+ + NO^- \longrightarrow NO_2 + H_2O$$

Step 3: Equalize the number of electrons transferred in the oxidation and reduction half-reactions.
The oxidation half-reaction transfers three electrons, but the reduction half-reaction transfers only one. Multiply the reduction half-reaction by three to equalize the number of electrons transferred.

$$3(e^- + 2H^+ + NO^- \longrightarrow NO_2 + H_2O)$$
$$3e^- + 6H^+ + 3NO_3^- \longrightarrow 3NO_2 + 3H_2O$$

Step 4. Add the half-reaction and cancel identical species which appear on both sides.

$$Au + 4Cl^- \longrightarrow AuCl_4^- + 3e^-$$
$$3e^- + 6H^+ + 3NO_3^- \longrightarrow 3NO_2 + 3H_2O$$

$$3\cancel{e^-} + 6H^+ + 3NO_3^- + Au + 4Cl^- \longrightarrow 3NO_2 + 3H_2O + AuCl_4^- + 3e^-$$

We can cancel the three electrons which appear on both sides, so the equation is

$$6H^+(aq) + 3NO_3^-(aq) + Au(s) + 4Cl^-(aq) \longrightarrow 3NO_2(g) + 3H_2O(l) + AuCl_4^-(aq)$$

Step 5. Check to be sure the elements and the charges balance.
Each side has three nitrogens, nine oxygens, one gold, and four chlorine atoms. The charge on each side is the same, 1-, so the equation is balanced.

	$3NO_3^- + 6H^+ + Au + 4Cl^- \rightarrow 3NO_2 + 3H_2O + AuCl_4^-$
elements	3N 9O 6H 1Au 4Cl $\rightarrow$ 3N 9O 6H 1Au 4Cl
charge	1- 1-

18.5 ELECTROCHEMISTRY--AN INTRODUCTION

Chemical and electrical energy can be interchanged. The study of the interchange of these two forms of energy is called **electrochemistry**. The interchange of energy forms can occur in either direction. You can convert chemical energy to electrical energy, or you can use an electrical current to produce a chemical reaction. Let's first consider how a chemical reaction can be used to produce electrical energy.

How Can a Chemical Reaction Produce an Electrical Current?

During a redox reaction, electrons are transferred whenever the reactants collide in solution. It is not possible to use the electron transfer to generate electrical energy under these circumstances. In order to harness the energy of a redox reaction, it is necessary to physically separate the two half-reactions in two separate containers which are connected by a wire. The electrons which are transferred between the oxidation half-reaction and the reduction half-reaction travel along the wire, producing an electrical current.

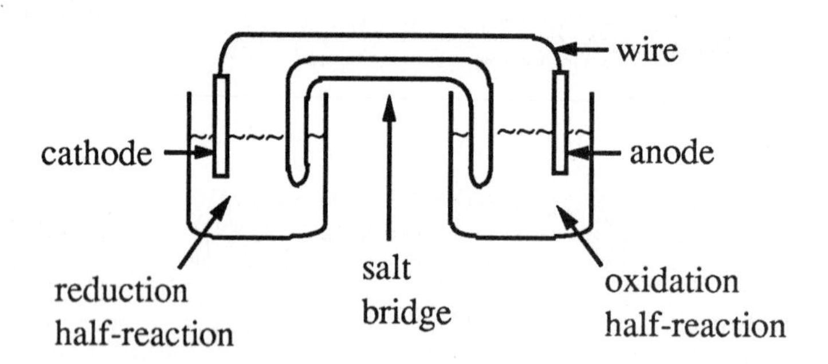

A reaction between two half-cells connected only by a wire will not happen unless there is another connection between the two containers which allows ions to flow freely back and forth. As electrons leave one container, and travel along the wire to the other container, differences in charges in the two containers would occur. The container with the oxidation half-reaction would build up a positive charge from the loss of electrons. The container with the reduction half-reaction would build up a negative charge from the gain of electrons. The extra connection which contains ions allows negative ions to travel to the container losing electrons, and positive ions to travel to the container gaining electrons, so the net charge in each container is zero. This connection is called a salt bridge. The current which is produced in a cell such as this can be used to do useful work, and is the principle upon which batteries are made. Cells powered by two separate half-reactions connected by a wire and by some type of connection to allow ion exchange are called **galvanic cells**. The electrode where electrons are lost is called the **anode**, and the electrode where electrons are gained is called the **cathode**.

18.6 BATTERIES

What Is a Battery?

A **battery** is a galvanic cell. Different batteries use different chemical reactions to generate a current.

How Does the Lead Storage Battery Work?

The **lead storage battery** is responsible for generating the current which starts automobiles. The half-reactions and the overall reaction of the lead storage battery are shown below.

$$Pb + H_2SO_4 \rightarrow PbSO_4 + 2H^+ + 2e^- \quad \text{oxidation}$$

$$PbO_2 + H_2SO_4 + 2e^- + 2H^+ \rightarrow PbSO_4 + 2H_2O \quad \text{reduction}$$

$$Pb(s) + PbO_2(s) + 2H_2SO_4(aq) \rightarrow 2PbSO_4(s) + 2H_2O(l) \quad \text{overall}$$

The solid Pb is cast into a metal grid, for easy contact with the liquid H_2SO_4, and the solid PbO_2 is coated onto a Pb grid, also for contact with H_2SO_4. Notice that H_2SO_4 participates in both the oxidation and the reduction half-reactions. Solid Pb, PbO_2 and H_2SO_4 are all used up in the reaction to generate a current.

What Are Dry Cell Batteries?

A dry cell battery is one which contains no liquid electrolytes. The components are either solids, or moist pastes. Dry cells are produced in both an acidic and a basic (alkaline) version. The half-reactions in both these cells are complex, but are presented below.

Acidic

$$Zn \rightarrow Zn^{2+} + 2e^- \quad \text{oxidation half-reaction}$$

$$2NH_4^+ + 2MnO_2 + 2e^- \rightarrow Mn_2O_3 + 2NH_3 + H_2O \quad \text{reduction half-reaction}$$

Basic

$$Zn + 2OH^- \rightarrow ZnO(s) + H_2O + 2e^- \quad \text{oxidation half-reaction}$$

$$2MnO_2 + H_2O + 2e^- \rightarrow Mn_2O_3 + 2OH^- \quad \text{reduction half-reaction}$$

18.7 CORROSION

What Is Corrosion and How Can It Be Prevented?

Corrosion is the oxidation of metals. The most common example is the rusting of iron in the presence of O_2. Oxidation of metals often results in weakening of the structural properties of the metals. Sometimes the metal oxide forms a coating on the outside of the metal which can protect the metal against further oxidation. This is true in the case of aluminum. The metal oxide coating of iron flakes off, exposing fresh metal to the rusting process. So rusted iron is not protected against further corrosion.

Corrosion of iron can be prevented by producing stainless steel, an alloy which contains chromium and nickel. The added metals form tough metal oxide coats which do not flake off. Metals can also be painted with protective paint, and surfaces can be plated with another metal which forms a stable oxide coat.

Another method of protecting iron from corrosion is cathodic protection. **Cathodic protection** can be achieved by attaching a piece of metal, such as magnesium, which oxidizes more readily than iron does, to the piece of iron. The magnesium is oxidized, leaving the iron intact.

18.8 ELECTROLYSIS

What Is Electrolysis?

Automobile lead storage batteries last several years. The reason we can reuse the battery to start our cars many times is that the battery is recharged while we drive. When we apply an electric current through the alternater of the car, we force the reaction to produce Pb, PbO_2 and H_2SO_4. This is just the opposite of the reaction which produces a current to start our cars. This is an example of **electrolysis**, the use of electrical energy to drive a chemical reaction that would not otherwise occur. By charging the battery, we force the reaction toward the left, and we keep the amount of Pb, PbO_2 and H_2SO_4 high enough so that the car will start when we want it to.

How Can Pure Aluminum Be Produced By Electrolysis?

Aluminum is found in nature as aluminum oxide. Production of aluminum from its ore proved to be difficult and expensive, because the oxide is a very stable molecule. For this reason, aluminum utensils were very expensive until a better method for producing pure aluminum was discovered. Pure aluminum can be produced from aluminum oxide by electrolysis. In the process, Al^{3+} gains three electrons to become Al.

LEARNING REVIEW

1. For each of the partial reactions, decide whether oxidation or reduction is occurring.

   a. $Li \rightarrow Li^+ + e^-$
   b. $Br_2 + 2e^- \rightarrow 2Br^-$
   c. $S^{2-} \rightarrow S + 2e^-$

2. For each reaction below, identify which element is oxidized and which is reduced.

   a. $Ca(s) + I_2(g) \rightarrow CaI_2(s)$
   b. $2K(s) + S(s) \rightarrow K_2S(s)$
   c. $6Na(s) + N_2(g) \rightarrow 2Na_3N(s)$

3. Determine the oxidation states for each element in the substances below.

   a. CH_4
   b. SO_4^{2-}
   c. $NaHCO_3$
   d. N_2O_5
   e. HIO_4

4. Determine the oxidation state for each element in the reactions below.

   a. $4Fe(s) + 3O_2(g) + 12HCl(aq) \rightarrow 4FeCl_3(aq) + 6H_2O(l)$
   b. $Zn(s) + 2AgNO_3(aq) \rightarrow Zn(NO_3)_2(aq) + 2Ag(s)$
   c. $MgCl_2(l) \rightarrow Mg(s) + Cl_2(g)$

5. During a redox reaction, does the reactant which is the reducing agent contain an element which is oxidized or reduced?

6. For each of the reactions below identify which atom is oxidized and which is reduced, and identify the oxidizing and reducing agents.

   a. $2C_2H_6(g) + 7O_2(g) \rightarrow 4CO_2(g) + 6H_2O(g)$
   b. $2KNO_3(l) \rightarrow 2KNO_2(l) + O_2(g)$
   c. $3CuO(s) + 2NH_3(aq) \rightarrow 3Cu(s) + N_2(g) + 3H_2O(l)$
   d. $K_2Cr_2O_7(aq) + 14HI(aq) \rightarrow 2CrI_3(s) + 2KI(aq) + 3I_2(s) + 7H_2O(l)$

7. Balance each of the reactions below by the half-reaction method.

a. $Zn(s) + Cu^{2+}(aq) \rightarrow Zn^{2+}(aq) + Cu(s)$
b. $Re^{5+}(aq) + Sb^{3+}(aq) \rightarrow Re^{4+}(aq) + Sb^{5+}(aq)$

8. Each of the reactions below occurs in acidic solution. Balance each one by the half-reaction method.

a. $H_2S(aq) + NO_3^-(aq) \rightarrow S(s) + NO(g)$
b. $H_5IO_6(aq) + I^-(aq) \rightarrow I_2(s)$
c. $Cr_2O_7^{2-}(aq) + Sn^{2+}(aq) \rightarrow Sn^{4+}(aq) + Cr^{3+}(aq)$
d. $I_2(s) + NO_3^-(aq) \rightarrow IO_3^-(aq) + NO_2(g)$
e. $Mn^{2+}(aq) + BiO_3(s) \rightarrow MnO_4^-(aq) + Bi^{3+}(aq)$

9. Normally, when a redox reaction occurs, no useful work is produced. How can a redox reaction be made to perform useful work?

10. Label what is needed to complete the electrical circuit and allow the redox reaction to proceed.

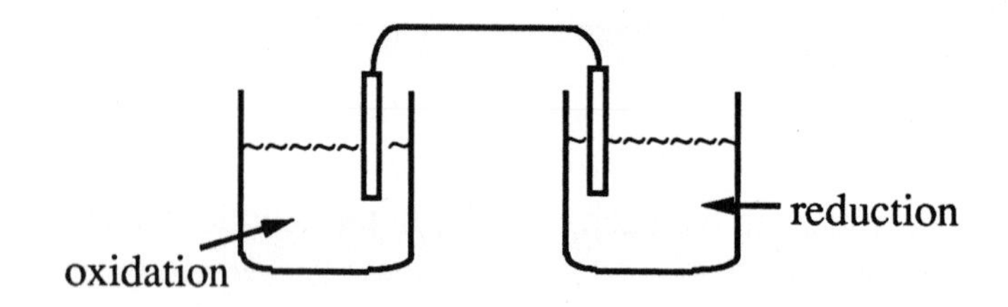

11. Based on the direction of electron flow, in which container does oxidation occur?

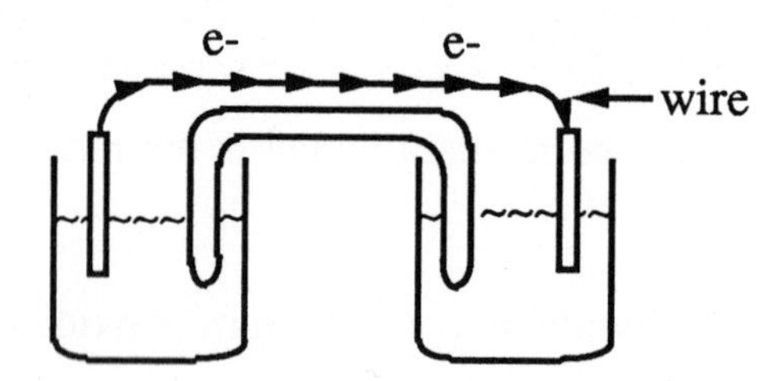

12. Briefly explain how the lead storage battery works.

13. What major differences are there between an acid dry cell battery and an alkaline dry cell?

14. List two methods for preventing the corrosion of metals.

15. Aluminum metal easily loses electrons to form Al_2O_3. How can aluminum metal be produced from its oxide?

ANSWERS TO LEARNING REVIEW

1. a. Lithium metal loses an electron. This is oxidation.
   b. Bromine gains an electron. This is reduction.
   c. The sulfide ion loses two electrons. This is oxidation.

2. a. Calcium is a metal from Group 2 and forms Ca^{2+} cations. Calcium metal loses two electrons. This is oxidation. Halogens such as I_2 form anions. Each atom in a molecule of I_2 gains one electron to form $2I^-$. This is reduction.
   b. Potassium metal from Group 1 forms K^+ cations. Potassium loses one electron so this is oxidation. Sulfur from Group 6 forms the S^{2-} anion. Sulfur gains two electrons, so it is reduced.
   c. Sodium metal forms Na^+ cations. Sodium loses electrons so it is oxidized. Nitrogen from Group 5 forms the N^{3-} anion. Each nitrogen atom gains three electrons, so it is reduced.

3. If you have trouble assigning oxidation states, look at the rules which are found in section 18.2 of your textbook and in this Study Guide.

   a. Rule 4 says that hydrogen bonded to a nonmetal such as carbon will have an oxidation state of 1+. There are four hydrogen atoms, so carbon must be 4- so that the sum of the charges is 0.

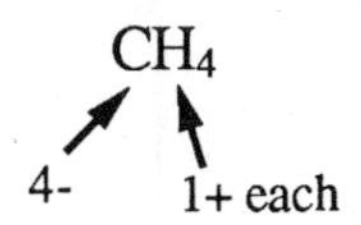

   b. Rule 3 says that oxygen is usually 2-. There are four oxygens, so sulfur must be 6+. The sum of the charges must be 2- because the charge on the sulfate ion is 2-.

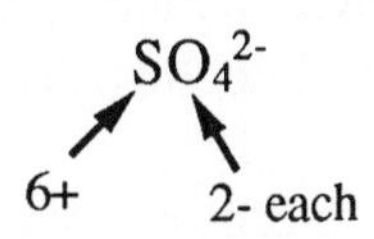

   c. Rule 2 says that the charge on Group 1 monatomic ions is 1+ so Na^+ is 1+. Rule 4 says that hydrogen is 1+ when covalently bonded to nonmetals. In this molecule, the hydrogen atom is covalently bonded to the CO_3 part of the molecule, so the oxidation state of hydrogen is 1+. Rule 3 says that oxygen is usually 2-. There are three of them. Carbon must be 6- (1 + 1) which is 4+.

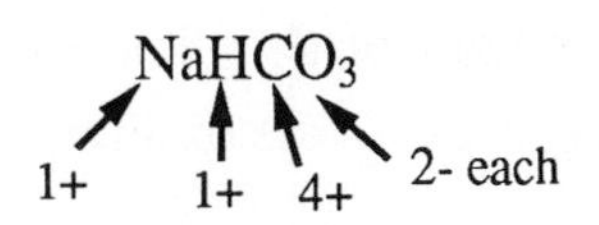

d. Rule 3 says that oxygen is usually 2-. There are five oxygen atoms. There are two nitrogen atoms, so each nitrogen atom must be 5+ to counterbalance the five oxygens.

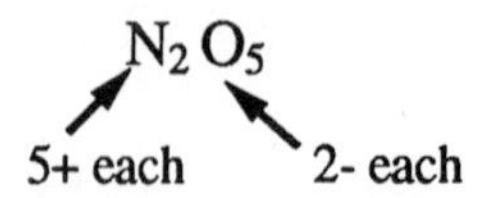

e. Rule 4 says that hydrogen is usually 1+. Rule 3 says that oxygen is usually 2-. If hydrogen is plus one and the four oxygen atoms are 2- each, then iodine must be 7+ so that the sum is zero.

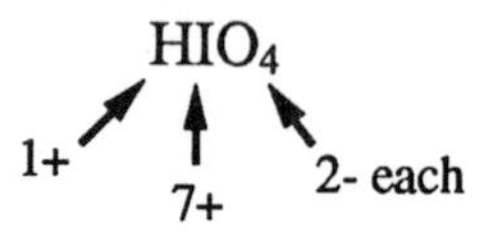

4. a.

$$4Fe(s) + 3O_2(g) + 12HCl(aq) \rightarrow 4FeCl_3(aq) + 6H_2O(l)$$

Fe: 0; O_2: 0; HCl: 1+ 1-; $FeCl_3$: 3+ 1- each; H_2O: 1+ each 2-

b.

$$Zn + 2AgNO_3 \rightarrow Zn(NO_3)_2 + 2Ag$$

Zn: 0; $AgNO_3$: 1+ 5+ 2- each; $Zn(NO_3)_2$: 2+ 5+ 2- each; Ag: 0

c.

$$MgCl_2 \rightarrow Mg + Cl_2$$

$MgCl_2$: 2+ 1- each; Mg: 0; Cl_2: 0

5. The reducing agent contains an element which is oxidized. So, the element which loses electrons during oxidation furnishes the electrons needed for reduction.

6. Use changes in oxidation state to determine which element is oxidized and which is reduced. Remember that the element which increases in oxidation state is oxidized. The oxidizing agent contains the element which is reduced, and the reducing agent contains the element which is oxidized.

a. Carbon in C_2H_6 has an oxidation state of 3-, while carbon in CO_2 has an oxidation state of 4+. The oxidation state increases so carbon is oxidized. C_2H_6 is the reducing agent. Oxygen in O_2 has an oxidation state of 0, and in CO_2 and in H_2O oxygen has an oxidation state of 2-. The oxidation state of oxygen decreases so oxygen is reduced. Oxygen gas is the oxidizing agent.

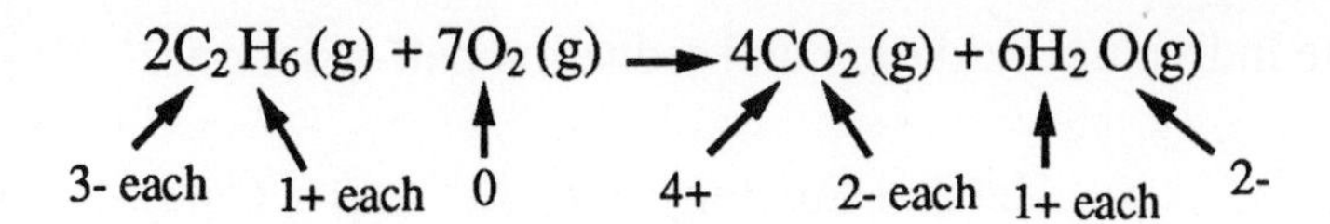

b. Nitrogen in KNO_3 has an oxidation state of 5+, while nitrogen in KNO_2 has an oxidation state of 3+. The oxidation state decreases from 5+ to 3+ so nitrogen is reduced. KNO_3 is the oxidizing agent. Oxygen in KNO_3 has an oxidation state of 2-, while molecular oxygen has an oxidation state of 0. The oxidation state of oxygen increases, so oxygen is oxidized. KNO_3 is the reducing agent.

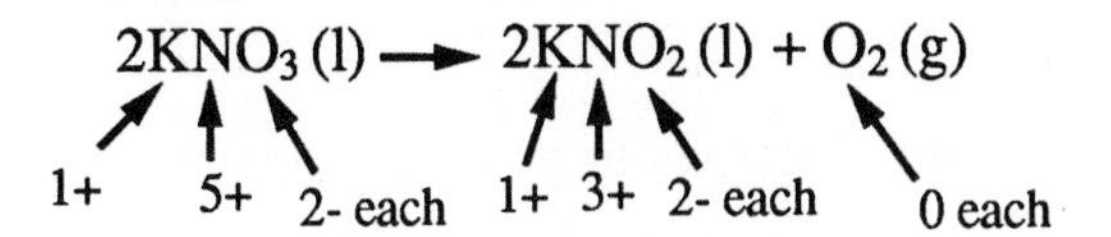

c. Copper in CuO has an oxidation state of 2+, while copper metal has an oxidation state of 0. The oxidation state decreases so copper is reduced. CuO is the oxidizing agent. Nitrogen in NH_3 has an oxidation state of 3-, while molecular nitrogen has an oxidation state of 0. The oxidation state increases, so nitrogen is oxidized. NH_3 is the reducing agent.

$$3CuO(s) + 2NH_3(l) \rightarrow 3Cu(s) + N_2(g) + 3H_2O(l)$$

2+ 2- 3- 1+ each 0 0 each 1+ each 2-

d. Chromium in $K_2Cr_2O_7$ has an oxidation state of 6+, while chromium in CrI_3 has an oxidation state of 3+. The oxidation state of chromium decreases so chromium is reduced. $K_2Cr_2O_7$ is the oxidizing agent. Iodine in HI has an oxidation state of 1- while iodine molecules have an oxidation state of 0. Note that some of the iodine does not change oxidation state. The iodine atoms in CrI_3 and KI both have oxidation states of 1-. Because the oxidation state of some of the iodine has increased, iodine is said to be oxidized. HI is the reducing agent.

$$K_2Cr_2O_7(aq) + 14HI(aq) \rightarrow 2CrI_3(s) + 3I_2(s) + 7H_2O(l) + 2KI$$

1+ each 6+ each 2- each 1+ 1- 3+ 1- each 0 each 1+ each 2- 1+ 1-

7. a. To balance redox reactions which do not occur in acid solution, follow the general steps given on page 599 of your textbook.

Write individual oxidation and reduction half-reactions.

$$Zn(s) + Cu^{2+}(aq) \rightarrow Zn^{2+}(aq) + Cu(s)$$

The oxidation state of zinc increases from 0 to 2+. Zinc is oxidized.

$$Zn \rightarrow Zn^{2+} \quad \text{oxidation half-reaction}$$

The oxidation state of copper decreases from 2^+ to 0, so copper is reduced.

$$Cu^{2+} \rightarrow Cu \quad \text{reduction half-reaction}$$

The Cu^{2+} ion gains electrons to produce copper metal. Two electrons are gained by the Cu^{2+} ion. To balance the 2+ charge on the left side of the reduction half-reaction, add two electrons to the left side.

$$2e^- + Cu^{2+} \rightarrow Cu$$

Zinc metal loses two electrons to become the Zn^{2+} ion. To balance the 2+ charge on the right side of the oxidation half-reaction, add two electrons to the right side.

$$Zn \rightarrow Zn^{2+} + 2e^-$$

Balance the number of atoms in each half-reaction. In this reaction the number of copper atoms is the same on both sides, so the coefficients of solid copper and Cu^{2+} ion do not need to be adjusted.

Zinc metal loses two electrons to become the Zn^{2+} ion. The number of zinc atoms is the same on both sides, so the coefficients of solid zinc and Zn^{2+} ion do not need to be adjusted.

In a balanced oxidation-reduction reaction the number of electrons lost must equal the number of electrons gained. In the oxidation half-reaction two electrons are lost and in the reduction half-reaction two electrons are gained. Because the number of electrons is the same in both half-reactions, no adjustment is needed to the number of electrons.

Add the two half-reactions together.

$$2e^- + Cu^{2+} \rightarrow Cu$$
$$Zn \rightarrow Zn^{2+} + 2e^-$$

$$2\cancel{e^-} + Cu^{2+} + Zn \rightarrow Cu + Zn^{2+} + 2\cancel{e^-}$$

Now cancel the electrons which appear on both sides to give the overall reaction.

$$Cu^{2+}(aq) + Zn(s) \rightarrow Cu(s) + Zn^{2+}(aq)$$

We can check our work to make sure the elements and charges are the same on both sides. There is one copper and one zinc on each side, and the charge is 2+ on each side, so the equation is balanced.

	$Cu^{2+}(aq) +$	$Zn(s)$	$\rightarrow$	$Cu(s) +$	$Zn^{2+}(aq)$
elements	1 Cu	1 Zn	$\rightarrow$	1 Cu	1 Zn
charge	2+		$\rightarrow$	2+	

b. Write the individual oxidation and reduction half-reactions.

$$Re^{5+}(aq) + Sb^{3+}(aq) \rightarrow Re^{4+}(aq) + Sb^{5+}(aq)$$

The oxidation state of rhenium decreases from 5+ to 4+. Rhenium is reduced.

$$Re^{5+} \rightarrow Re^{4+} \quad \text{reduction half-reaction}$$

The oxidation state of antimony increases from 3+ to 5+. Antimony is oxidized.

$$Sb^{3+} \rightarrow Sb^{5+} \quad \text{oxidation half-reaction}$$

The Re^{5+} ion gains an electron to become the Re^{4+} ion. To balance the 1+ charge on the left side of the reduction half-reaction, add one electron to the left side.

$$e^- + Re^{5+} \rightarrow Re^{4+}$$

The Sb^{3+} ion loses two electrons to become the Sb^{5+} ion. To balance the 2+ charge on the right side of the oxidation half-reaction, add two electrons to the right side.

$$Sb^{3+}(aq) \rightarrow Sb^{5+}(aq) + 2e^-$$

The number of rhenium atoms and antimony atoms is the same on both sides, so the coefficients of Re^{5+}, Re^{4+}, Sb^{3+}, and Sb^{5+} do not need to be adjusted.

In a balanced redox reaction the number of electrons gained and lost must be equal, so multiply the reduction half-reaction by two so that both half-reactions transfer two electrons.

$$2(e^- + Re^{5+} \rightarrow Re^{4+})$$
$$2e^- + 2Re^{5+} \rightarrow 2Re^{4+}$$

Now, add the two half-reactions together.

$$2e^- + 2Re^{5+} \rightarrow 2Re^{4+}$$
$$Sb^{3+} \rightarrow Sb^{5+} + 2e^-$$

$$\cancel{2e^-} + 2Re^{5+} + Sb^{3+} \rightarrow 2Re^{4+} + Sb^{5+} + \cancel{2e^-}$$

Cancel the electrons which appear on both sides of the equation to give the overall balanced reaction.

$$2Re^{5+}(aq) + Sb^{3+}(aq) \rightarrow 2Re^{4+}(aq) + Sb^{5+}(aq)$$

We can check our work to make sure the elements and charges are the same on both sides. There are two rheniums and one antimony on both sides, and the charge is 13+ on each side, so the equation is balanced.

	$2Re^{5+}(aq)$	$+ Sb^{3+}(aq)$	$\rightarrow$	$2Re^{4+}(aq)$	$+ Sb^{5+}(aq)$
elements	2 Re	1 Sb	$\rightarrow$	2Re	1 Sb
charge	13+		$\rightarrow$	13+	

8. When balancing oxidation-reduction reactions in acidic solution, use the five steps given in section 18.4 of your textbook.

a. Step 1: Write equations for the oxidation and reduction half-reactions.

$$H_2S(aq) + NO_3^-(aq) \rightarrow S_8(s) + NO(g)$$

Oxidation states: H: 1+ each; S (in H_2S): 2-; N: 5+; O (in NO_3^-): 2- each; S (in S_8): 0 each; N (in NO): 2+; O (in NO): 2-

Sulfur in H_2S loses two electrons to become elemental sulfur, so sulfur is oxidized. The oxidation half-reaction is

$$H_2S \rightarrow S_8 \qquad \text{oxidation half-reaction}$$

Nitrogen in NO_3^- gains three electrons to become NO, so nitrogen is reduced. The reduction half-reaction is

$$NO_3^- \rightarrow NO \qquad \text{reduction half-reaction}$$

Step 2a: Balance all elements except hydrogen and oxygen.
The right side of the oxidation half-reaction has eight sulfur atoms so we will need eight on the left.

$$8H_2S \rightarrow S_8$$

The reduction half-reaction contains one nitrogen atom on both sides so no adjustment is needed.

Step 2b: Balance the oxygen atoms using H_2O.
The oxidation half-reaction contains no oxygen atoms. The reduction half-reaction has three oxygen atoms on the left and only one on the right, so add two molecules of H_2O to the right side.

$$NO_3^- \rightarrow NO + 2H_2O$$

Step 2c: Balance hydrogens using H^+.
The oxidation half-reaction has sixteen hydrogens on the left, and none on the right, so add $16H^+$ to the right side.

$$8H_2S \rightarrow S_8 + 16H^+$$

The reduction half-reaction has four hydrogens on the right, and none on the left, so add $4H^+$ to the left side.

$$4H^+ + NO_3^- \rightarrow NO + 2H_2O$$

Step 2d: Balance the charge using electrons.
To balance the sixteen positive charges on the right side of the reduction half-reaction, add sixteen electrons to the right side.

	$8H_2S$	$\rightarrow$	$S_8 + 16H^+$
charge	0	$\rightarrow$	16+

$$8H_2S \rightarrow S_8 + 16H^+ + 16e^-$$

To balance the three positive charges on the left side of the oxidation half-reaction, add three electrons to the left side.

	$4H^+ + NO_3^-$	$\rightarrow$	$NO + 2H_2O$
charge	3+	$\rightarrow$	0

$$3e^- + 4H^+ + NO_3^- \rightarrow NO + 2H_2O$$

Step 3: The oxidation half-reaction transfers sixteen electrons and the reduction half-reaction transfers three electrons, so we must equalize the number of electrons transferred. Multiply the oxidation half-reaction by three and the reduction half-reaction by sixteen.

$$16(3e^- + 4H^+ + NO_3^- \longrightarrow NO + 2H_2O)$$
$$48e^- + 64H^+ + 16NO_3^- \longrightarrow 16NO + 32H_2O$$

$$3(8H_2S \longrightarrow S_8 + 16H^+ + 16e^-)$$
$$24H_2S \longrightarrow 3S_8 + 48H^+ + 48e^-$$

Step 4: Now, add the half-reactions together and cancel species which appear on both sides. The forty-eight electrons appear on both sides and cancel. All forty-eight hydrogen ions cancel on the right, and forty-eight cancel on the left, leaving sixteen hydrogen ions on the left.

$$24H_2S \longrightarrow 3S_8 + 48H^+ + 48e^-$$
$$48e^- + 64H^+ + 16NO_3^- \longrightarrow 16NO + 32H_2O$$

$$48\cancel{e^-} + 64\cancel{H^+} + 16NO_3^- + 24H_2S \longrightarrow 16NO + 32H_2O + 3S_8 + 48\cancel{H^+} + 48\cancel{e^-}$$

The equation becomes

$$16H^+(aq) + 16NO_3^-(aq) + 24H_2S(aq) \longrightarrow 16NO(g) + 32H_2O(l) + 3S_8(s)$$

Step 5: Let's check the elements and the charges on each side. There are sixty-four hydrogen atoms, sixteen nitrogen atoms, forty-eight oxygen atoms, and twenty-four sulfur atoms on each side. The charge on each side is zero, so the equation is balanced.

$$16H^+(aq) + 16NO_3^-(aq) + 24H_2S(aq) \longrightarrow 16NO(g) + 32H_2O(l) + 3S_8(s)$$

elements	64 H	16 N	48 O	24 S	→	64 H	16 N	48 O	24 S
charge		0			→			0	

b. Step 1: Write the equations for the oxidation and reduction half-reactions.

$$H_5IO_6(aq) + I^-(aq) \longrightarrow I_2(s)$$

(oxidation numbers: H 1+ each; I 7+; O 2- each; I^- 1-; I_2 0)

Iodine in H_5IO_6 gains seven electrons to become I^-. Iodine is reduced. The reduction half-reaction is

$$H_5IO_6 \rightarrow I_2 \quad \text{reduction half-reaction}$$

Iodide ion loses an electron to become I_2. Iodine is oxidized. The oxidation half-reaction is

$$I^- \rightarrow I_2 \quad \text{oxidation half-reaction}$$

In this oxidation-reduction reaction, iodine is both the species oxidized and the species reduced.

Step 2a: Let's balance all the elements except oxygen and hydrogen. For the reduction half-reaction there is one iodine atom on the left, and two on the right. Change the coefficient of H_5IO_6 from one to two so that the number of iodine atoms is the same on each side.

$$2H_5IO_6 \rightarrow I_2$$

For the oxidation half-reaction, put a coefficient of two on the left side to balance the two iodine atoms on the right.

$$2I^- \rightarrow I_2$$

Step 2b: There are twelve oxygen atoms on the left side of the reduction half-reaction, so put twelve molecules of H_2O on the right to balance the oxygen atoms.

$$2H_5IO_6 \rightarrow I_2 + 12H_2O$$

The reduction half-reaction contains no oxygen atoms.

Step 2c: There are twenty-four hydrogen atoms on the right side of the reduction half-reaction, but only ten on the left. Add fourteen hydrogen ions to the left side.

$$14H^+ + 2H_5IO_6 \rightarrow I_2 + 12H_2O$$

The oxidation half-reaction contains no hydrogen atoms.

Step 2d: To balance the 14+ charge on the left side of the reduction half-reaction, add fourteen electrons to the left side.

$$14H^+ + 2H_5IO_6 \rightarrow I_2 + 12H_2O$$

charge 14+ → 0

$$14e^- + 14H^+ + 2H_5IO_6 \rightarrow I_2 + 12H_2O$$

The reduction half-reaction is now balanced.

To balance the 2- charge on the left side of the oxidation half-reaction, add two electrons to the right side.

$$2I^- \rightarrow I_2$$

charge 2- → 0

$$2I^- \rightarrow I_2 + 2e^-$$

Step 3: Because the reduction half-reaction transfers fourteen electrons and the oxidation half-reaction transfers two electrons, multiply the oxidation half-reaction by seven.

$$7(2I^- \rightarrow I_2 + 2e^-)$$

$$14I^- \rightarrow 7I_2 + 14e^-$$

Step 4: Now add the half-reactions together.

$$14e^- + 14H^+ + 2H_5IO_6 \rightarrow I_2 + 12H_2O$$

$$14\,I^- \rightarrow 7I_2 + 14e^-$$

$$\cancel{14e^-} + 14H^+ + 2H_5IO_6 + 14I^- \rightarrow I_2 + 12H_2O + 7I_2 + \cancel{14e^-}$$

The electrons on both sides cancel and the number of iodine molecules can be combined to simplify the equation.

$$14H^+ + 2H_5IO_6 + 14I^- \rightarrow 8I_2 + 12H_2O$$

Step 5: Let's check the elements and charges on each side. There are twenty-four hydrogen atoms, sixteen iodine atoms and twelve oxygen atoms on each side. The charge on both sides is zero, so the equation is balanced.

$$14H^+(aq) + 2H_5IO_6(aq) + 14I^-(aq) \rightarrow 8I_2(s) + 12H_2O(l)$$

elements 24 H 16 I 12 O → 48H 16 I 12 O

charge 0 → 0

c. Step 1: Write equations for the oxidation and reduction half-reactions.

$$Cr_2O_7^{2-}(aq) + Sn^{2+}(aq) \rightarrow Sn^{4+}(aq) + Cr^{3+}(aq)$$

6+ each 2- each 2+ 4+ 3+

Chromium in $Cr_2O_7^{2-}$ gains three electrons to become Cr^{3+}. Chromium is reduced. The reduction half-reaction is

$$Cr_2O_7^{2-} \rightarrow Cr^{3+} \quad \text{reduction half-reaction}$$

The Sn^{2+} ion loses two electrons to become Sn^{4+}. Tin is oxidized. The oxidation half-reaction is

$$Sn^{2+} \rightarrow Sn^{4+} \quad \text{oxidation half-reaction}$$

Step 2a: The reduction half-reaction has two chromium atoms on the left and one on the right, so change the coefficient of Cr^{3+} to two.

$$Cr_2O_7^{2-} \rightarrow 2Cr^{3+}$$

The oxidation half-reaction has one tin on each side. The coefficients need no adjustment.

Step 2b: In the reduction half-reaction there are seven oxygen atoms on the left and none on the right, so add seven molecules of H_2O to the right to balance the oygens.

$$Cr_2O_7^{2-} \rightarrow 2Cr^{3+} + 7H_2O$$

The oxidation half-reaction contains no oxygen atoms.

Step 2c: The reduction half-reaction has fourteen hydrogens on the right and none on the left, so add fourteen hydrogen ions to the left side.

$$14H^+ + Cr_2O_7^{2-} \rightarrow 2Cr^{3+} + 7H_2O$$

The oxidation half-reaction needs no adjustments for hydrogens.

Step 2d: To balance the 6+ charge on the left side of the reduction half-reaction, add six electrons to the left side.

$$14H^+ + Cr_2O_7^{2-} \rightarrow 2Cr^{3+} + 7H_2O$$

charge 12+ → 6+

$$6e^- + 14H^+ + Cr_2O_7^{2-} \rightarrow 2Cr^{3+} + 7H_2O$$

The oxidation half-reaction has a 2+ charge on the left, and 4+ charge on the right, so add two electrons to the right to balance the charge.

$$Sn^{2+} \rightarrow Sn^{4+}$$

charge 2+ → 4+

$$Sn^{2+} \rightarrow Sn^{4+} + 2e^-$$

Step 3: In the reduction half-reaction six electrons are gained, and in the oxidation half-reaction two electrons are lost. Multiply the oxidation half-reaction times three to equalize the number of electrons transferred.

$$3(Sn^{2+} \rightarrow Sn^{4+} + 2e^-)$$
$$3Sn^{2+} \rightarrow 3Sn^{4+} + 6e^-$$

Step 4: Now add the two half-reactions together.

$$6e^- + 14H^+ + Cr_2O_7^{2-} \rightarrow 2Cr^{3+} + 7H_2O$$
$$3Sn^{2+} \rightarrow 3Sn^{4+} + 6e^-$$

$$\cancel{6e^-} + 14H^+ + 2Cr_2O_7^{2-} + 3Sn^{2+} \rightarrow 4Cr^{3+} + 7H_2O + 3Sn^{4+} + \cancel{6e^-}$$

The electrons cancel and the equation becomes

$$14H^+ + Cr_2O_7^{2-} + 3Sn^{2+} \rightarrow 2Cr^{3+} + 7H_2O + 3Sn^{4+}$$

Step 5: Let's check the elements and charges on both sides. There are fourteen hydrogen atoms, two chromium atoms, seven oxygen atoms and three tin atoms on each side. The charge is 18+ on both sides; the equation is balanced.

$$14H^+(aq) + Cr_2O_7^{2-}(aq) + 3Sn^{2+}(aq) \rightarrow 2Cr^{3+}(aq) + 7H_2O(l) + 3Sn^{4+}(aq)$$

elements 14H 2 Cr 7 O 3 Sn → 14 H 2 Cr 7 O 3 Sn

charge 18+ → 18+

d. Step 1: Write equations for the oxidation and reduction half-reactions.

$$I_2\,(s) + NO_3^-\,(aq) \rightarrow IO_3^-\,(aq) + NO_2\,(g)$$

0 (I in I_2); 5+, 2- each (N, O in NO_3^-); 5+, 2- each (I, O in IO_3^-); 4+, 2- each (N, O in NO_2)

Each atom in molecular iodine loses five electrons to become IO_3^-. Iodine is oxidized. The oxidation half-reaction is

$$I_2 \rightarrow IO_3^- \quad \text{oxidation half-reaction}$$

Nitrogen in NO_3^- gains one electron to become NO_2. Nitrogen is reduced. The reduction half-reaction is

$$NO_3^- \rightarrow NO_2 \quad \text{reduction half-reaction}$$

Step 2a: There are two iodine atoms on the left side of the oxidation half-reaction and one on the right. Change the coefficient of IO_3^- to two.

$$I_2 \rightarrow 2IO_3^-$$

The reduction half-reaction has one nitrogen atom on each side.

Step 2b: The oxidation half reaction has six oxygen atoms on the right, so add six molecules of H_2O to the left.

$$6H_2O + I_2 \rightarrow 2IO_3^-$$

The reduction half-reaction has three oxygen atoms on the left, but only two on the right, so add one molecule of H_2O to the right.

$$NO_3^- \rightarrow NO_2 + H_2O$$

Step 2c: The oxidation half-reaction has twelve hydrogens on the left, so add twelve hydrogen ions to the right.

$$6H_2O + I_2 \rightarrow 2IO_3^- + 12H^+$$

The reduction half-reaction has two hydrogen atoms on the right, so add two hydrogen ions to the left.

$$2H^+ + NO_3^- \rightarrow NO_2 + H_2O$$

Step 2d: To balance the 10+ charge on the right side of the oxidation half-reaction, add ten electrons to the right side.

$$6H_2O + I_2 \rightarrow 2IO_3^- + 12H^+$$

charge $\quad 0 \rightarrow 10+$

$$6H_2O + I_2 \rightarrow 2IO_3^- + 12H^+ + 10e^-$$

To balance the 1+ charge on the left side of the reduction half-reaction, add one electron to the left side.

$$2H^+ + NO_3^- \rightarrow NO_2 + H_2O$$

charge $\quad 1+ \rightarrow 0$

$$e^- + 2H^+ + NO_3^- \rightarrow NO_2 + H_2O$$

Step 3: In the oxidation half-reaction ten electrons are lost, while one electron is gained during reduction. Multiply the reduction half-reaction by ten to equalize the electrons transferred.

$$10(e^- + 2H^+ + NO_3^- \rightarrow NO_2 + H_2O)$$

$$10e^- + 20H^+ + 10NO_3^- \rightarrow 10NO_2 + 10H_2O$$

Step 4: Now add the two half-reactions together.

$$6H_2O + I_2 \rightarrow 2IO_3^- + 12H^+ + 10e^-$$
$$10e^- + 20H^+ + 10NO_3^- \rightarrow 10NO_2 + 10H_2O$$

$$10\cancel{e^-} + 20\cancel{H^+} + 10NO_3^- + 6H_2O + I_2 \rightarrow 10NO_2 + 10H_2O + 2IO_3^- + 12\cancel{H^+} + 10\cancel{e^-}$$

The ten electrons on each side cancel. There are twenty hydrogen ions on the left and twelve on the right, so cancel the twelve hydrogen ions on the right and leave eight on the left. There are six water molecules on the left and ten on the right so cancel the six molecules on the left, which leaves four on the right. The equation becomes

$$8H^+(aq) + 10NO_3^-(aq) + I_2(s) \rightarrow 10NO_2(g) + 4H_2O(l) + 2IO_3^-(aq)$$

Step 5: Let's check the elements and charges on each side. There are eight hydrogen atoms, ten nitrogen atoms, thirty oxygen atoms and two iodine atoms on each side. The charge on each side is 2-, so the equation is balanced.

$$8H^+(aq) + 10NO_3^-(aq) + I_2(s) \rightarrow 10NO_2(g) + 4H_2O(l) + 2IO_3^-(aq)$$

elements $\quad$ 8 H $\quad$ 10 N $\quad$ 30 O $\quad$ 2 I $\rightarrow$ 8 H $\quad$ 10 N $\quad$ 30 O $\quad$ 2 I

charge $\quad$ 2- $\rightarrow$ 2-

e. Step 1: Write the equations for the oxidation and reduction half-reactions.

$$Mn^{2+}(aq) + BiO_3(s) \longrightarrow MnO_4^-(aq) + Bi^{3+}(aq)$$

2+ 6+ 2- each 7+ 2- each 3+

The Mn^{2+} ion loses five electrons to become the permanganate ion. Manganese is oxidized. The oxidation half-reaction is

$$Mn^{2+} \longrightarrow MnO_4^- \quad \text{oxidation half-reaction}$$

Bismuth in BiO_3 gains three electrons to become the Bi^{3+} ion. Bismuth is reduced. The reduction half-reaction is

$$BiO_3 \longrightarrow Bi^{3+} \quad \text{reduction half-reaction}$$

Step 2a: For both the oxidation and reduction half-reactions, all elements other than oxygen are balanced.

Step 2b: The oxidation half-reaction has four oxygen atoms on the right and none on the left, so add four molecules of water to the left.

$$4H_2O + Mn^{2+} \longrightarrow MnO_4^-$$

The reduction half-reaction has three oxygen atoms on the left and none on the right, so add three water molecules to the right.

$$BiO_3 \longrightarrow Bi^{3+} + 3H_2O$$

Step 2c: The oxidation half-reaction has eight hydrogens on the left, so add $8H^+$ to the right.

$$4H_2O + Mn^{2+} \longrightarrow MnO_4^- + 8H^+$$

The reduction half-reaction has six hydrogens on the right, so add $6H^+$ to the left.

$$6H^+ + BiO_3 \longrightarrow Bi^{3+} + 3H_2O$$

Step 2d: The oxidation half-reaction has an excess of 5+ charge on the right side, so add $5e^-$ to the right side to balance the charge.

$$4H_2O + Mn^{2+} \longrightarrow MnO_4^- + 8H^+$$

charge $\quad 2+ \longrightarrow 7+$

$$4H_2O + Mn^{2+} \longrightarrow MnO_4^- + 8H^+ + 5e^-$$

The reduction half-reaction has an excess of 3+ charge on the left side, so add $3e^-$ to the left side to balance the charge.

$$6H^+ + BiO_3 \longrightarrow Bi^{3+} + 3H_2O$$

charge $\quad 6+ \longrightarrow 3+$

$$3e^- + 6H^+ + BiO_3 \longrightarrow Bi^{3+} + 3H_2O$$

Step 3: In the oxidation half-reaction five electrons are transferred, and in the reduction half-reaction three electrons are transferred. Multiply the oxidation half-reaction by three and the reduction half-reaction by five to balance the number of electrons transferred.

$$3(4H_2O + Mn^{2+} \longrightarrow MnO_4^- + 8H^+ + 5e^-)$$
$$12H_2O + 3Mn^{2+} \longrightarrow 3MnO_4^- + 24H^+ + 15e^-$$

$$5(3e^- + 6H^+ + BiO_3 \longrightarrow Bi^{3+} + 3H_2O)$$
$$15e^- + 30H^+ + 5BiO_3 \longrightarrow 5Bi^{3+} + 15H_2O$$

Step 4: Now add the two half-reactions together.

$$12H_2O + 3Mn^{2+} \longrightarrow 3MnO_4^- + 24H^+ + 15e^-$$
$$15e^- + 30H^+ + 5BiO_3 \longrightarrow 5Bi^{3+} + 15H_2O$$

$$15\not{e}^- + 30\not{H}^+ + 5BiO_3 + 12H_2\not{O} + 3Mn^{2+} \longrightarrow 5Bi^{3+} + 15H_2\not{O} + 3MnO_4^- + 24\not{H}^+ + 15\not{e}^-$$

The fifteen electrons on each side cancel, and the number of hydrogen ions and water molecules can be reduced. There are twenty-four hydrogen ions on the right and thirty on the left side. Cancel the twenty-four on each side and we are left with six hydrogen ions on the left. Cancel twelve water molecules on each side and we are left with three water molecules on the right. The equation becomes

$$6H^+ + 5BiO_3 + 3Mn^{2+} \longrightarrow 5Bi^{3+} + 3MnO_4^- + 3H_2O$$

Step 5: Let's check the elements and the charges on each side. There are six hydrogen atoms, five bismuth atoms, fifteen oxygen atoms and three manganese atoms on each side. There is a 12+ charge on each side so the equation is balanced.

$$6H^+(aq) + 5BiO_3(s) + 3Mn^{2+}(aq) \rightarrow 5Bi^{3+}(aq) + 3MnO_4^-(aq) + 3H_2O(l)$$

elements	6 H	5 Bi	15 O	3 Mn	→	6 H	5 Bi	15 O	3 Mn
charge		12+			→			12+	

9. Usually redox reactions occur in a single container. If we separate the oxidation half-reaction from the reduction half-reaction but use a salt bridge to allow ions to flow, we can require the electron transfer to occur through a wire. The current produced can be used to do work.

10. The two solutions must be connected by a salt bridge so that ions can flow between the two containers. When electrons flow in one direction, the salt bridge allows negative charge to be balanced by the flow of cations.

11. During oxidation, electrons are lost, so oxidation occurs in the left container and reduction in the right container.

12. In the lead storage battery there are lead grids connected by a metal bar. Lead is oxidized to form Pb^{2+}, which combines with SO_4^{2-} from sulfuric acid (battery acid) to form solid $PbSO_4$. The substance which gains electrons and is reduced is PbO_2, which is coated on the lead grids. Pb^{4+} in PbO_2 is reduced to Pb^{2+}, which combines with SO_4^{2-} from H_2SO_4 to form $PbSO_4$. So the product of both oxidation and reduction is $PbSO_4$. The oxidation and reduction half-reactions are separated so that useful work, such as starting your car, can be accomplished.

13. Both acid and alkaline dry cells contain zinc, which is oxidized to Zn^{2+} with a loss of two electrons. However, in the alkaline cell, hydroxide ions are present instead of ammonium chloride.

$$Zn \rightarrow Zn^{2+} + 2e^- \quad \text{acid dry cell oxidation}$$

$$Zn + 2OH^- \rightarrow ZnO(s) + H_2O + 2e^- \quad \text{alkaline dry cell oxidation}$$

In both the acid and alkaline dry cell versions, MnO_2 is reduced to Mn_2O_3, but the alkaline version contains OH^- ions.

$$2e^- + 2NH_4^+ + 2MnO_2 \rightarrow Mn_2O_3 + 2NH_3 + H_2O \quad \text{acid dry cell reduction}$$

$$2e^- + 2MnO_2 + H_2O \rightarrow Mn_2O_3 + 2OH^- \quad \text{alkaline dry cell reduction}$$

14. The most common method for preventing corrosion is to coat the metal to be protected with paint or with a metal plating. The metal plating keeps oxygen necessary for corrosion away from the metal underneath. Cathodic protection is used to prevent corrosion of buried fuel tanks and pipelines. A metal such as magnesium, which is very easily

oxidized, is connected by a wire to the metal which is to be protected. The magnesium is preferentially oxidized and the metal tank is left unharmed. The magnesium must be replaced periodically as it is oxidized to Mg^{2+} ions. Iron is often alloyed with metals such as chromium and nickel, both of which form protective oxide coats.

15. Aluminum metal can be produced by electrolysis of aluminum oxide. Electrolysis is the process of forcing a current through an electrochemical cell to cause a chemical change which would not occur naturally. During electrolysis, aluminum is reduced.

$$2Al_2O_3(s) \rightarrow 4Al(s) + 3O_2(g)$$

PRACTICE EXAM

1. In the reaction below, which element is oxidized and which is reduced?

$$Mg(s) + Ni(NO_3)_2(aq) \rightarrow Ni(s) + Mg(NO_3)_2(aq)$$

a. Magnesium is oxidized and nitrogen is reduced.
b. Nitrogen is oxidized and magnesium is reduced.
c. Oxygen is oxidized and nitrogen is reduced.
d. Magnesium is oxidized and nickel is reduced.
e. Nickel is oxidized and magnesium is reduced.

2. Which of the oxidation states for the atoms in the reaction below is completely correct?

$$NaSeO_3 + Cl_2 + 2NaOH \rightarrow NaSeO_4 + 2NaCl + H_2O$$

a. $NaSeO_3 + Cl_2 + 2NaOH \rightarrow NaSeO_4 + 2NaCl + H_2O$
2- 5+ 2- 1- 1+ 2- 1+ 3- 7+ 4- 2+ 2- 1+ 2-

b. $NaSeO_3 + Cl_2 + 2NaOH \rightarrow NaSeO_4 + 2NaCl + H_2O$
1+ 5+ 6- 2- 1+ 2- 1+ 1+ 7+ 8- 2+ 1- 2+ 2-

c. $NaSeO_3 + Cl_2 + 2NaOH \rightarrow NaSeO_4 + 2NaCl + H_2O$
1+ 5+ 2- 0 1+ 2- 1+ 1+ 7+ 2- 1+ 1- 1+ 2-

d. $NaSeO_3 + Cl_2 + 2NaOH \rightarrow NaSeO_4 + 2NaCl + H_2O$
1+ 1+ 2- 0 1+ 2- 1+ 1+ 1+ 2- 1+ 1- 1+ 2-

e. $NaSeO_3 + Cl_2 + 2NaOH \rightarrow NaSeO_4 + 2NaCl + H_2O$
1+ 2+ 3- 0 2+ 1- 1- 2+ 2+ 4- 1+ 1- 2+ 2-

3. In this reaction:

$$Bi(OH)_3 + K_2SnO_2 \rightarrow Bi + K_2SnO_3 + H_2O$$

a. Bismuth is oxidized and $Bi(OH)_3$ is the oxidizing agent.
b. Tin is oxidized and K_2SnO_2 is the oxidizing agent.
c. Tin is oxidized and $Bi(OH)_3$ is the oxidizing agent.
d. Bismuth is oxidized and $Bi(OH)_3$ is the oxidizing agent.
e. Bismuth is oxidized and K_2SnO_2 is the oxidizing agent.

4. What is the complete balanced equation for the reaction below?

$$Fe^{3+}(aq) + Sn^{2+}(aq) \rightarrow Fe^{2+}(aq) + Sn^{4+}(aq)$$

a. $2Fe^{3+}(aq) + Sn^{2+}(aq) \rightarrow Sn^{4+}(aq) + 2Fe^{2+}(aq)$

b. $Fe^{3+}(aq) + 2Sn^{2+}(aq) \rightarrow 2Sn^{4+}(aq) + Fe^{2+}(aq)$

c. $Fe^{3+}(aq) + Sn^{2+}(aq) \rightarrow Sn^{4+}(aq) + Fe^{2+}(aq)$

d. $3Fe^{3+}(aq) + 2Sn^{2+}(aq) \rightarrow 2Sn^{4+}(aq) + 3Fe^{2+}(aq)$

e. $Fe^{3+}(aq) + 4Sn^{2+}(aq) \rightarrow 4Sn^{4+}(aq) + Fe^{2+}(aq)$

5. What is the complete balanced equation for the reaction below, which occurs in acidic solution?

$$H_2S(aq) + Cr_2O_7^{2-}(aq) \rightarrow Cr^{3+}(aq) + S(s)$$

a. $H_2S(s) + Cr_2O_7^{2-}(aq) \rightarrow 2Cr^{3+}(aq) + S(s) + 2H^+(aq)$

b. $8H^+(aq) + 3H_2S(aq) + Cr_2O_7^{2-}(aq) \rightarrow 2Cr^{3+}(aq) + 3S(s) + 7H_2O(l)$

c. $3H_2S(aq) + Cr_2O_7^{2-}(aq) \rightarrow 2Cr^{3+}(aq) + 3S(s) + 6H_2O(l)$

d. $28H^+(aq) + 3H_2S(aq) + 2Cr_2O_7^{2-}(aq) \rightarrow 4Cr^{3+}(aq) + 3S(s) + 14H_2O(l) + 6H^+(aq)$

e. $10H^+(aq) + 2H_2S(s) + Cr_2O_7^{2-}(aq) \rightarrow 2S(s) + 2Cr^{3+}(aq) + 7H_2O(l)$

6. Which statement about the galvanic cell below is <u>not</u> correct?

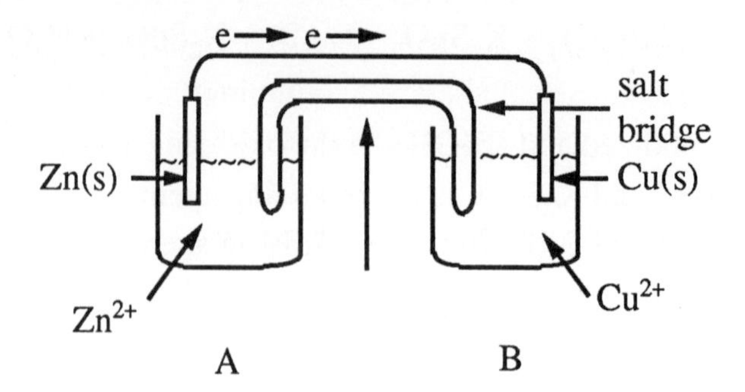

a. The zinc electrode is the cathode.
b. The salt bridge allows ion transfer between compartments.
c. In compartment A, oxidation occurs.
d. Copper is the oxidizing agent.
e. The reduction half-reaction is $Cu^{2+}(aq) + 2e^{-} \rightarrow Cu(s)$.

7. Which statement about the lead storage battery is true?

a. PbO_2 is the cathode.
b. The electrolyte solution is NHO_3.
c. Pb is reduced.
d. Pb is the cathode.
e. The product of both the oxidation and reduction half-reactions is PbO_2.

8. Which is <u>not</u> a feature of an acid dry cell battery?

a. The inner zinc liner is oxidized to Zn^{2+}.
b. The carbon rod is packed in a paste of MnO_2 and NH_4Cl.
c. The carbon rod down the center of the battery acts as the cathode.
d. An acid dry cell produces a potential of 12 volts.
e. The zinc inner liner is the anode.

9. Which statement about corrosion is <u>not</u> true?

a. Attaching a metal which does not oxidize easily to the metal which needs protecting is called cathodic protection.
b. One product of the corrosion of metal is the metal oxide.
c. Painting and plating metals are common methods for preventing corrosion.
d. Some corrosion products actually protect the metal from further corrosion.
e. The tarnishing of silver is a form of corrosion.

10. Which statement about electrolysis is true?

   a. Electrolysis speeds up a process which occurs naturally.
   b. During the electrolysis of aluminum, aluminum is oxidized.
   c. When electrolysis occurs, electrolytes are not needed to balance the charges on ions.
   d. Electrolysis of water produces H^+ and OH^-.
   e. The lead storage battery can be recharged by electrolysis.

PRACTICE EXAM ANSWERS

1. d (18.1)
2. c (18.2)
3. c (18.3)
4. a (18.4)
5. b (18.4)
6. a (18.5)
7. a (18.6)
8. d (18.6)
9. a (18.7)
10. e (18.8)

CHAPTER 19: RADIOACTIVITY AND NUCLEAR ENERGY

INTRODUCTION

Most chemical properties depend on the arrangement of electrons, and many chemical reactions involve the transfer of electrons from one atom to another. But the events and reactions described in this chapter depend on the properties of the nucleus of an atom. The best known nuclear reactions produce energy in nuclear reactors and in nuclear explosions. You will learn about these reactions and other processes in this chapter.

AIMS FOR THIS CHAPTER

1. Know how to classify radioactive decay, and be able to balance a nuclear equation. (Section 19.1)
2. Understand how elements can undergo nuclear transformation when bombarded with small particles. (Section 19.2)
3. Know the common methods for detecting radioactivity. (Section 19.3)
4. Know the definition of half-life, and how half-life can be used to calculate changes in the concentration of radionuclides over time. (Section 19.3)
5. Know how carbon-14 dating can be used to determine the approximate age of some artifacts. (Section 19.4)
6. Know how radiotracers are used for disease diagnosis in medicine. (Section 19.5)
7. Become familiar with the terms fission and fusion. (Section 19.6)
8. Know how nuclear fission occurs. (Section 19.7)
9. Understand how a fission nuclear reactor works. (Section 19.8)
10. Know how nuclear fusion occurs. (Section 19.9)
11. Know the factors which affect the damage done by radiation. (Section 19.10)

QUICK DEFINITIONS

Radioactive — The term used to describe a nucleus which breaks down spontaneously to produce another nucleus and one or more particles. (Section 19.1)

beta particle — Also called a β-particle. Beta particles are electrons, and are symbolized in nuclear equations as ${}^{0}_{-1}e$. (Section 19.1)

alpha particle — Also called an α-particle. Alpha particles are helium nuclei, with four neutrons and two protons. They are symbolized in nuclear equations as ${}^{4}_{2}He$. (Section 19.1)

β-particle production A nuclear decay process which is accompanied by the loss of a β-particle. β-particle production has the effect of changing a neutron to a proton. (Section 19.1)

gamma ray Also called a γ-ray. Gamma rays are high-energy photons of light. Gamma ray production is a way for the nucleus to get rid of excess energy. The production of gamma rays often occurs at the same time as nuclear decay, which produces other particles. (Section 19.1)

Positron A particle with low mass, like an electron, but with a positive charge. It is symbolized in nuclear equations as $^{0}_{1}e$. (Section 19.1)

Positron production A nuclear decay process which is accompanied by the loss of a positron. Positron production has the effect of changing a proton to a neutron. (Section 19.1)

Electron capture Electrons can be captured by the nucleus, as well as emitted during nuclear decay. The electron which is captured is an inner orbital electron. (Section 19.1)

Decay series Some radioactive nuclei must decay several times before producing a nucleus which does not decay further. The steps which result in a stable nucleus are called a decay series. (Section 19.1)

Nuclear transformation The change of one element into another by bombarding the element with a nuclear particle. (Section 19.2)

Particle accelerators Chambers which can cause particles to move at very high speeds. (Section 19.2)

Transuranium elements All elements with atomic numbers from 93 to 109. (Section 19.2)

Geiger counter Also called a Geiger-Müller counter. An instrument for measuring radioactivity. It works by measuring the current produced when high energy decay particles knock an electron off a neutral Ar atom. (Section 19.3)

Scintillation counter An instrument for measuring radioactivity. It works by counting the number of flashes of light produced when high energy decay particles hit a substance such as sodium iodide. The number of light flashes is an indication of the amount of radioactivity. (Section 19.3)

Half-life The amount of time required for the decay of exactly one-half of the nuclei in a radioactive sample. (Section 19.3)

Radiocarbon dating A technique used to date artifacts made from plants (wood or cloth). The technique depends upon measuring the amount of carbon-14 still present in the artifact, and using the known half-life of carbon-14 to estimate the number of years since the artifact was produced. Also called carbon-14 dating. (Section 19.4)

Radiotracers Radionuclides suitable for introduction into living systems. Their accumulation in different tissues is observed and measured. Radiotracers can help with the diagnosis of many diseases. (Section 19.5)

Fusion Combining two light nuclei to make a heavier one. This process gives off a large amount of energy. (Section 19.6)

Fission Breaking apart a heavy nucleus into two lighter ones. This process gives off a large amount of energy. (Section 19.6)

Chain reaction A process which can keep itself going. (Section 19.7)

Critical mass A mass of fissionable material sufficient to produce a chain reaction. (Section 19.7)

Reactor core The part of a nuclear reactor where the uranium is located and where the fission reaction takes place. (Section 19.8)

Moderator A sheathing to slow down neutrons so that they have a better chance of causing uranium atoms to split. (Section 19.8)

Control rods Rod-shaped substances which can be raised and lowered between the uranium fuel rods to control the rate of fission. The control rods work by absorbing neutrons. (Section 19.8)

Breeder reactors — Reactors which produce energy, and which also produce the fissionable fuel, ${}^{239}_{94}Pu$. (Section 19.8)

Somatic damage — Damage to tissues from exposure to radiation, causing sickness or death. (Section 19.10)

Genetic damage — Damage to the genetic machinery of reproductive cells in parents which causes birth defects and other disorders in offspring. (Section 19.10)

Rem — A unit of radiation exposure whose amount can be correlated with the danger of illness or death. (Section 19.10)

CONTENT REVIEW

19.1 RADIOACTIVE DECAY

What Is Radioactive Decay?

Not all nuclei are stable. Many decay spontaneously, producing a new nucleus, and in addition, some type of nuclear particle. The nucleus of cobalt-60 is unstable. It spontaneously decays to produce nickel-60 and an electron. This reaction can be written

$$ {}^{60}_{27}Co \rightarrow {}^{60}_{28}Ni + {}^{0}_{-1}e $$

When writing nuclear decay reactions, always show the atomic mass, A, and the atomic number, Z, for each element and each particle. Nuclear equations must be balanced. The sum of the atomic masses must be the same on both sides of the equation, and the sum of the atomic numbers must be the same on both sides. In the equation above, the atomic mass is sixty on one side, and sixty plus zero on the other. The atomic number is twenty-seven on one side and twenty-eight minus one, or twenty-seven, on the other. Both sides are balanced.

What Types of Radioactive Decay Are There?

Different types of radioactive decay are defined by the type of particle produced in the reaction. One type of decay produces an **alpha particle** (α-particle), which is a helium nucleus, ${}^{4}_{2}He$. The decay of bismuth-211 produces an alpha particle and an isotope of thallium.

$$ {}^{211}_{83}Bi \rightarrow {}^{4}_{2}He + {}^{207}_{81}Tl $$

The total atomic mass on each side is the same, 211 on the left side and 207 plus four on the right. The total atomic number is also the same on both sides, eighty-three on the left and eighty-one plus two on the right. When an α-particle is lost, the net result for the new nucleus is a loss of four in mass number, and a loss of two in atomic number.

Beta particles (β-particles) are often produced during nuclear decay. A **beta particle** is an electron, symbolized ${}^{0}_{-1}e$. The decay of berillium-10 produces a β-particle.

$${}^{10}_{4}Be \rightarrow {}^{0}_{-1}e + {}^{10}_{5}B$$

The net effect of β-particle production is to change a neutron to a proton. The atomic number in the new nucleus increases by one. The mass number does not change.

Sometimes gamma rays (γ-rays) are produced during nuclear decay, usually accompanied by another particle. A **γ-ray** is a high energy photon of light with no mass or atomic number and is symbolized ${}^{0}_{0}\gamma$. The decay of radium-226 produces radon and both a γ-ray and an α-particle.

$${}^{226}_{88}Ra \rightarrow {}^{222}_{86}Rn + {}^{4}_{2}He + {}^{0}_{0}\gamma$$

The production of a γ-ray does not change either the atomic mass or the atomic number.

A **positron** has very little mass and a positive charge, and is symbolized ${}^{0}_{1}e$. Oxygen-15 decays to produce a positron and nitrogen.

$${}^{15}_{8}O \rightarrow {}^{15}_{7}N + {}^{0}_{1}e$$

The production of a positron does not change the mass number, and decreases the atomic number by one.

Sometimes, the nucleus can capture one of its own inner orbital electrons. This process is called **electron capture**. The result is a new nucleus. Argon-37 can capture an electron to become chlorine-37. The capture of an electron decreases the atomic number by one.

$${}^{37}_{18}Ar + {}^{0}_{-1}e \rightarrow {}^{37}_{17}Cl$$

19.2 NUCLEAR TRANSFORMATIONS

What Are Nuclear Transformations?

Nuclear transformation is the change of one element to another. While nuclear decay and electron capture are natural occurences, a nuclear transformation results from the bombardment of a nucleus with high speed particles while inside a **particle accelerator**. The most common particles used to bombard nuclei are α-particles and neutrons.

19.3 DETECTION OF RADIOACTIVITY AND HALF-LIFE

How Can Radioactivity Be Measured?

Two common instruments for measuring radioactivity are the **Geiger counter** and the **scintillation counter**. The Geiger counter has a tube or probe filled with argon gas. The high energy particles released during nuclear decay pass through the walls of the tube and hit some of the argon atoms. Argon atoms which have been hit by high energy particles lose an electron.

$$\text{Ar} + \text{high energy particle} \longrightarrow \text{Ar}^{+} + \text{e-}$$

The argon ions conduct a current which can be measured. A high amount of radioactive decay means a large current and a large number of clicks from the speaker of the Geiger counter.

When a decay particle hits a substance such as sodium iodide in a scintillation counter, light is emitted. A sensor detects the flash of light. The more light flashes there are, the more radioactivity is present in the sample being tested.

What Is the Half-life Of a Sample of Radioactive Nuclei?

The **half-life** of a sample of radioactive nuclei is the time it takes for one-half of the radioactive nuclei to decay. Half-lives are expressed in units of time; minutes, hours, seconds, or sometimes years. Each different radioactive nuclide has a different half-life. The half-life of sodium-24 is 15.0 hours. If you begin with a 10.0 g sample of sodium-24, after 15.0 hours there will remain 5.0 g of sodium-24, and after 30.0 hours, 2.5 g.

19.4 DATING BY RADIOACTIVITY

How Can Radioactive Nuclei Be Used to Date Artifacts?

The concentration of certain radioactive nuclei present in artifacts, coupled with a knowledge of their half-lives, allows scientists to calculate the age of some artifacts. Carbon-14 is commonly used for this purpose. Carbon-14 is found in the atmosphere, usually as part of a carbon dioxide molecule. Plants of all kinds use the CO_2 from the atmosphere, along with sunlight, to produce the plant's structural molecules. As the plant grows, carbon-14 becomes part of the plant's tissues. When the plant dies, or is harvested for use, no more carbon-14 is incorporated into the plant's tissues, and as carbon-14 decays, it is no longer replaced. Carbon-14 has a long half-life, 5730 years. By knowing how many carbon-14 nuclei are left in a piece of old wood, and knowing how many carbon-14 nuclei would have been present originally, and by using the half-life, it is possible to date a piece of wood. An old wood tool which contains one fourth of the original amount of carbon-14 would be approximately 11,460 years old. By 5730 years, the tool would have half the original amount of carbon-14. After an additional 5730 years, the current amount of carbon-14 would only be one fourth of the original amount.

19.5 MEDICAL APPLICATIONS OF RADIOACTIVITY

How Are Radioactive Nuclei Used in Medicine?

Radiotracers are radioactive nuclei that are useful for diagnosis or treatment of certain diseases. Whether or not a certain organ incorporates the radiotracer as part of its tissue can help determine whether the tissue is healthy or diseased.

19.6 NUCLEAR ENERGY

What Are the Ways Nuclear Energy Can Be Produced?

Nuclear energy can be produced by fusion or by fission. **Fission** is the splitting of a relatively heavy nucleus into two lighter nuclei. **Fusion** is the combining of two relatively light nuclei to produce a heavier one. Both processes produce a large amount of energy.

19.7 NUCLEAR FISSION

How Does Nuclear Fission Occur?

During fission, the relatively heavy uranium nucleus is bombarded with neutrons and is split into two lighter nuclei. Uranium-235 can split into many different small nuclei, among them, barium-142 and krypton-91.

$$^{1}_{0}n + ^{235}_{92}U \rightarrow ^{142}_{56}Ba + ^{91}_{36}Kr + 3\,^{1}_{0}n$$

The neutrons produced during fission can collide with more uranium-235 atoms, causing fission and the production of more neutrons. A **chain reaction** can occur, as the neutrons from one fission reaction can cause more fission reactions to occur.

19.8 NUCLEAR REACTORS

How Can Fission Reactions Be Used To Produce Electrical Energy?

A **reactor core** contains uranium-235 fuel and is the site of the fission reactions. The number of fissions which occurs can be controlled in the reactor by rods which absorb some of the neutrons, and prevent the reaction from proceeding too fast. Tremendous amounts of heat energy are produced during fission. The heat is extracted with a liquid, often water, which is circulated through pipes to produce steam, which operates electrical generators.

19.9 NUCLEAR FUSION

How Can Nuclear Fusion Occur?

Fusion is the combining of light nuclei to form a heavier one. The process of fusion produces more energy per mole of reactant than does fission. Fusion reactions produce the sun's energy. It is thought that several reactions occur before a final product is produced. One possible series of reactions begins with protons, ${}^{1}_{1}H$, and ends with helium, ${}^{4}_{2}He$.

$${}^{1}_{1}H + {}^{1}_{1}H \rightarrow {}^{2}_{1}H + {}^{0}_{1}e + \text{energy}$$

$${}^{1}_{1}H + {}^{2}_{1}H \rightarrow {}^{3}_{2}He + \text{energy}$$

$${}^{3}_{2}He + {}^{3}_{2}He \rightarrow {}^{4}_{2}He + 2\,{}^{1}_{1}H + \text{energy}$$

Scientists would like to use fusion reactions to produce energy on earth. Fusion reactions require collisions between protons, which repel each other because they have like charges. A great amount of heat energy must be used to overcome the repulsions. As yet, scientists have not been able to produce conditions under which fusion will occur.

19.10 EFFECTS OF RADIATION

How Does Nuclear Radiation Damage Human Tissues?

Energy in many forms, including nuclear radiation, can cause damage to human tissues. There are two kinds of damage, somatic and genetic. **Somatic damage** results when the tissues themselves are injured. The damage can cause sickness or death. Sometimes the damage shows up years later in the form of cancer. **Genetic damage** occurs when the genetic material in reproductive cells is damaged, causing death or birth defects to children.

What Factors Control the Amount of Damage to Tissues?

1. The energy level of the radiation. Radiation with a high energy level does more damage than radiation with a low energy level.

2. The ability of the radiation to penetrate body tissues. The higher the penetrating ability, the more damage can occur. γ-rays can penetrate readily into tissues, β-particles can penetrate 1 cm, and α-particles do not penetrate the skin.

3. The ionizing ability of radiation. Some radioactive particles can cause ions to form in the body. Ions produced from neutral molecules do not function the same way as the neutral molecules do, so body functions can be affected. γ-rays penetrate deeply, but do not cause much ionization. α-particles, while they do not penetrate past the skin, cause a great deal of ionization.

4. The chemical properties of the radiation source. Radionuclides can be ingested with contaminated food. Some of the radionuclides can react chemically with molecules in the body to form compounds that remain in the body for long periods. The chances of damage to tissues is greater if the radionuclides stay in the body than if they pass through the body without reacting.

LEARNING REVIEW

1. In a balanced nuclear equation, which two quantities must be the same on both sides of the equation?

2. What is the atomic number and the mass number of each of the particles below?

   a. gamma ray
   b. positron
   c. alpha particle
   d. beta particle

3. Write balanced nuclear equations for the decay of the radioactive particles below.

   a. $^{226}_{86}Rn$ decays to produce an α-particle and a γ-ray.
   b. $^{70}_{31}Ga$ decays to produce a β-particle.
   c. $^{144}_{60}Nd$ decays to produce a β-particle.
   d. $^{234}_{92}U$ decays to produce an α-particle and a γ-ray.

4. Complete and balance these nuclear equations.

a. $^{161}_{67}Ho + ? \rightarrow ^{161}_{66}Dy$
b. $^{10}_{4}Be \rightarrow ? + ^{0}_{-1}e$
c. $? + ^{0}_{-1}e \rightarrow ^{44}_{21}Sc$
d. $^{253}_{99}Es + ^{4}_{2}He \rightarrow ^{1}_{1}H + ?$
e. $^{59}_{29}Cu \rightarrow ? + ^{59}_{28}Ni$
f. $^{55}_{25}Mn + ^{1}_{1}H \rightarrow ^{1}_{0}n + ?$

5. Show the products formed when each of the nuclides below is bombarded with a smaller nuclide.

a. $^{238}_{92}U + ^{12}_{6}C \rightarrow ? + 4\,^{1}_{0}n$
b. $^{238}_{92}U + ^{14}_{7}N \rightarrow ? + 5\,^{1}_{0}n$

6. When the particles below are emitted during radioactive decay, what is the effect on the atomic mass and on the atomic number of the nuclide undergoing decay?

a. β-particle
b. α-particle
c. positron

7. Two instruments for detecting radioactivity are the Geiger counter and the scintillation counter. Briefly explain how each one works.

8. In a sample of the nuclides below, which would exhibit the highest number of decay events during a fixed period of time?

	name	**half-life**
a.	potassium-42	12.4 hours
b.	hydrogen-3	12.5 years
c.	plutonium-239	2.44×10^4 years

9. If a sample of 5.0×10^{20} iodine-131 atoms with a half-life of 8 days is allowed to decay for 48 days, how many iodine-131 atoms will remain?

10. A sample of polonium-218 with a half-life of 3.05 minutes was allowed to decay for 21.36 minutes. If there are 2.0×10^{11} atoms of polonium-218 remaining, how many atoms were in the original sample?

11. A wooden post from an ancient village has 25% of the carbon-14 found in living trees. How old is the wooden post? The half-life of carbon-14 is 5730 years.

12. Why do you think that most nuclides used in medicine as radiotracers have short half-lives?

13. Why is the mass of fissionable material important when initiating a fission reaction?

14. What safety features would prevent a nuclear explosion in case of a serious malfunction of a nuclear reactor?

15. What radionuclide is used as a fissionable fuel in nuclear reactors?

16. What features of breeder reactors make them an attractive potential source of energy?

17. Why do you think that the fusion process would supplant fission if the technology were available?

18. What differences exist between genetic and somatic damage caused by radioactivity?

19. a. Why is the ionizing ability of a radiation source important in determining the biological effects of radiation?
    b. What type of radioactive particles causes the most ionization?

ANSWERS TO LEARNING REVIEW

1. The sum of the atomic numbers (Z) and the sum of the mass numbers (A) must be the same on both sides of a nuclear equation.

2. a. A gamma ray has a mass number of zero and an atomic number of zero.
   b. A positron has a mass number of zero and an atomic number of 1+.
   c. An alpha particle has a mass number of four and an atomic number of 2+.
   d. A beta particle has a mass number of zero and an atomic number of 1-.

3. a. When $^{226}_{86}Rn$ decays to produce an alpha particle and a gamma particle, the mass number of the new nuclide is decreased by four to 222. The atomic number decreases by two to eighty-four. The new nuclide would have a mass number of 222 and an atomic number of eighty-four. The element with atomic number of eighty-four is polonium, so the new nuclide is $^{222}_{84}Po$.

$$^{226}_{86}Rn \longrightarrow {}^{4}_{2}He + {}^{0}_{0}\gamma + {}^{222}_{84}Po$$

b. When $^{70}_{31}Ga$ decays to produce a beta particle, the mass number of the new nuclide does not change. The atomic number increases by one to thirty-two. The new nuclide would have a mass number of seventy, and an atomic number of thirty-two. The element with an atomic number of thirty-two is germanium, so the new nuclide is $^{70}_{32}Ge$.

$$^{70}_{31}Ga \rightarrow {}^{0}_{-1}e + {}^{70}_{32}Ge$$

c. When $^{144}_{60}Nd$ decays to produce a beta particle, the mass number does not change. The atomic number increases by one to sixty-one. The new nuclide would have a mass number of 144 and an atomic number of sixty-one. The element with an atomic number of sixty-one is promethium, so the new nuclide is $^{144}_{61}Pm$.

$$^{144}_{60}Nd \rightarrow {}^{0}_{-1}e + {}^{144}_{61}Pm$$

d. When $^{234}_{92}U$ decays to produce an alpha particle, the mass number decreases by four to 230. The atomic number decreases by two to ninety. The new nuclide would have a mass number of 230 and an atomic number of ninety. The element with an atomic number of ninety is thorium, so the new nuclide is $^{230}_{90}Th$.

$$^{234}_{92}U \rightarrow {}^{4}_{2}He + {}^{230}_{90}Th$$

4. a. This problem provides the nuclides before and after decay and asks for the identity of an unknown decay particle. Because the mass number does not change on either side, the mass number of the particle is zero. The atomic number decreases by one on the right side, so the atomic number of the unknown particle is 1- so that the sum of atomic numbers on each is the same. The unknown particle has a mass number of zero and an atomic number of -1. It is a β-particle. This is an example of electron capture.

$$^{161}_{67}Ho + {}^{0}_{-1}e \rightarrow {}^{161}_{66}Dy$$

b. This problem provides a nuclide on the left that decays to an unknown nuclide and a β-particle. Because the mass number of the β-particle is zero, the mass number of the unknown nuclide must be ten to balance the left side. The atomic number increases by one to become five to balance the four on the left side. The element with an atomic number of five is boron, so the nuclide is $^{10}_{5}B$.

$$^{10}_{4}Be \rightarrow {}^{10}_{5}B + {}^{0}_{-1}e$$

c. This problem provides the identity of a particle that combines with an unknown nuclide to produce the nuclide $^{44}_{21}Sc$. The β-particle has a mass number of zero, so the mass number of the unknown nuclide must be forty-four. The β-particle has an atomic number of 1-, so the atomic number of the unknown nuclide must be twenty-two so that the sum of the atomic numbers is the same on each side. The element with atomic number twenty-two is titanium, so the unknown nuclide is $^{44}_{22}Ti$.

$$^{44}_{22}Ti + ^{0}_{-1}e \rightarrow ^{44}_{21}Sc$$

d. This problem provides a nuclide which reacts with an alpha particle to produce a proton and an unknown nuclide. The total mass number on the left side is 257. On the right, the proton has a mass number of one, so the unknown nuclide must have a mass number of 256 so that both sides are balanced. The total atomic number on the left side is 101. On the right, the proton has an atomic number of one, so that the atomic number of the unknown nuclide must be 100. The element with an atomic number of 100 is fermium, so the nuclide is $^{256}_{100}Fm$.

$$^{253}_{99}Es + ^{4}_{2}He \rightarrow ^{1}_{1}H + ^{256}_{100}Fm$$

e. This problem provides a nuclide which decays to an unknown particle and to a nuclide of nickel. The mass number on both sides is fifty-nine, so the mass number of the unknown particle must be zero. The atomic number of the nuclide on the left is twenty-nine and the atomic number of the nuclide on the right is twenty-eight. The unknown particle has an atomic number of 1+. The particle with a mass number of zero and an atomic number of 1+ is a positron, $^{0}_{1}e$.

$$^{59}_{29}Cu \rightarrow ^{0}_{1}e + ^{59}_{28}Ni$$

f. The problem provides a nuclide of manganese reacting with a proton ($^{1}_{1}H$) to produce a neutron ($^{1}_{0}n$) and an unknown nuclide. The total mass number on the left is fifty-six and on the right the mass number is one, so the mass number of the unknown nuclide is fifty-five. The total atomic number is twenty-six on the left and zero on the right. The atomic number of the unknown nuclide is twenty-six. The element which has an atomic number of twenty-six is iron, so the nuclide is $^{55}_{26}Fe$.

$$^{55}_{25}Mn + ^{1}_{1}H \rightarrow ^{1}_{0}n + ^{55}_{26}Fe$$

5. The nuclides below are bombarded with much smaller nuclei. The products are a new nuclide and a subatomic particle.

   a. The problem provides a nuclide of uranium that is bombarded with a smaller carbon nucleus to produce an unknown nuclide and four neutrons. The total mass number on the left is 238 plus twelve, which is 250. Each neutron on the right has a mass number of one, so the total mass number of the neutrons is four. The mass number of the new nuclide is 246. The total atomic number on the left is ninety-two plus six, which is ninety-eight. Each of the four neutrons on the right has an atomic number of zero, so the atomic number of the new nuclide is ninety-eight. The element with an atomic number of ninety-eight is californium, Cf. The new nuclide is $^{246}_{98}Cf$.

$$^{238}_{92}U + ^{12}_{6}C \rightarrow ^{246}_{98}Cf + 4\,^{1}_{0}n$$

   b. The problem provides a nuclide of uranium that is bombarded with a nitrogen nucleus to produce an unknown nuclide and five neutrons. The total mass number on the left is 252. Each of the five neutrons on the right has a mass number of one, so the total mass number of the neutrons is five. The mass number of the nuclide is 247. The neutrons on the right each have an atomic number of zero, so the unknown nuclide has an atomic number of ninety-nine. The element with atomic number ninety-nine is einsteinium, Es. The unknown nuclide is $^{247}_{99}Es$.

$$^{238}_{92}U + ^{14}_{7}N \rightarrow ^{247}_{99}Es + 5\,^{1}_{0}n$$

6. a. Emission of a β-particle, $^{0}_{-1}e$, does not change the mass number of the new nuclide compared with the original nuclide. Because a β-particle has an atomic number of 1-, the effect of β-particle decay is to increase the atomic number by one.

   b. Emission of an alpha particle, $^{4}_{2}He$, decreases the mass number of the nuclide by four. An alpha particle has an atomic number of two, so the effect of alpha particle decay is to decrease the atomic number of the nuclide by two.

   c. Emission of a positron, $^{0}_{1}e$, does not change the mass number of the nuclide. A positron has an atomic number of 1+ so the effect of positron decay is to decrease the atomic number of the nuclide by one.

7. The Geiger-Müller counter, or Geiger counter, has a probe which is placed close to the source of radioactivity. The probe contains atoms of argon gas which lose an electron when hit by a high-speed subatomic particle. The argon cation and accompanying electron produce a momentary pulse of electrical current which is detected by the Geiger counter. The amount of radioactive material is directly related to the number of pulses detected.

Radioactivity can also be detected with a scintillation counter. High-speed decay particles collide with a substance inside the scintillation counter such as sodium iodide. The sodium iodide emits a flash of light when struck. Each flash of light is counted and the number of flashes is directly related to the amount of radioactivity.

8. The half-life of potassium-42 is 12.4 hours, which means that fifty percent of a sample of potassium-42 would decay in 12.4 hours. Plutonium-239 has a half-life of 24,400 years, which means it would take 24,400 years for fifty percent of a plutonium-239 sample to decay. The shorter the half-life, the quicker a nuclide decays, so the nuclide with the smallest half-life produces the most decay events over time. Of the three nuclides, potassium-42 would produce the most decay events in any fixed amount of time.

9. Because the half-life of iodine-131 is eight days, the number of iodine-131 atoms in any sample will decrease by fifty percent after eight days. So, after eight days there will be $5.0 \times 10^{20}/2 = 2.5 \times 10^{20}$ iodine-131 atoms left. After another eight days (for a total of sixteen days) there would be $2.5 \times 10^{20}/2 = 1.3 \times 10^{20}$ iodine-131 atoms left. After another eight days (for a total of twenty-four days) there would be $1.3 \times 10^{20}/2 = 6.3 \times 10^{19}$ iodine-131 atoms left. After three more eight-day periods (for a total of forty-eight days) there would be 7.8×10^{18} iodine-131 atoms left.

10. To find the number of polonium-218 atoms initially present in a sample, we need to know how many complete half-life periods occurred. The total decay time is 21.36 minutes and the half-life of plutonium-218 is 3.05 minutes. Divide the total decay period by the half-life to give the total number of half-life events which have occurred.

$$\frac{21.36 \text{ minutes}}{3.05 \text{ minute/half-life}} = 7.00 \text{ half-lives}$$

7.00 half-life decays occurred to produce 2.0×10^{11} atoms of plutonium-218. By working backwards we can calculate the number of atoms originally present. After six half-lives, twice as much plutonium 218 would have been present as at seven half-lives. So after six half-lives 4.0×10^{11} atoms of plutonium-218 would have been present. At five half-lives, 8.0×10^{11} atoms of plutonium-218 would have been present. At four half-lives, 1.6×10^{12} atoms of plutonium-218 would have been present, and before any decay had occurred, there would have been 2.6×10^{13} atoms of plutonium-218.

11. A piece of wood which contains 25% of the carbon-14 found in freshly cut wood has undergone two half-life decays. The first half-life would decrease the carbon-14 from 100% to 50%, and the second half-life would decrease the carbon-14 content from 50% to 25%. So, a piece of wood which has undergone two half-life decays would be 2 times 5730, or 11,460, years old.

12. Using any radiotracer inside the human body poses some risk of damage by the high-speed decay particles. Radiotracers with a short half-life will rapidly decay and produce many decay particles in a short period of time. Doctors can use small amounts of radiotracer and still detect their presence because the numbers of decay particles are high at first. Because the half-life is short, most of the radiotracer usually decays quickly.

13. Each fission event produces neutrons in addition to producing nuclides. Neutrons themselves collide with uranium-235 atoms and cause fission. If the mass of uranium-235 is too small, not enough neutrons are produced to sustain fission, and the process dies out. A certain mass, called the critical mass, produces just enough neutrons to keep fission going, but too large a mass causes a fission chain reaction which leads to overheating and a violent explosion.

14. Nuclear reactors have many safety features, including control rods which are made of substances which absorb neutrons. The control rods can be raised or lowered between the fuel rods to control how fast the nuclear reaction occurs. If a serious problem occurs with the reactor, the control rods automatically lower into the core so that the fission process slows down. The amount of fissionable fuel present in any nuclear reactor is below the critical mass, so that even in the worst possible case a nuclear explosion would not occur.

15. Uranium-235, ${}^{235}_{92}U$, is used as a fuel in nuclear reactors. Natural uranium contains only 0.7% of this nuclide. It is enriched to 3% for use as a fuel.

16. The supply of fissionable uranium-235 will run out eventually. Breeder reactors themselves produce a fissionable fuel, ${}^{239}_{94}Pu$, which can be collected and used to fuel another reactor.

17. Fusion would quickly supplant fission because fuel for fusion is readily available in sea water. Fusion reactors would produce helium as an end product, and not the wide variety of radionuclides produced from fission. Safe disposal of the nuclear waste from fission is a concern which does not occur with fusion reactors.

18. Somatic damage is the damage done directly to the tissues of the organism. Somatic damage usually occurs soon after exposure to the radiation source. Genetic damage is the kind of damage done to the reproductive machinery of the human body. Genetic damage occurs at the time of exposure but may not show up until the birth of offspring.

19. a. When biomolecules are ionized by a radiation source, they no longer perform their functions in the body.
    b. Alpha particles, although they do not penetrate deeply into tissue, cause a large amount of ionization.

PRACTICE EXAM

1. Which statement about radioactive decay is not true?

   a. Loss of a γ-ray results in a decrease by one in the mass number.
   b. Loss of a β-particle results in an increase by one in the atomic number.
   c. Loss of a positron results in no change in the mass number.
   d. Loss of an α-particle results in a loss of four in mass number.
   e. Loss of a β-particle results in no change in the mass number.

2. What is the correct balanced nuclear reaction for the emission of a β-particle from a nuclide of silver, ${}^{113}_{47}Ag$?

   a. ${}^{113}_{47}Ag \rightarrow {}^{-1}_{0}e + {}^{114}_{47}Ag$
   b. ${}^{113}_{47}Ag \rightarrow {}^{4}_{2}e + {}^{109}_{45}Rh$
   c. ${}^{113}_{47}Ag \rightarrow {}^{0}_{1}e + {}^{113}_{46}Pd$
   d. ${}^{113}_{47}Ag \rightarrow {}^{0}_{-1}e + {}^{113}_{48}Cd$
   e. ${}^{113}_{47}Ag \rightarrow {}^{0}_{-1}e + {}^{113}_{46}Pd$

3. Which is the correct balanced nuclear equation for the process below?

$$ {}^{253}_{99}Es + {}^{4}_{2}He \rightarrow \; ? + {}^{1}_{0}n $$

   a. ${}^{253}_{99}Es + {}^{4}_{2}He \rightarrow {}^{250}_{97}Bk + {}^{1}_{0}n$
   b. ${}^{253}_{99}Es + {}^{4}_{2}He \rightarrow {}^{256}_{101}Md + {}^{1}_{0}n$
   c. ${}^{253}_{99}Es + {}^{4}_{2}He \rightarrow {}^{257}_{101}Md + {}^{1}_{0}n$
   d. ${}^{253}_{99}Es + {}^{4}_{2}He \rightarrow {}^{258}_{101}Md + {}^{1}_{0}n$
   e. ${}^{253}_{99}Es + {}^{4}_{2}He \rightarrow {}^{249}_{97}Bk + {}^{1}_{0}n$

4. A nuclide of radium, Ra, has a half-life of 3.6 days. If a sample of radium-223 begins with 8.5×10^{20} atoms, how many atoms will be left after 18.0 days?

   a. 2.1×10^{20}
   b. 1.1×10^{20}
   c. 5.3×10^{19}
   d. 2.7×10^{19}
   e. 1.3×10^{19}

5. Which statement about radiocarbon dating is <u>not</u> true?

   a. Carbon-14 is continuously produced in the earth's atmosphere.
   b. Carbon-14 is produced in the atmosphere when $^{14}_{7}N$ captures a neutron.
   c. When plants die, the amount of carbon-14 they contain begins to decline.
   d. After 5730 years, a cut tree would contain 25% of the original amount of carbon-14.
   e. A living plant inceases the amount of carbon-14 in its tissues over time.

6. Which nuclide would be least likely to be used as a radiotracer in medicine?

   a. ^{24}Na
   b. ^{131}I
   c. ^{59}Fe
   d. ^{32}P
   e. ^{235}U

7. Nuclear fission produces large amounts of energy. Which statement about fission is <u>not</u> true?

   a. The amount of energy released from the fission of a mole of nuclear fuel is around 26 million times the amount of energy produced from a mole of methane gas.
   b. Alpha particles are needed to initiate fission.
   c. The nuclide which undergoes fission is $^{235}_{92}U$.
   d. Fission produces neutrons and many smaller nuclides.
   e. A critical mass of fissionable material is needed to produce a fission bomb.

8. Nuclear reactors produce electrical energy. Which statement about nuclear reactors is true?

   a. Breeder reactors use small nuclides from sea water as a fuel.
   b. The moderator keeps neutrons from colliding with uranium atoms.
   c. The amount of uranium fuel present in a nuclear reactor could cause a nuclear explosion if improperly handled.
   d. Breeder reactors produce a fuel other than uranium-235.
   e. Control rods reflect neutrons back into the uranium so that the chain reaction continues.

9. Which statement about fusion is <u>not</u> true?

   a. For nuclear particles to fuse, they must first get very close to each other.
   b. Fusion is more difficult to initiate than fission.
   c. No known examples of fusion are yet available for chemists to study.
   d. Fusion can occur when light nuclei, such as nuclides of hydrogen and helium, fuse together.
   e. Fusion produces more energy per mole than fission does.

10. Which is <u>not</u> a factor in determining the biological damage caused by a source of radiation?

   a. The molar mass of the radionuclide.
   b. The chemical properties of the radiation source.
   c. The penetrating ability of the radiation.
   d. The energy of the radiation.
   e. The ionizing ability of the radiation.

PRACTICE EXAM ANSWERS

1. a (19.1)
2. d (19.1)
3. b (19.2)
4. d (19.3)
5. d (19.4)
6. e (19.5)
7. b (19.7)
8. d (19.8)
9. c (19.9)
10. a (19.10)

9. Which statement about fusion is not true?

a. For nuclear particles to fuse, they must first get very close to each other.
b. Fusion is more difficult to initiate than fission.
c. [illegible] examples of fusion are yet [illegible] for [illegible]
d. Fusion can occur when light nuclei, such as isotopes of hydrogen and helium, fuse together.
e. Fusion gives off more energy per [illegible] than fission does.

10. What [illegible] radiation?

a. [illegible]
b. The [illegible] of the radioactive [illegible]
c. The [illegible] particles in the radioactive [illegible]
d. The [illegible] of the [illegible]
e. [illegible]

ESSAY TEST [illegible]

[illegible]

CHAPTER 20: ORGANIC CHEMISTRY

INTRODUCTION

There is a whole field of chemistry, called organic chemistry, that is devoted to the study of compounds and reactions of the element carbon. No other element forms as many different compounds as carbon. The compounds range from the simple molecule methane, which we burn as fuel, to the complex molecules which carry genetic information. This chapter will help you learn the language of organic chemistry by introducing you to how carbon-containing molecules are formed, how they function, and how they are named.

AIMS FOR THIS CHAPTER

1. Know what kinds of bonds are formed when a carbon atom bonds to one or more atoms. (Section 20.1)
2. Know the names and formulas for the first ten alkanes. (Section 20.2)
3. Know how to write the formulas of structural isomers of the alkanes. (Section 20.3)
4. Know how to name alkanes systematically and how to write correct structures from names. (Section 20.4)
5. Know the major components of petroleum and how they are used. (Section 20.5)
6. Know what reactions the alkanes undergo. (Section 20.6)
7. Know how to name alkenes and alkynes, and how to describe addition reactions. (Section 20.7)
8. Know how to recognize and draw a common feature of aromatic hydrocarbons and draw the structural formula of the aromatic hydrocarbon benzene. (Section 20.8)
9. Know how to name the aromatic hydrocarbons. (Section 20.9)
10. Be able to recognize each of the organic functional groups. (Section 20.10)
11. Know how to name and how to classify the alcohols. (Section 20.11)
12. Be familiar with the properties and uses of the common alcohols. (Section 20.12)
13. Be familiar with the properties and the uses of common aldehydes and ketones. (Section 20.13)
14. Know how to name aldehydes and ketones. (Section 20.14)
15. Be familiar with the structures and properties of the common carboxylic acids and esters, and be able to name them. (Section 20.15)
16. Know the two basic ways that polymers are made from monomers, and be able to draw a section of polymer, given the monomers. (Section 20.16)

QUICK DEFINITIONS

Hydrocarbons Compounds which are composed of only carbon and hydrogen. (Section 20.2)

Saturated — Carbon compounds in which all of the carbons are bonded to four other atoms. Each carbon atom has formed the maximum number of bonds. (Section 20.2)

Alkanes — Saturated hydrocarbons. All carbon atoms are bonded to four other atoms. (Section 20.2)

Normal alkanes — Alkanes whose carbon atoms are found in a single unbranched chain. Normal alkanes are also called straight-chain alkanes, or unbranched hydrocarbons. (Section 20.2)

Straight-chain alkanes — See the entry under normal alkanes.

Unbranched hydrocarbons — See the entry under normal alkanes.

Structural isomerism — Structural isomers have the same numbers and kinds of atoms, but the atoms are attached to each other in a different order. (Section 20.3)

Substituent — A group or a branch which substitutes for a hydrogen. (Section 20.4)

Alkyl group — A substituent formed from an alkane by removing a hydrogen. (Section 20.4)

Petroleum — A thick dark liquid usually found beneath the earth's surface and used as a source of hydrocarbons for gasoline and many other important organic molecules. (Section 20.5)

Natural gas — Often found with petroleum. Composed mainly of methane, with some ethane, propane and butane. (Section 20.5)

Kerosene fraction — The part of petroleum which contains molecules with ten to eighteen carbons. (Section 20.5)

Gasoline fraction — The part of petroleum which contains molecules with five to ten carbons. (Section 20.5)

Pyrolytic cracking — A process where long molecules in the kerosene fraction are broken into smaller molecules which can be used as gasoline. (Section 20.5)

Combustion reaction	The type of reaction represented when a hydrocarbon reacts with O_2 gas to produce CO_2 and H_2O. (Section 20.6)
Substitution reactions	Reactions in which one or more hydrogens of an alkane are substituted with a different atom. (Section 20.6)
Dehydrogenation reactions	The removal of two hydrogens from a hydrocarbon to produce an unsaturated hydrocarbon. (Section 20.6)
Alkenes	A hydrocarbon with one or more double carbon-carbon bonds. (Section 20.7)
Alkynes	A hydrocarbon with one or more triple carbon-carbon bonds. (Section 20.7)
Addition reactions	New atoms add to the carbons involved in double or triple bonds, producing new single bonds on the carbons. (Section 20.7)
Hydrogenation reactions	A type of addition reaction where two hydrogen atoms add to the carbons involved in a double bond, creating two new carbon-hydrogen bonds. (Section 20.7)
Halogenation	The addition of halogen atoms to the carbons involved in a double bond, creating new single bonds between the carbons and the halogen atoms. (Section 20.7)
Polymerization	The joining of small molecules together to form a large molecule. (Section 20.7)
Benzene	A hydrocarbon with six carbons joined in a ring, and six hydrogens, one attached to each carbon atom. At room temperature, benzene is a liquid. (Section 20.8)
Phenyl group	A benzene ring used as a substituent. (Section 20.9)
Hydrocarbon derivatives	Hydrocarbons which contain other atoms besides carbon and hydrogen. (Section 20.10)
Functional groups	Groups of atoms which are commonly found together and which modify the chemistry of a hydrocarbon. (Section 20.10)

Alcohols — A hydrocarbon derivative which contains the hydroxyl functional group, -OH. (Section 20.11)

Carbonyl group — A functional group with no OH group where an oxygen atom is attached to a carbon by a double bond. (Section 20.13)

$$-\overset{\overset{\displaystyle O}{\|}}{C}-$$

Ketone — A hydrocarbon derivative where the carbon of a carbonyl group is attached to an R group on each side. (Section 20.13)

$$R-\underset{\underset{\displaystyle O}{\|}}{C}-R'$$

Aldehyde — A hydrocarbon derivative where the carbon of a carbonyl group is attached to an R group on one side, and a hydrogen on the other side. (Section 20.13)

$$R-\underset{\underset{\displaystyle O}{\|}}{C}-H$$

Carboxyl group — A functional group containing a carbonyl and a hydroxyl group. (Section 20.15)

$$-\overset{\overset{\displaystyle O}{\|}}{C}-OH$$

Carboxylic acids — Hydrocarbon derivatives which contain one or more carboxyl groups and have the general formula RCOOH. (Section 20.15)

Ester — A functional group which is formed when a carboxylic acid reacts with an alcohol. A molecule of water is lost in the process. (Section 20.15)

$$-\overset{\overset{\displaystyle O}{/\!/}}{C}-O-H + HO-R \rightarrow -\overset{\overset{\displaystyle O}{/\!/}}{C}-OR + H_2O$$

Polymers — Long molecules made by joining together many small molecules. (Section 20.16)

Monomer Small molecules from which polymers are made. (Section 20.16)

Addition polymerization Polymers made by adding together many monomers without losing any atoms from the monomers. (Section 20.16)

Condensation polymerization Polymers in which a small molecule such as water is produced each time a monomer is joined to the end of the growing chain. (Section 20.16)

Copolymer A polymer made from more than one kind of monomer. (Section 20.16)

Homopolymer A polymer made from only one kind of monomer. (Section 20.16)

Dimer A molecule made from two monomers joined together. (Section 20.16)

Polyester A condensation polymer made from a dialcohol and a dicarboxylic acid. (Section 20.16)

CONTENT REVIEW

20.1 CARBON BONDING

What Kinds of Bonds Does Carbon Form With Other Atoms?

A carbon atom can form a maximum of four covalent bonds with other atoms. The VSEPR model tells us that electron pairs try to spread out as far as possible, to minimize repulsions from the negatively charged electrons. The shape assumed by a carbon atom bonded to four atoms is a tetrahedron. In this shape the four electron pairs are as far apart as possible. Any time carbon is bonded to four other atoms, the shape will be a tetrahedron. Carbon can bond to fewer than four other atoms when it forms a double or a triple bond. When carbon forms a double bond with an atom, two electron pairs are used to make the double bond. When carbon forms a triple bond, three electron pairs are used to form the triple bond. Carbons with either double or triple bonds do not have a tetrahedral shape.

20.2 ALKANES

What Are the Alkanes?

Hydrocarbons are compounds composed entirely of carbon and hydrogen. Hydrocarbons are either saturated or unsaturated. A **saturated hydrocarbon** is one where all of the carbons are bonded to four atoms. An **unsaturated hydrocarbon** has at least one carbon atom bonded to less than four atoms, that is, it has at least one double or triple bond. **Alkanes** are saturated hydrocarbons.

What Are the Structures and Formulas For Some of the Alkanes?

The simplest alkane contains one carbon atom bonded to four hydrogens and is called methane. The formula for methane is CH_4. All alkanes except methane have more than one carbon atom bonded together. Hydrogens are bonded to each carbon so that each carbon atom forms four bonds. Ethane has two carbons and a structure of CH_3CH_3. The alkane with three carbons is propane, with a structure of $CH_3CH_2CH_3$. Some alkanes have all the carbon atoms in a chain, with no branches or side chains. These are the **normal**, or **straight-chain** alkanes, also called **unbranched hydrocarbons**. You should learn the names of the structures of the first ten alkanes given in Table 20.1 of your textbook. All of the organic chemistry in this chapter uses the names given in this table, so if you cannot remember the names and structures of the alkanes, you will have difficulty completing the material in this chapter.

20.3 STRUCTURAL FORMULAS AND ISOMERISM

What Is Structural Isomerism?

When we write a formula such as C_4H_{10}, we can tell from the formula that this hydrocarbon contains four carbon atoms and ten hydrogen atoms. But we cannot tell how the carbon atoms are arranged. Are all four carbons in a single chain, or is there a branch? We can draw two different **structural formulas** of C_4H_{10}.

```
                                  CH3 CHCH3
        CH3 CH2 CH2 CH3    and        |
                                     CH3
```

Structural formulas tell us the order of the bonding. Two molecules with the same formula but a different order to the bonding are called **structural isomers**. Each structural isomer is a different molecule, with a different name and different chemical properties. Structural formulas are very useful in organic chemistry because they help us to identify a unique structural isomer.

20.4 NAMING ALKANES

How Are Alkanes Assigned Individual Names?

Common names are used widely in organic chemistry, but common names do not tell us anything about the structure of the compound. Chemists devised a systematic naming scheme so that anyone who learned the system could write the structural formula of an organic compound from the name.

Before we look at the rules for naming organic compounds, we need to look at hydrocarbon substituents, or branches, and see how they are formed and named. Many organic compounds have side branches on the main carbon chain. For example, the molecule below has several branches.

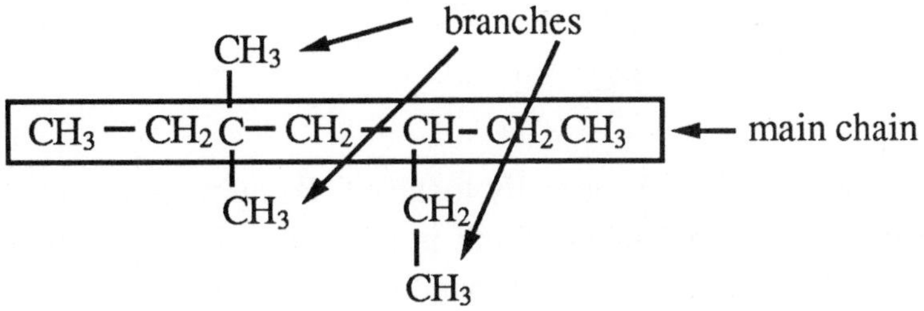

```
                     branches
         CH3  <----
          |
 | CH3 - CH2 C - CH2 - CH - CH2 CH3 |  <-- main chain
          |           |
         CH3         CH2
                      |
                     CH3
```

The side branches are formed from alkanes by removing a hydrogen atom. CH_3CH_3 can be attached to a carbon chain by removing a hydrogen from one of the carbons to form the group CH_3CH_2–. CH_4 can form a branch by losing a hydrogen to become CH_3–. Any alkane can become a branch in the same way, by removing a hydrogen from one of the carbons. The groups are given a name derived from the alkane name. When CH_4 loses a hydrogen to become CH_3–, the name changes from meth**ane** to meth**yl**. When CH_3CH_3 loses a hydrogen to become CH_3CH_2–, the name changes from eth**ane** to eth**yl**. When alkanes become substituents, the substituents are named by dropping the -ane ending, and adding -yl. The most common alkyl substituents are given in Table 20.2 of your text. You should learn these names in order to be able to easily name organic compounds.

The rules for naming organic compounds are summarized below.

1. Find the longest chain of carbons. The chain can be bent, as long as the carbons are in a continuous chain. This chain is given the alkane name corresponding to the number of carbons in the chain.
2. Number each carbon in the longest chain. Begin numbering from the end which is closest to a substituent. If starting from either end gives a substituent on the same numbered carbon, then continue numbering from either end until you come to the next substituent. The end which locates all substituents with the lowest numbers is the correct end to begin numbering from.
3. Name each substituent, and locate its position with a number. Separate the number and the name of the substituent with a hyphen.
4. If there are two or more identical alkyl groups, use a prefix such as di or tri to indicate how many alkyl groups of a given type there are. Use numbers to locate **each** alkyl group. Separate each number with a comma.
5. Assemble all the parts of the name. The alkyl groups with their numbers should come first, followed by the alkane name. If there is more than one substituent, put them in alphabetical order based on the substituent name. Do not use the prefixes di, tri, and so on, when assigning alphabetical order, only the actual alkyl name.

If the structure you are to name is given in condensed form, expand the formula so that you can see each bond attachment before you begin to name. $CH_3CHCH_3CH_2CH_2CH_2CHCH_3CH_3$ should be expanded to

```
CH3 CHCH2 CH2 CH2 CHCH3
     |             |
    CH3           CH3
```

Now you can see where each branch is actually located. To help locate the longest continuous carbon chain, draw a box around what you think is the longest chain, then count and number each carbon in the box. Try drawing boxes around each potential longest chain until you are certain that you have identified the longest continuous chain. Everything outside of the box is a substituent, so drawing a box around the chain makes it easier to identify and locate substituents. Do not be discouraged because the molecule looks complicated. Organic molecules are built from small familiar pieces, and once the pieces are named, you can put the names together in the proper order.

Example:
Name the molecule below.

```
CH3 CH2 CH2 CHCH2 CH2
             |     |
            CH2   CH3
             |
            CH3
```

Draw a box around what appears to be the longest carbon chain until you find the longest.

```
|CH3 CH2 CH2 CHCH2 CH2|      |CH3 CH2 CH2 CH-|-CH2 CH2      |CH3 CH2 CH2 CHCH2 CH2|
             |     |                    |        |                       |     |
            CH2   CH3                  |CH2|    CH3                     CH2   |CH3|
             |                          |                                |
            CH3                        |CH3|                            CH3

  6 carbons                     6 carbons                        7 carbons
                                                                 longest
```

The longest chain has seven carbons, so this molecule is a heptane. Now number the chain from the end closest to a substituent. In this case, the only substituent is on the middle carbon so we can number from either end.

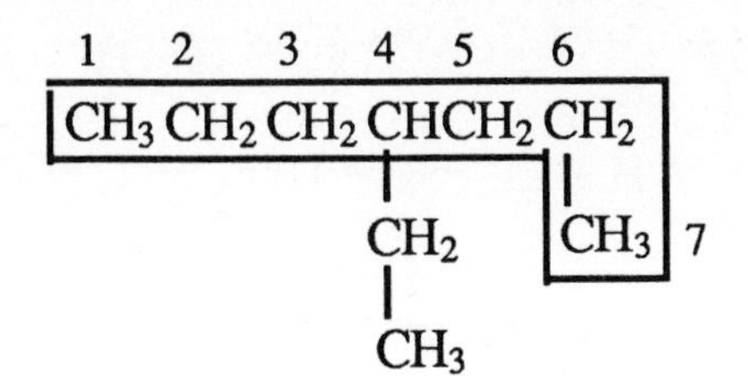

There is one substituent on carbon four, an ethyl group, called 4-ethyl. When we assemble the parts of the name, we have 4-ethylheptane.

20.5 PETROLEUM

What Is Petroleum and How Is It Used?

Petroleum is a thick liquid composed mainly of hydrocarbons and is believed to have been formed from decayed plant remains. The hydrocarbons in petroleum are from five to more than twenty-five carbons in length. Natural gas is often found with petroleum and consists of alkanes from one to four carbons in length, but most of the natural gas is methane, CH_4.

Petroleum is not very useful until the hydrocarbons have been separated into fractions based on the length of the carbon chain. The separation is performed by heating the petroleum until the low molecular weight molecules with low boiling points boil, leaving the rest of the petroleum behind. By raising the boiling point, more molecules boil, until finally, the largest molecules are left behind.

20.6 REACTIONS OF ALKANES

What Kinds of Reactions Do the Alkanes Undergo?

The only kinds of bonds the alkanes have are carbon-carbon and carbon-hydrogen covalent bonds. Both these bonds are relatively stable, that is, they do not break very easily, so alkanes do not react with many substances. They do undergo combustion. In these **combustion reactions,** alkanes react with O_2 to produce CO_2 and H_2O.

Example:

$$2C_3H_8(g) + 10O_2(g) \rightarrow 6CO_2(g) + 8H_2O(g)$$

propane

Alkanes also undergo substitution reactions. In a **substitution reaction**, one or more hydrogen atoms are substituted by different atoms, in this case, by halogen atoms. Ultraviolet light, $h\nu$, is required to supply energy for the reaction.

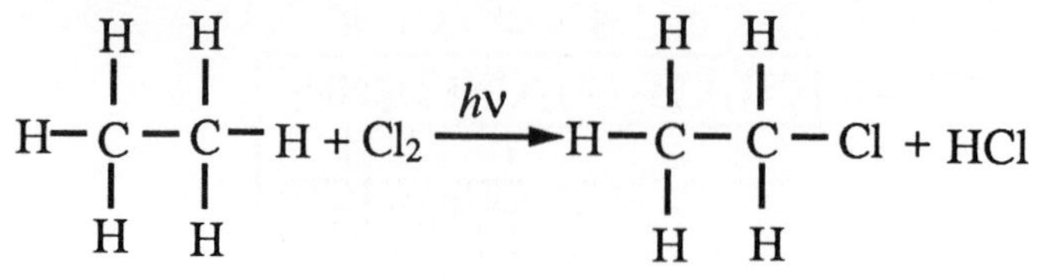

$$H_3C{-}CH_3 + Cl_2 \xrightarrow{h\nu} H_3C{-}CH_2Cl + HCl$$

$$\begin{array}{c} \text{H}\ \ \text{H} \\ |\ \ \ \ | \\ \text{H}-\text{C}-\text{C}-\text{H} \\ |\ \ \ \ | \\ \text{H}\ \ \text{H} \end{array} + Cl_2 \xrightarrow{h\nu} \begin{array}{c} \text{H}\ \ \text{H} \\ |\ \ \ \ | \\ \text{H}-\text{C}-\text{C}-\text{Cl} \\ |\ \ \ \ | \\ \text{H}\ \ \text{H} \end{array} + HCl$$

ethane 1-chloroethane

We need to name molecules which have halogen substituents. The chlorine atom becomes a chloro substituent, and a bromine atom becomes a bromo substituent. Halogen substituents are given location numbers, just as alkyl substituents are. $CH_2ClCH_2CH_2CH_2Cl$ is named 1,4-dichlorobutane.

Alkanes also undergo **dehydrogenation reactions**, which means the removal of hydrogen atoms. Under certain conditions, a hydrogen atom can be removed from adjacent carbons, resulting in a double bond between the two carbons.

$$CH_3CH_2CH_3 \xrightarrow[Cr_2O_3]{500°} CH_2 = CHCH_3 + H_2$$

propane propene

20.7 ALKENES AND ALKYNES

How Can We Name Organic Compounds Which Contain Double Or Triple Bonds?

Alkenes are hydrocarbons which contain at least one carbon-carbon double bond. **Alkynes** are hydrocarbons which contain at least one carbon-carbon triple bond. The rules for naming these compounds are similar to those for the alkanes, with some modifications for the multiple bonds.

1. Find the longest continuous chain containing the double or triple bond.
2. Determine what the correct alkane name would be, based on the number of carbons in the chain. Drop the **-ane** ending, and add **-ene** if a double bond is present, or **-yne** if a triple bond is present.
3. Number the chain beginning at the end closest to the multiple bond. Use a number to show where in the chain the multiple bond occurs.
4. Name and locate the substituents as you did for the alkanes. Assemble the parts into a complete name.

Example:

$$CH_3CH_2C{\equiv}CCH_2\underset{\displaystyle CH_3}{\underset{|}{C}}HCH_2CH_3$$

The longest chain which includes the triple bond has eight carbons.

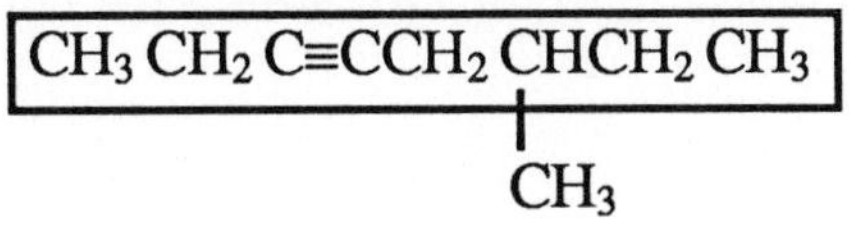

$$\boxed{CH_3\,CH_2\,C{\equiv}CCH_2\,\underset{\displaystyle CH_3}{\underset{|}{C}}HCH_2\,CH_3}$$

The alkane name for an eight-carbon chain is octane. Because there is a triple bond, drop the -ane ending and add -yne to produce octyne. Begin numbering the chain from the left to give the triple bond the lowest number.

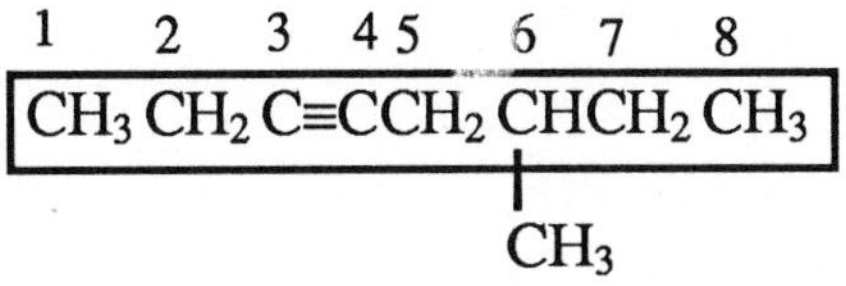

$$\overset{1}{CH_3}\,\overset{2}{CH_2}\,\overset{3}{C}{\equiv}\overset{4}{C}\overset{5}{CH_2}\,\underset{\displaystyle CH_3}{\underset{|}{\overset{6}{C}}}H\overset{7}{CH_2}\,\overset{8}{CH_3}$$

The triple bond begins on carbon three, so this would be 3-octyne. There is one substituent, a methyl group, located on carbon six, so the entire name is 6-methyl-3-octyne.

What Reactions Do Alkenes Undergo?

Alkenes undergo different addition reactions. In an **addition reaction**, the double bond is broken, and new single bonds are formed between the two carbon atoms at the site of the double bond where the atoms are added. In a **hydrogenation reaction**, hydrogen is added to a double bond.

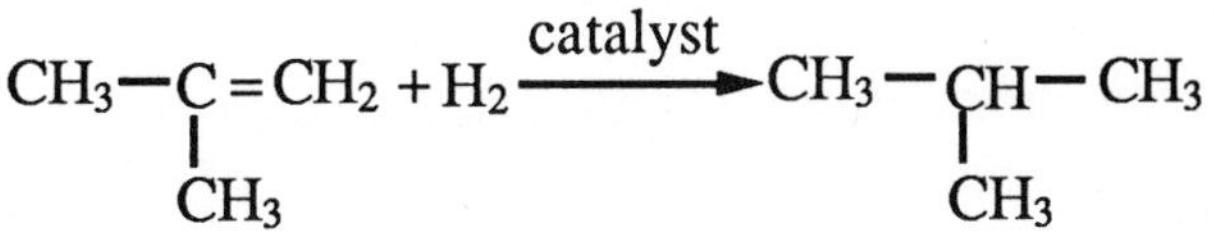

$$CH_3-\underset{\displaystyle CH_3}{\underset{|}{C}}=CH_2 + H_2 \xrightarrow{\text{catalyst}} CH_3-\underset{\displaystyle CH_3}{\underset{|}{C}}H-CH_3$$

2-methylpropene   2-methylpropane

In a **halogenation reaction**, halogen atoms are added to unsaturated hydrocarbons.

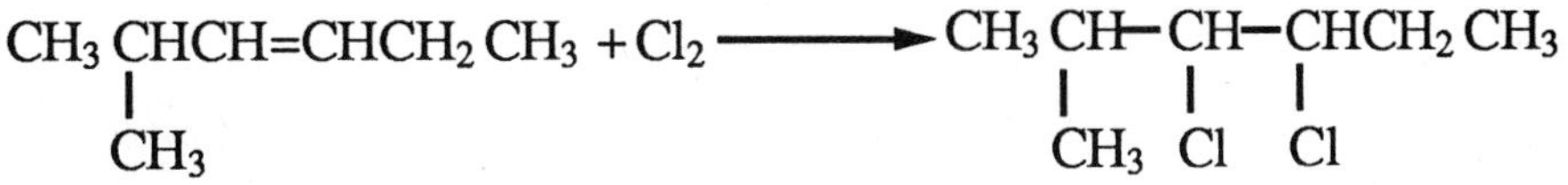

$$CH_3\,\underset{\displaystyle CH_3}{\underset{|}{C}}HCH{=}CHCH_2\,CH_3 + Cl_2 \longrightarrow CH_3\,\underset{\displaystyle CH_3}{\underset{|}{C}}H-\underset{\displaystyle Cl}{\underset{|}{C}}H-\underset{\displaystyle Cl}{\underset{|}{C}}HCH_2\,CH_3$$

2-methyl-3-hexene   3,4-dichloro-2-methylhexane

20.8 AROMATIC HYDROCARBONS

What Are Aromatic Hydrocarbons?

Some of the hydrocarbons in coal and petroleum were found to have pleasant odors. These were called **aromatic hydrocarbons**. The fragrant molecules all have as part of their structures a ring made of six carbons and six hydrogens called a **benzene** ring. Benzene is a flat molecule whose actual bonding is a hybrid of the bonding in the two structures below.

and

Because it is difficult to draw one structure which accurately shows the bonding of benzene, the structure of benzene is often indicated by a ring with a circle inside. The carbons are understood to be at the corners of the hexagon, and the hydrogens are not shown.

20.9 NAMING AROMATIC COMPOUNDS

How Are the Aromatic Hydrocarbons Named?

Hydrogens can be removed from alkanes and alkenes and be replaced by substituents. Hydrogens on benzene can also be replaced with substituents of various kinds. Benzene with one substituent is called a **monosubstituted** benzene. The rules for naming monosubstituted benzenes tell us to name the substituent as a prefix, followed by the word benzene. The molecule below would be named propylbenzene.

$CH_2\,CH_2\,CH_3$

Do not put a space between the prefix and the word benzene. A few of the monosubstituted benzenes are not named by the rules, but have accepted common names. Among these are toluene and phenol.

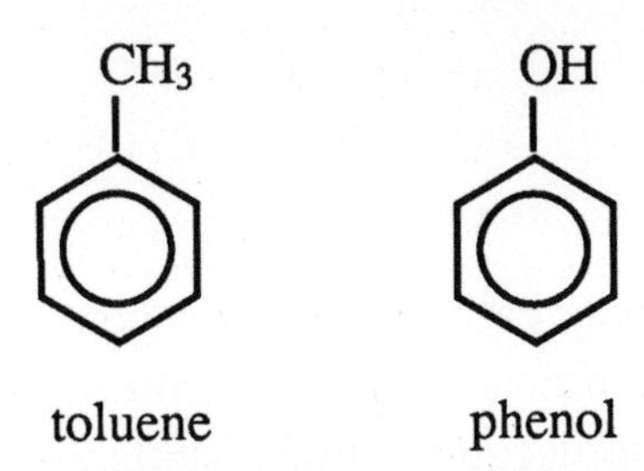

toluene phenol

When the substituent on a benzene ring is long or complex, it is often easier to name the compound by making the benzene ring the substituent. Benzene becomes a substituent by removing a hydrogen from one of the carbons. The substituent is called phenyl.

For example, the molecule below is called 2,2-dimethyl-4-phenylhexane.

CH_3

$CH_3-C-CH_2-CH-CH_2-CH_3$

CH_3

For the disubstituted benzenes, it is necessary to use a number to locate both of the substituents. For example, the structures below are 1,2-dinitrobenzene, 1,3-dinitrobenzene, and 1,4-dinitrobenzene.

NO_2 NO_2 NO_2 NO_2 NO_2 NO_2 NO_2 NO_2 NO_2 NO_2

1,2-dinitrobenzene	1,3-dinitrobenzene	1,4-dinitrobenzene	1,3-dinitrobenzene not 1,5-dinitrobenzene	1,2-dinitrobenzene not 1,6-dinitrobenzene

There is no 1,5 or 1,6-dinitrobenzene, because you should always use the lowest numbers. 1,5-dinitrobenzene should be named 1,3-dinitrobenzene, and 1,6-dinitrobenzene should be named 1,2-dinitrobenzene. Another way of locating the substituents uses the prefixes *ortho-* (*o*-), *meta-* (*m*-) and *para-* (*p*-) to show the relationship between the two substituents. 1,2-dinitrobenzene is called *o*-dinitrobenzene, 1,3-dinitrobenzene is called *m*-dinitrobenzene and 1,4-dinitrobenzene is called *p*-dinitrobenzene. You should be able to name disubstituted benzenes using either numbers or prefixes.

20.10 FUNCTIONAL GROUPS

What Are Functional Groups?

Not all organic molecules contain only carbon and hydrogen. Many contain other atoms, such as oxygen and nitrogen. These atoms often exist as **functional groups**, groups of atoms that impart special chemical characteristics to molecules. Study Table 20.5 in your textbook, which lists the structures and classes of the common functional groups. Make sure you can correctly identify them when you see them as part of larger molecules.

20.11 ALCOHOLS

How Are Alcohols Classified and Named?

Alcohols all contain the hydroxyl functional group (-OH). To name the simple alcohols, drop the **-e** from the alkane name, and add **-ol.** Methane becomes methanol, and ethane becomes ethanol. Use a number to locate the hydroxyl only if necessary. Ethanol does not need a number because the hydroxyl can only be on carbon number 1. Propanol needs a number, because the hydroxyl can be placed on either an end carbon or a middle carbon. The molecule below is named 3-hexanol, not 4-hexanol. Always begin numbering from the end which gives the smallest numbers to the functional groups.

$$\overset{6}{CH_3}-\overset{5}{CH_2}-\overset{4}{CH_2}-\underset{\displaystyle OH}{\overset{3}{CH}}-\overset{2}{CH_2}-\overset{1}{CH_3}$$

Alcohols are classified by the number of hydrocarbon fragments attached to the carbon the hydroxyl is bonded to. If the carbon with the hydroxyl is bonded to one hydrocarbon fragment and two hydrogen atoms, the alcohol is a **primary alcohol.** If the carbon with the hydroxyl is bonded to two hydrocarbon fragments and one hydrogen atom, the alcohol is a **secondary alcohol.** If the carbon with the hydroxyl is bonded to three hydrocarbon fragments, and no hydrogen atoms, the alcohol is a **tertiary alcohol.**

$$CH_3CH_2\underset{\displaystyle OH}{CH}CH_3$$

is a secondary alcohol. The carbon the -OH is attached to is bonded to two carbons.

$$CH_3-\underset{\displaystyle CH_3}{\overset{\displaystyle CH_3}{C}}-CH_2OH$$

is a primary alcohol. The carbon the -OH is attached to is bonded to one carbon.

$$\begin{array}{c} CH_3 \\ | \\ CH_3-C-CH_2CH_3 \\ | \\ OH \end{array}$$

is a tertiary alcohol. The carbon the -OH is attached to is bonded to three carbons.

20.12 PROPERTIES AND USES OF ALCOHOLS

How Are Alcohols Prepared and Used?

The two smallest alcohols, methanol and ethanol, are used in large quantities by industry. Methanol is prepared from carbon monoxide by a hydrogenation reaction.

$$CO + 2H_2 \xrightarrow[ZnO/Cr_2O_3]{400°} CH_3OH$$

It is used as a fuel, and as a starting material for many other products.

Ethanol is produced biologically and industrially. The biological production of ethanol begins with a sugar, and uses biological catalysts (enzymes) found in yeast to produce the ethanol.

$$C_6H_{12}O_6 \xrightarrow{\text{yeast}} 2CH_3CH_2OH + 2CO_2$$

Ethanol produced this way is used for beverages such as beer or wine. Ethanol can also be produced for use in industry. Ethylene undergoes an addition reaction with water to produce ethyl alcohol.

$$CH_2{=}CH_2 + H_2O \xrightarrow{\text{acid catalyst}} CH_3CH_2OH$$

20.13 ALDEHYDES AND KETONES

What Are the Structures of Aldehydes and Ketones and How Are They Made?

Aldehydes and ketones both contain the carbonyl group,

$$\begin{array}{c} -C- \\ \| \\ O \end{array}$$

In a **ketone** the carbonyl group is bonded to two carbons. Ketones have the carbonyl group on a middle carbon.

The general formula for a ketone is

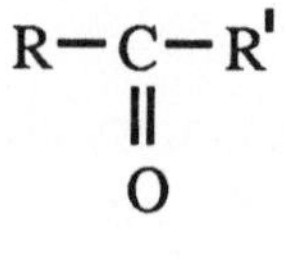

$$\begin{array}{c} R-\underset{\displaystyle \overset{\|}{O}}{C}-R' \end{array}$$

When the carbonyl group is bonded to a hydrogen and one carbon, the molecule is an **aldehyde.** Aldehydes have the carbonyl group on the end of the molecule. The general formula for an aldehyde is

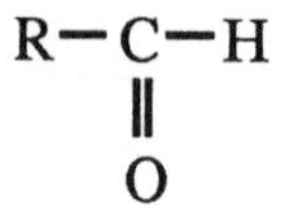

$$\begin{array}{c} R-\underset{\displaystyle \overset{\|}{O}}{C}-H \end{array}$$

Both aldehydes and ketones can be produced by oxidation of alcohols. The oxidation of a primary alcohol produces an aldehyde, while the oxidation of a secondary alcohol produces a ketone.

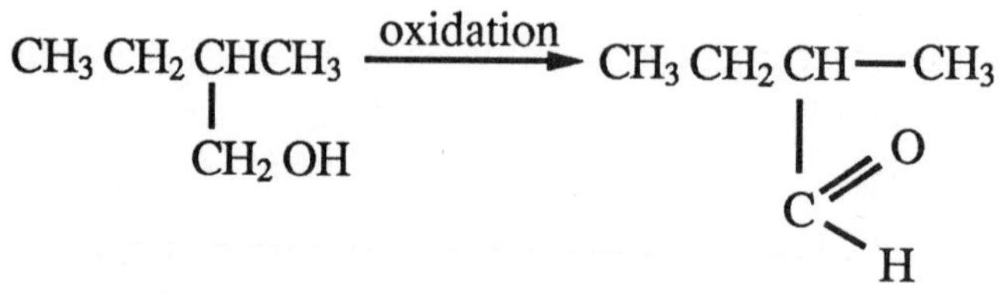

$$CH_3CH_2\underset{\displaystyle CH_2OH}{\underset{|}{C}}HCH_3 \xrightarrow{\text{oxidation}} CH_3CH_2\underset{\displaystyle C(=O)H}{\underset{|}{C}}H-CH_3$$

primary alcohol → aldehyde

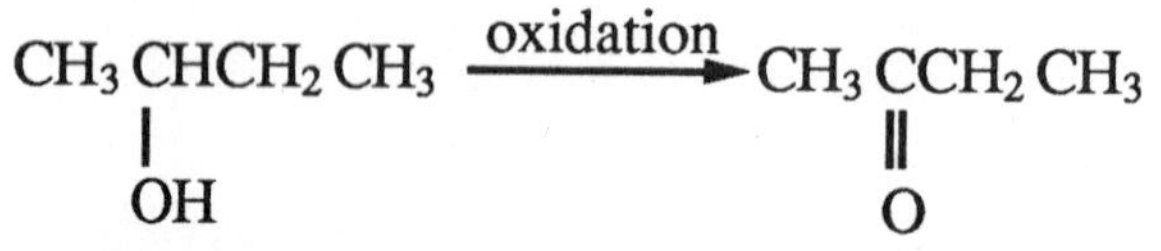

$$CH_3\underset{\displaystyle OH}{\underset{|}{C}}HCH_2CH_3 \xrightarrow{\text{oxidation}} CH_3\underset{\displaystyle O}{\underset{\|}{C}}CH_2CH_3$$

secondary alcohol → ketone

20.14 NAMING ALDEHYDES AND KETONES

How Are Aldehydes and Ketones Named?

Determine the correct alkane name, drop the -**e** and add an -**al** ending for aldehydes, and an -**one** ending for ketones. You never need a number to locate the aldehyde group. It is always on the end, so it is always on carbon 1. Use a number to locate the carbonyl group in the ketones.

Example:

$$CH_3CH_2CH(CH_2CH_3)CH_2C(=O)H$$ is 3-ethylpentanal, and

$$CH_3CH_2CH(CH_3)-CH(CH_3)C(=O)CH_3$$ is 3,4-dimethyl-2-hexanone

There are a few common names whose structures you should know. Methanal, $HC(=O)H$, is also called formaldehyde. Ethanal, $CH_3C(=O)H$, is called acetaldehyde. Propanone, $CH_3C(=O)CH_3$, is called acetone. A benzene ring with a carbonyl group, $C_6H_5C(=O)H$, is called benzaldehyde.

A method sometimes used to name ketones names the alkyl groups on each side of the carbonyl group followed by the word ketone.

Example:

$$CH_3C(=O)CH_2CH_2CH_3$$

would be named methylpropylketone.

20.15 CARBOXYLIC ACIDS AND ESTERS

What Are the Structures and Names of the Common Carboxylic Acids?

Carboxylic acids contain the carboxyl group

$$-\overset{\overset{\Large O}{\|}}{C}-OH$$

Carboxylic acids are all weak acids, they donate a proton by the reaction

$$RCOOH(aq) + H_2O(l) \rightleftharpoons H_3O^+(aq) + RCOO^-(aq)$$

The dissociation equilibrium for this reaction is to the left. Most of the acid is undissociated. Carboxylic acids are named by droping the **-e** of the alkane name and adding **-oic acid**. Heptane would become heptanoic acid, and pentane would become pentanoic acid. The carboxyl group is always on the end of a molecule, so you do not need to assign the carboxyl functional group a number. You should learn the common names for the carboxylic acids given in Table 20.7 of your textbook. The common names are often used in place of the systematic names.

Carboxylic acids can react with an alcohol to produce an **ester** and water.

Example:

Propanoic acid reacts with methanol as shown below.

$$CH_3CH_2\overset{\overset{O}{/\!/}}{C}-\boxed{OH + H}\,OCH_3 \rightarrow CH_3CH_2\overset{\overset{O}{/\!/}}{C}-OCH_3 + H_2O$$

propanoic acid methanol methyl propanoate

Common names for the esters are often used and have two parts, the part from the alcohol, and the part from the carboxylic acid. The part of the ester which came from the alcohol is named as an alkyl group, and the part of the ester which came from the carboxylic acid is named by replacing the **-ic** ending with **-ate**. The ester made from propanoic acid and methanol would be named methyl propanoate.

20.16 POLYMERS

What Are Polymers and How Are They Made?

Polymers are large molecules made from smaller molecules joined end to end. Polymers can contain thousands of small molecules, each of which is called a **monomer**. One way polymers are produced is by addition polymerization. During **addition polymerization,** the monomers join together, often by breaking a double bond, and form new single bonds with the lengthening chain. No atoms are lost during addition polymerization. All the atoms present in the monomers are found in the polymer. The best known example is polyethylene, produced from ethylene monomers.

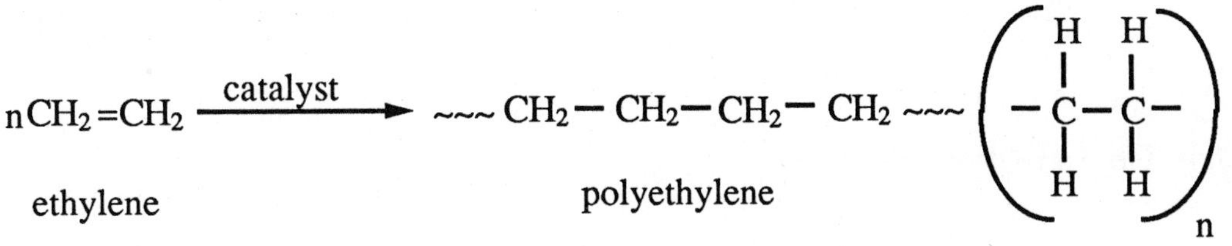

$$nCH_2{=}CH_2 \xrightarrow{\text{catalyst}} \sim\sim\sim CH_2{-}CH_2{-}CH_2{-}CH_2 \sim\sim\sim \quad \left(\begin{array}{c} H\ \ H \\ |\ \ \ | \\ -C-C- \\ |\ \ \ | \\ H\ \ H \end{array} \right)_n$$

ethylene polyethylene

Another way polymers can be produced is by a condensation reaction. In a **condensation reaction,** a small molecule such as H_2O or HCl is produced when the monomers react. Nylon is produced by a condensation reaction between two different monomers, hexamethylenediamine and adipic acid. A water molecule is also produced.

LEARNING REVIEW

1. Match the term below with the correct definition.

a.	hydrocarbon	contains one or more double or triple bonds
b.	alkane	an unbranched molecule
c.	normal	contains only carbon-carbon single bonds
d.	unsaturated	composed of carbon and hydrogen

2. What is the shape of a carbon tetrachloride molecule, CCl_4?

3. Write the condensed formula for an unbranched alkane with five carbons.

4. A branched alkane with the formula below has seven carbons. Can this alkane be represented by the general formula C_nH_{2n+2}?

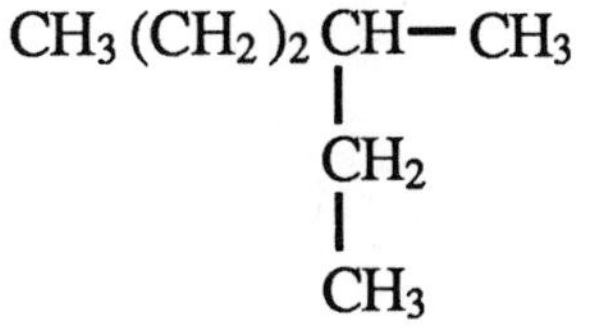

$$\begin{array}{l} CH_3(CH_2)_2\,CH{-}CH_3 \\ \qquad\qquad\quad | \\ \qquad\qquad\ CH_2 \\ \qquad\qquad\quad | \\ \qquad\qquad\ CH_3 \end{array}$$

5. What are structural isomers?

6. Write all structural isomers of hexane, C_6H_{14}.

7. Name the alkanes below.

a.
```
CH3 CH—CH—CH2 CH—CH3
    |   |     |
    CH3 CH3   CH3
```

b.
```
CH3CH2CH—CH2CH3
      |
      CH—CH3
      |
      CH3
```

c.
```
CH3 CH—CH—CH3
    |  |
    CH2 CH3
    |
    CH3
```

d.
```
                 CH3
                 |
                 CH2
                 |
CH3CH2CH—CH—CH—CH2CH3
      |  |
      CH2 CH3
      |
      CH3
```

e.
```
      CH3
      |
CH3—C—CH3
      |
      CH3
```

f.
```
                          CH3
                          |
CH3—CH—CH2—CH—C—CH3
      |        |  |
      CH3      CH2 CH3
               |
               CH2—CH2—CH3
```

8. Are these molecules structural isomers?

a.

$$\begin{array}{l} CH_3\,CH-CH-CH_3 \\ \quad\;\; | \qquad | \\ \quad CH_3 \;\; CH_2-CH_3 \end{array}$$

b.

$$\begin{array}{l} \qquad\quad CH_3 \\ \qquad\quad\; | \\ CH_3-CH \quad CH_3 \\ \qquad\quad | \qquad\; | \\ \qquad CH_2-CH \\ \qquad\qquad\quad\; | \\ \qquad\qquad\;\; CH_3 \end{array}$$

9. Write structural formulas for the molecules below.

a. 2-methyl-4-*sec*-butyloctane
b. 2-methylpropane
c. 3-ethylpentane
d. 2,2,4-trimethylhexane

10. Each of the alkane names below is **incorrect**. Write the structural formula for each molecule. Then, name it correctly.

a. 2-ethylbutane
b. 1-methylpentane
c. 4,4-dimethylhexane
d. 2-ethyl-2-methylheptane

11. Petroleum itself is not very useful. It must be separated into fractions to be useful. How is petroleum separated into the fractions which have useful properties?

12. When chlorine gas reacts with pentane, one of the hydrogens is substituted by a chlorine atom. Several structural isomers are possible depending on which hydrogen atom is substituted. Show all possible structural isomers which could be produced when chlorine reacts with pentane in the presence of ultraviolet light.

13. Write the structures for the products of the reactions below.

a. $$\begin{array}{l} CH_3\,CHCH_3 + Cl_2 \xrightarrow{h\nu} \\ \qquad | \\ \quad\; CH_3 \end{array}$$

b. $CH_4 + O_2 \longrightarrow$

c. $CH_3CH_2CH_3 \xrightarrow[500^\circ C]{CrO_3}$

14. Name each of the molecules below.

a. 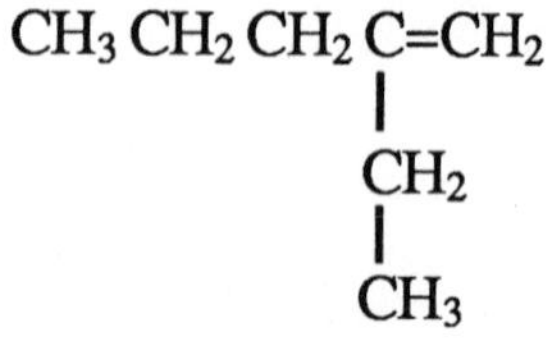 $CH_3CH_2CH_2C{=}CH_2$ (with CH_2CH_3 attached to the C)

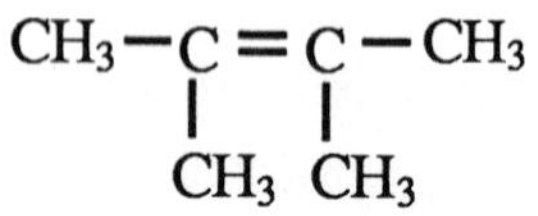

$$\begin{array}{l} CH_3CH_2CH_2\underset{\displaystyle\begin{array}{c}|\\CH_2\\|\\CH_3\end{array}}{C}{=}CH_2 \end{array}$$

b. $$CH_3\underset{\displaystyle\begin{array}{c}|\\CH_3\end{array}}{CH}-C{\equiv}C-CH_2\underset{\displaystyle\begin{array}{c}|\\CH_3\end{array}}{CH}-CH_3$$

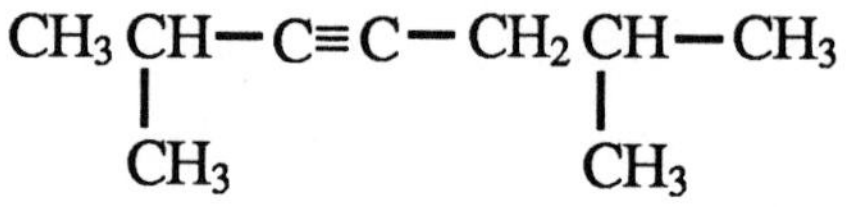

c. $$CH_3-\underset{\displaystyle\begin{array}{c}|\\CH_3\end{array}}{C}=\underset{\displaystyle\begin{array}{c}|\\CH_3\end{array}}{C}-CH_3$$

d. 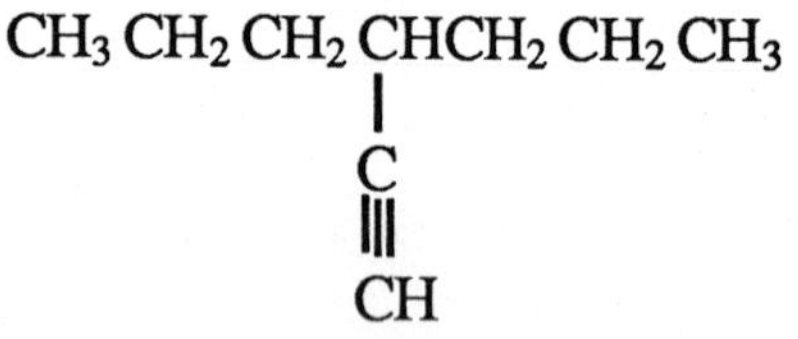 $$CH_3CH_2CH_2\underset{\displaystyle\begin{array}{c}|\\C\\|||\\CH\end{array}}{CH}CH_2CH_2CH_3$$

15. Write structural formulas for the products of the reactions below.

a. 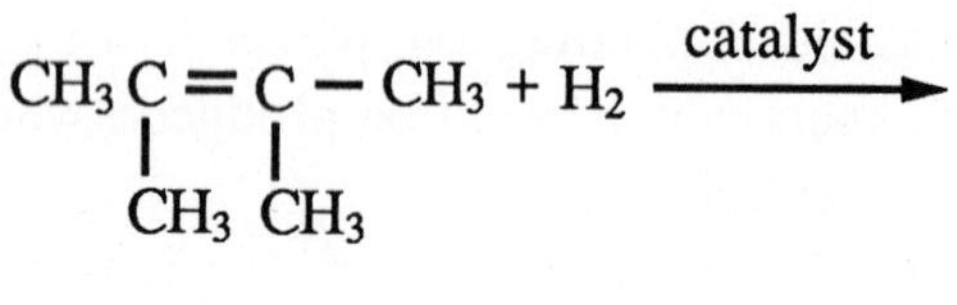 $$CH_3\underset{\displaystyle\begin{array}{c}|\\CH_3\end{array}}{C}=\underset{\displaystyle\begin{array}{c}|\\CH_3\end{array}}{C}-CH_3 + H_2 \xrightarrow{\text{catalyst}}$$

b. $CH_3CH_2CH_2CH{=}CH_2 + Cl_2 \longrightarrow$

c. 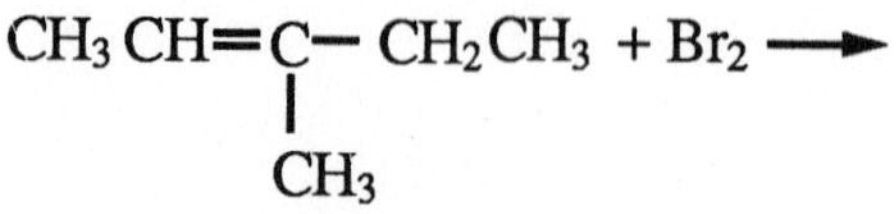 $$CH_3CH=\underset{\displaystyle\begin{array}{c}|\\CH_3\end{array}}{C}-CH_2CH_3 + Br_2 \longrightarrow$$

16. Name the aromatic hydrocarbons below.

a.

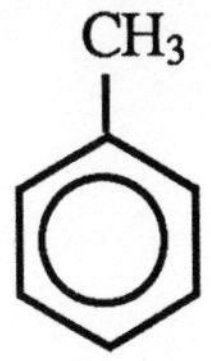
CH3

b.

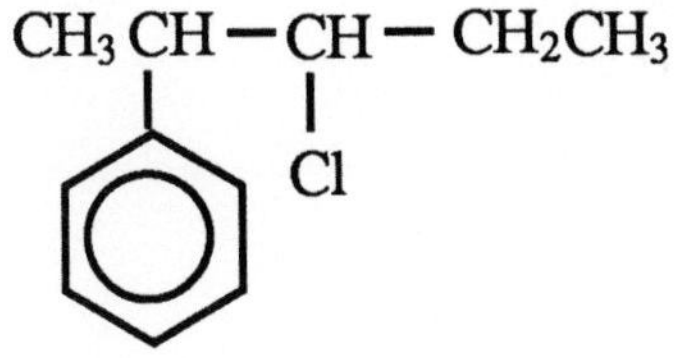
CH3 CH—CH— CH2CH3
Cl

c.

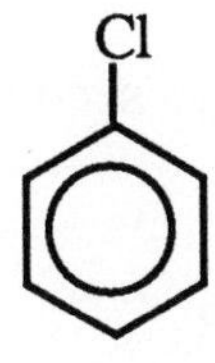
Cl

d.

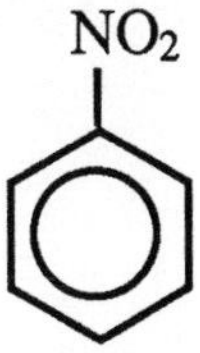
NO2

17. Name the disubstituted aromatic hydrocarbons below.

a.

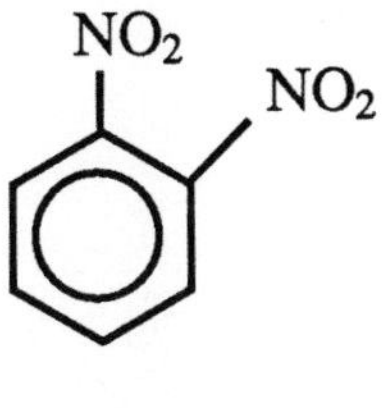
NO2
NO2

b.

OH
Cl

c. CH_3 (benzene ring) CH_2CH_3

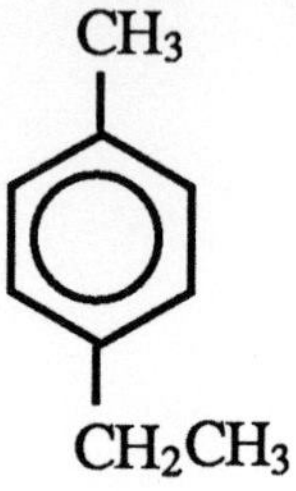

d. CH_3 (benzene ring) Br

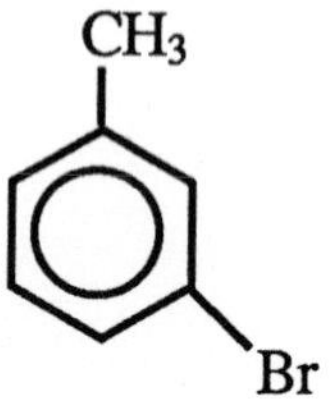

18. Name each of the alcohols below and decide whether they are primary, secondary, or tertiary alcohols.

a.

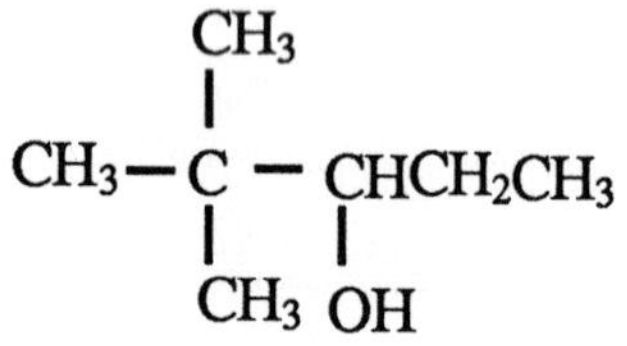

$$\begin{array}{c} CH_3 \\ | \\ CH_3-C-CHCH_2CH_3 \\ |\quad\ \ | \qquad\quad \\ CH_3\ OH \qquad\quad \end{array}$$

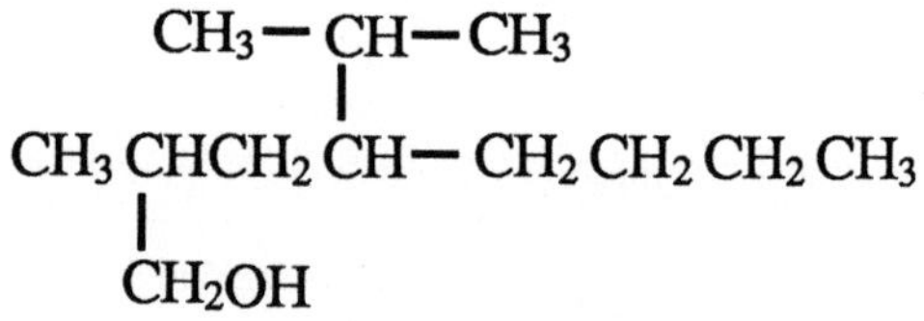

b.

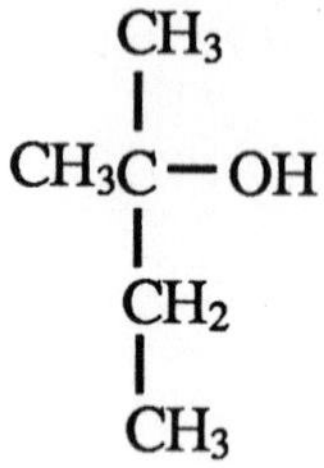

$$\begin{array}{c} CH_3 \\ | \\ CH_3C-OH \\ | \\ CH_2 \\ | \\ CH_3 \end{array}$$

c.

$$\begin{array}{c} CH_3 \\ | \\ CH_3CH_2C-CH-CH_2CH_2OH \\ |\quad\ \ | \\ Br\ \ CH_3 \end{array}$$

d.

$$\begin{array}{c} CH_3-CH-CH_3 \\ | \\ CH_3CHCH_2CH-CH_2CH_2CH_2CH_3 \\ | \\ CH_2OH \end{array}$$

e. CH_3OH

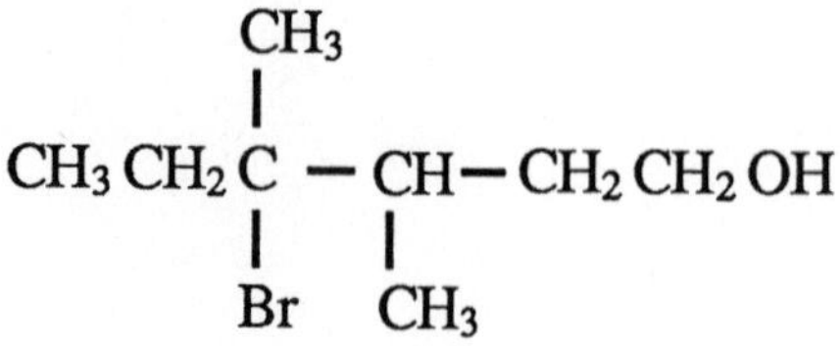

f.

$$\begin{array}{c} CH_3CH_2CHCH_2CH_3 \\ | \\ CH_2CH_2CHCH_2CHCH_3 \\ |\qquad\ \ | \\ CH_2\quad\ OH \\ | \\ CH_3 \end{array}$$

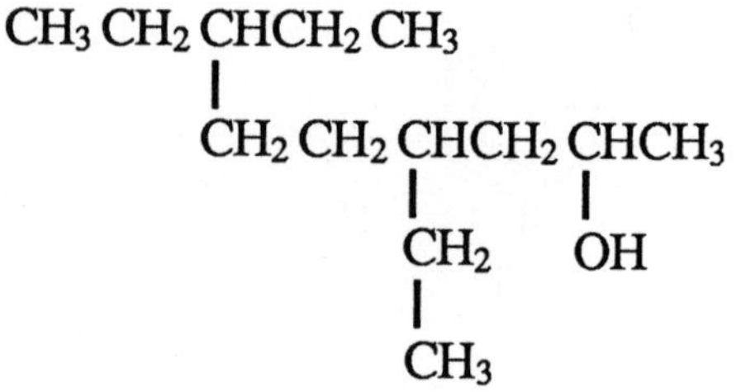

19. Match the alcohols below with the appropriate description.

a.	methanol	used in antifreeze
b.	phenol	found in alcoholic beverages
c.	ethanol	commonly known as wood alcohol
d.	ethylene glycol	used in the production of plastics

20. Show the structures of the aldehydes and ketones which would be produced from the oxidation of the following alcohols.

a. $C_6H_5CH_2OH \xrightarrow{\text{oxidation}}$

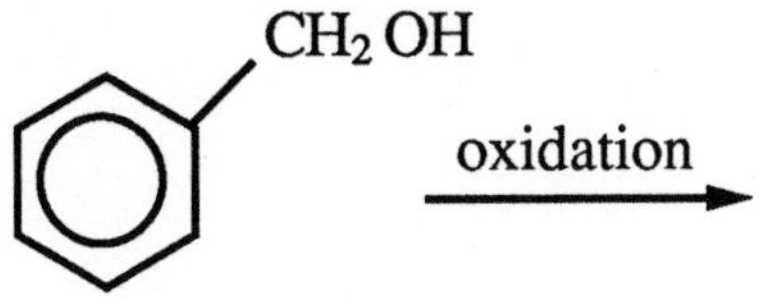

b. $CH_3CH(CH_3)CH_2OH \xrightarrow{\text{oxidation}}$

$$\begin{array}{l} CH_3\,\underset{\displaystyle CH_3}{\underset{|}{C}H}CH_2\,OH \end{array}$$

c. $CH_3CH_2CH(OH)CH_2CH_3 \xrightarrow{\text{oxidation}}$

$$CH_3\,CH_2\,\underset{\displaystyle OH}{\underset{|}{C}H}CH_2\,CH_3$$

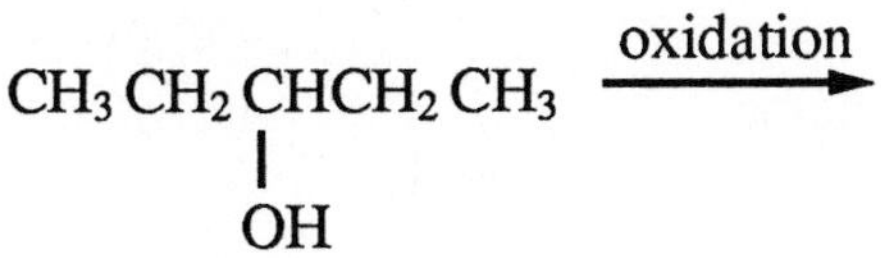

21. Name the aldehydes and ketones below.

a. $CH_3C(CH_3)_2-CH(CH_3)-CHO$

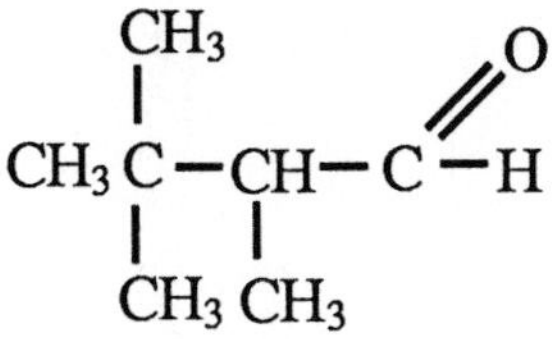

$$CH_3\,\underset{\displaystyle CH_3}{\underset{|}{\overset{\displaystyle CH_3}{\overset{|}{C}}}}-\underset{\displaystyle CH_3}{\underset{|}{CH}}-\overset{\displaystyle O}{\overset{\|}{C}}-H$$

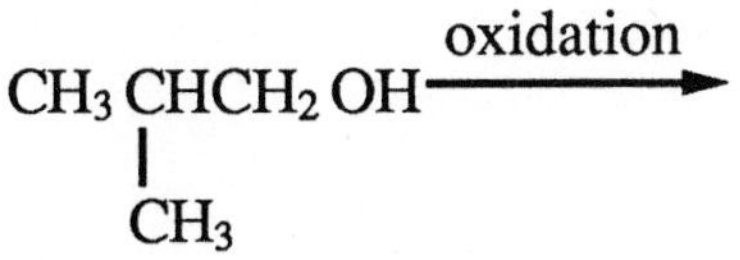

b. $CH_3\overset{\displaystyle O}{\overset{\|}{C}}-CH_2\,\overset{\displaystyle CH_2CH_3}{\overset{|}{C}}HCH_2CH_3$

c. $H-\overset{\displaystyle O}{\overset{\|}{C}}-\underset{\displaystyle Br}{\underset{|}{C}}HCH_2\,CH_2\,CH_2\,CH_2\,CH_3$

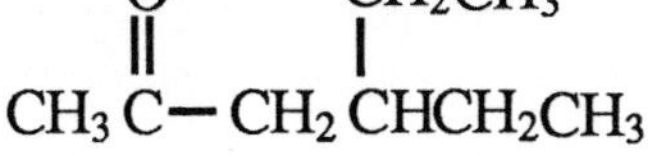

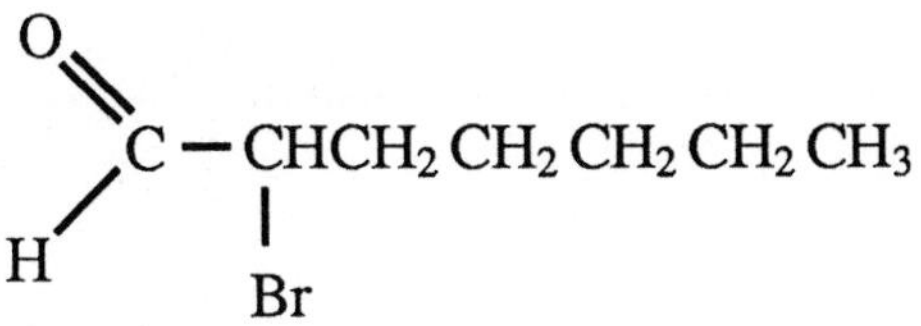

d. $CH_3CH_2CH(CH_3)–CH(CH_2CH_3)–C(=O)–CH_3$

e. $C_6H_5–CH_2CH(CH_3)–C(=O)–CH_3$

f. $CH_3C(CH_3)(CHO)–CCl_2CH_2CH_3$

22. Match the structure with the functional group name.

a. $HC(=O)H$	ketone
b. $CH_3C(=O)OH$	ester
c. CH_3CH_2OH	aldehyde
d. $CH_3C(=O)CH_3$	carboxylic acid
e. $CH_3C(=O)OCH_3$	alcohol

23. Show the structure of the carboxylic acid which is produced by oxidizing primary alcohols.

a. 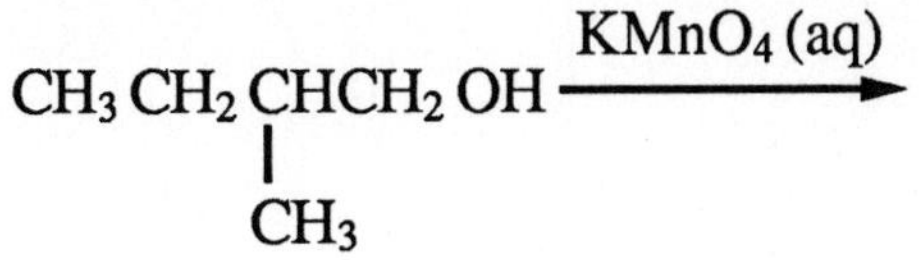 $CH_3CH_2CH(CH_3)CH_2OH \xrightarrow{KMnO_4(aq)}$

b. 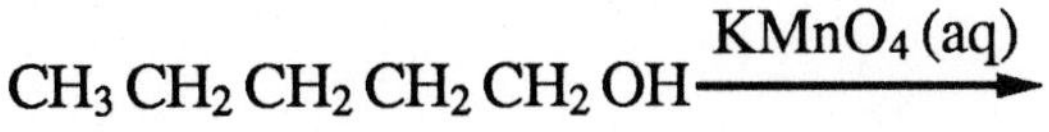 $CH_3CH_2CH_2CH_2CH_2OH \xrightarrow{KMnO_4(aq)}$

24. a. Show the structure of the ester formed when benzoic acid reacts with CH_3OH.

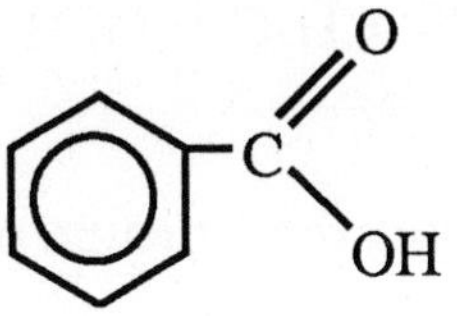

$C_6H_5-C(=O)-OH$

b. What is the name of the ester?

25. Name the carboxylic acids below.

a. 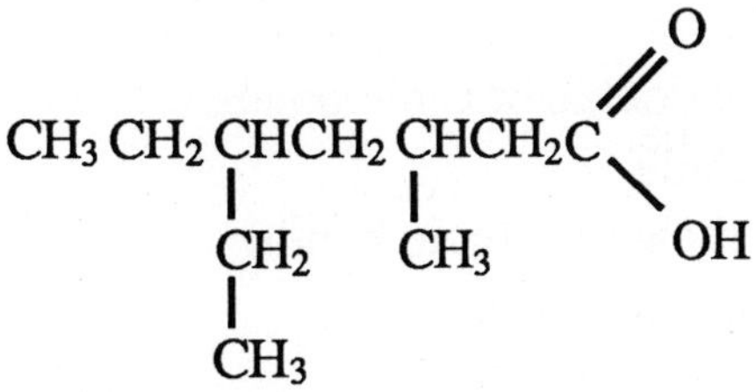 $CH_3CH_2CH(CH_2CH_3)CH_2CH(CH_3)CH_2C(=O)OH$

b. 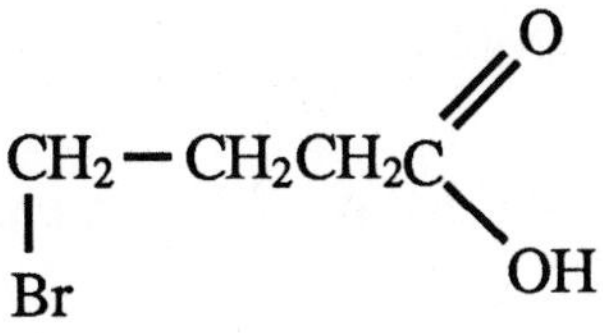 $CH_2(Br)-CH_2CH_2C(=O)OH$

c. 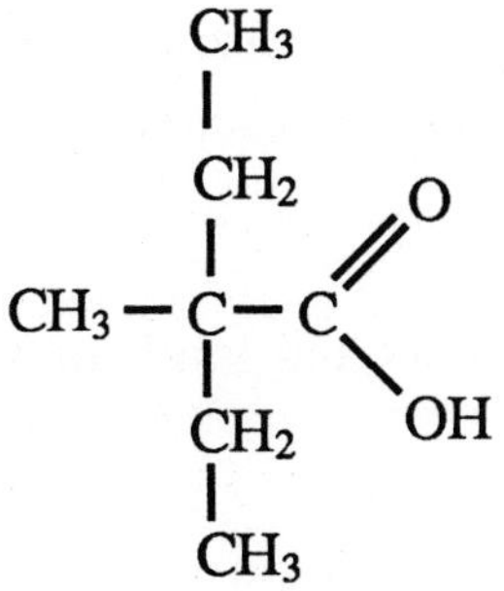 $CH_3-C(CH_2CH_3)(CH_2CH_3)-C(=O)OH$

26. Show the structure of the carboxylic acid and the alcohol which reacted to make the esters below.

a. 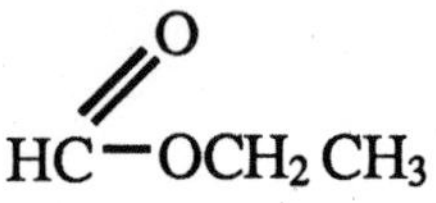 $HC(=O)-OCH_2CH_3$

b. 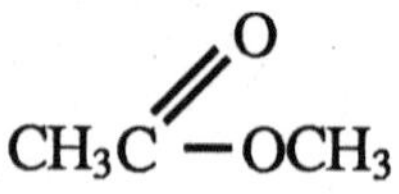 $CH_3C(=O)-OCH_3$

27. Show the structure of the polymer which would be produced from the monomer below.

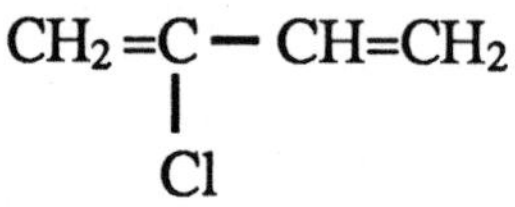

$$CH_2{=}C(Cl)-CH{=}CH_2$$

ANSWERS TO LEARNING REVIEW

1.

a. hydrocarbon — composed of carbon and hydrogen
b. alkane — contains only carbon-carbon single bonds
c. normal — an unbranched molecule
d. unsaturated — contains one or more double or triple bonds

2. In carbon tetrachloride carbon, is bonded to four other atoms. From the VSEPR model four pairs of bonding electrons spread out to form a tetrahedron.

3. The condensed structure of an alkane with five carbons will have a CH_3 on each end, and three CH_2, or methylene, units. The structure is $CH_3(CH_2)_3CH_3$.

4. This alkane has seven carbons and sixteen hydrogens, so it is represented by the general formula C_nH_{2n+2}. Both straight chain and normal alkanes are represented by the general formula C_nH_{2n+2}.

5. Structural isomers are molecules which have the same numbers and kinds of atoms, but a different arrangement of bonds.

6. One way to write all the structural isomers of a particular formula is to use the same system each time. Hexane has six carbons in a chain, so first write the structure of hexane itself.

```
    H  H  H  H  H  H
    |  |  |  |  |  |
 H— C— C— C— C— C— C—H
   1| 2| 3| 4| 5| 6|
    H  H  H  H  H  H
```

Then, remove a $-CH_3$ from the end of hexane, which leaves five carbons in the chain. Remove an -H from successive carbons on the chain and replace with the $-CH_3$. If we remove an -H from the end carbon, and replace with the $-CH_3$, we have not made a structural isomer, because the new molecule is just like hexane.

```
    H  H  H  H  H              H
    |  |  |  |  |              |
 H— C— C— C— C— C—           —C—H
    |1 |2 |3 |4 |5             |
    H  H  H  H  H              H
```

Put the $-CH_3$ on carbon number two. The $-CH_3$ is now located on the second carbon from the end. This is a new structural isomer.

```
    H  H  H  H  H
   1| 2| 3| 4| 5|
 H— C— C— C— C— C—H
    |  |  |  |  |
    H  |  H  H  H
    H— C—H
       |
       H
```

Now, move the $-CH_3$ to carbon number three.

```
    H  H  H  H  H
    |1 |2 |3 |4 |5
 H— C— C— C— C— C—H
    |  |  |  |  |
    H  H  |  H  H
       H— C—H
          |
          H
```

The $-CH_3$ is now located on the middle carbon, the third carbon from the end. We can move the $-CH_3$ to carbon number four, but this does not create a molecule with a different order. The carbons have the same order as when the $-CH_3$ is on the second carbon from the left. Both structures have $-CH_3$ on the second carbon from the end carbon.

```
   H  H  H  H  H                              H  H  H  H  H
   1| 2| 3| 4| 5|                             5| 4| 3| 2| 1|
H- C- C- C- C- C -H       same as          H- C- C- C- C- C -H
   |  |  |  |  |                              |  |  |  |  |
   H  |  H  H  H                              H  H  H  |  H
   H- C -H                                          H- C -H
      |                                                |
      H                                                H
```

We have exhausted the structural isomers which can be made by moving one $-CH_3$ from one location to another. Let's now remove another carbon from the chain and see what isomers can be made from a chain with four carbons and two $-CH_3$ groups.

```
   H  H  H  H       H       H
   |  |  |  |       |       |
H- C- C- C- C     - C -   - C -H
   1| 2| 3| 4|      5|      6|
   H  H  H  H       H       H
```

Substitute one $-CH_3$ on the second carbon, and one on the third carbon.

```
    H    H         H    H
    |1   |2        |3   |4
H---C----C---------C----C---H
    |    |         |    |
    H    |         |    H
      H- C -H   H- C -H
         |         |
         H         H
```

Now, substitute both $-CH_3$ groups on the second carbon.

```
           H
           |
        H- C -H
     H     |  H  H
     |1    |2 |3 |4
  H- C --- C- C- C -H
     |     |  |  |
     H     |  H  H
        H- C -H
           |
           H
```

If we put both $-CH_3$ groups on the third carbon, we have not created a structure with a new order of atoms. It is the same order as putting both $-CH_3$ groups on carbon number two.

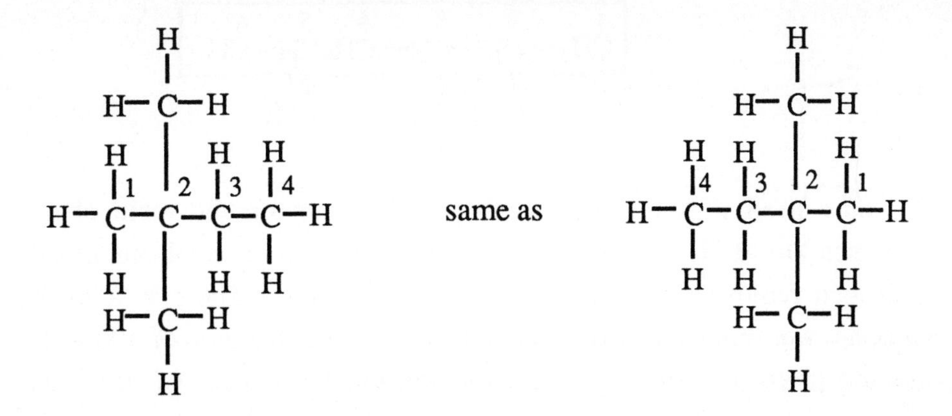

We can make no more different structural isomers with a four-carbon chain and two -CH_3 groups. Take another carbon from the chain.

```
    H  H  H        H      H        H
    |  |  |        |      |        |
 H—C—C—C—       —C     —C—      —C—H
   1| 2| 3|        |      |        |
    H  H  H        H      H        H
```

With a three-carbon chain and three -CH_3 groups, there are no new structural isomers. All the combinations we can make have the same order as ones we have already made. Thus there are five structural isomers of C_6H_{14}.

7. You can name alkanes by applying the rules in Section 20.4 of your textbook.

   a. The longest continuous chain of carbon atoms has six carbons so the parent alkane name is hexane.

```
 ┌─────────────────────────────────┐
 │CH3 CH — CH—CH2—CH—CH3           │
 └─────────────────────────────────┘
        |     |         |
       CH3   CH3       CH3
```

   We need to number the parent chain, assigning number one to the end which is closest to the first branch. In this case, the first branch, a -CH_3 group, would be on carbon two regardlesss of which end we begin numbering from. But, beginning the numbering from the left would locate the next branch, also a -CH_3, on carbon three. If we began from the right, the second substituent would be on carbon four.

$$\begin{array}{llllll} 1 & 2 & 3 & 4 & 5 & 6 \\ CH_3- & CH- & CH- & CH_2 & CH- & CH_3 \\ & | & | & & | & \\ & CH_3 & CH_3 & & CH_3 & \end{array}$$

We now need to name and number each group on the main chain. All three branches are $-CH_3$ groups. These are equivalent to methane molecules with a hydrogen removed, and are called methyl groups. There is a methyl group on carbons two, three, and five. When there is more than one of a kind of substituent, use the prefix di, tri, and so on, to indicate the number of times the substituent appears. These substituents would be named 2,3,5-trimethyl. Because there is only one kind of substituent, we do not have to list them in alphabetical order. The entire name is 2,3,5-trimethylhexane.

b. The longest continuous chain of carbon atoms has five carbons so the parent alkane name is pentane.

$$\begin{array}{l} CH_3\,CH_2\,CHCH_2\,CH_3 \\ \qquad\quad\;\; | \\ \qquad\quad\;\; CH-CH_3 \\ \qquad\quad\;\; | \\ \qquad\quad\;\; CH_3 \end{array}$$

We can number this alkane from either end. There is only one substituent, and it is on the middle carbon.

$$\begin{array}{l} 1 \quad\; 2 \quad\;\; 3 \;\; 4 \quad\; 5 \\ CH_3\,CH_2\,CHCH_2\,CH_3 \\ \qquad\quad\;\; | \\ \qquad\quad\;\; CH-CH_3 \\ \qquad\quad\;\; | \\ \qquad\quad\;\; CH_3 \end{array}$$

The group on carbon three is an isopropyl group, so this molecule is called 3-isopropylpentane.

c. The longest continuous chain of carbon atoms has five carbon atoms so the parent alkane name is pentane. Remember that the longest chain will not always be drawn horizontally on the page.

$$\begin{array}{llll} CH_3- & CH- & CH- & CH_3 \\ & | & | & \\ & CH_2 & CH_3 & \\ & | & & \\ & CH_3 & & \end{array}$$

Number the chain beginning with the end on the right side in order to give the first substituent the lowest possible number.

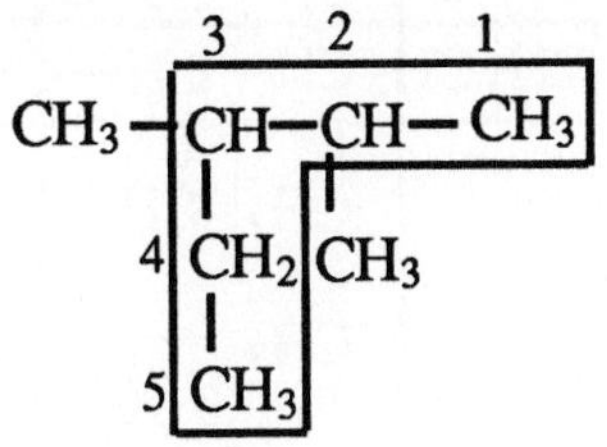

```
          3    2    1
CH3 — CH — CH — CH3
           |      |
       4  CH2  CH3
           |
       5  CH3
```

There are two substituents, both methyl groups, on carbons two and three. The substituents are named 2,3-dimethyl. The whole name is 2,3-dimethylpentane.

d. The longest continuous chain of carbon atoms has seven carbons, so the parent alkane name is heptane.

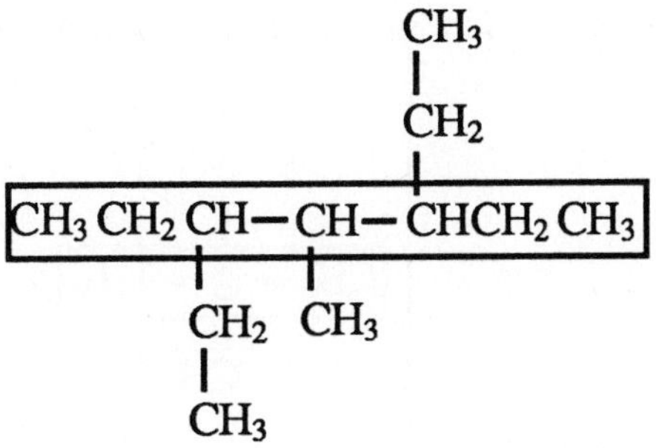

```
                              CH3
                               |
                              CH2
                               |
CH3 CH2 CH — CH — CHCH2 CH3
             |      |
            CH2   CH3
             |
            CH3
```

You can also draw the box another way, which gives a chain of equal length.

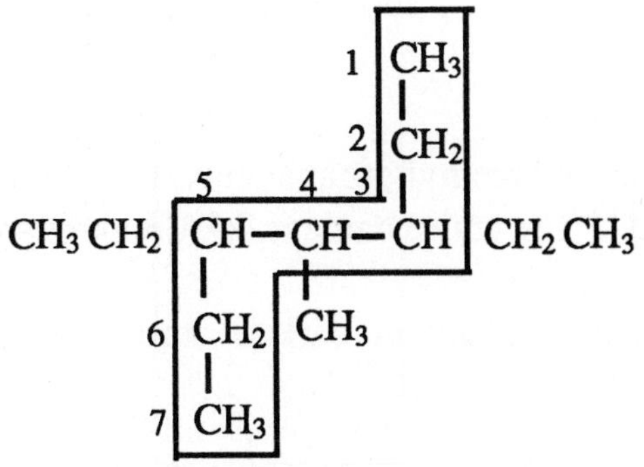

```
                            1  CH3
                                |
                            2  CH2
              5        4   3    |
CH3 CH2  CH — CH — CH  CH2 CH3
              |      |
          6  CH2   CH3
              |
          7  CH3
```

Number the chain from either end.

```
                            CH3
                             |
                            CH2
1   2    3     4     5|  6   7
CH3 CH2 CH — CH — CHCH2 CH3
         |     |
        CH2   CH3
         |
        CH3
```

The substituents are a $-CH_3$, or methyl group, on carbon four, and two CH_3CH_2-, or ethyl groups, on carbons three and five. The two ethyl groups are named 3,5-diethyl. When assembling the substituent names, list ethyl before methyl because the groups must be listed alphabetically. The group names are 3,5-diethyl-4-methyl. The entire name is 3,5-diethyl-4-methylheptane.

e. The longest continuous chain of carbons has three carbons, so the parent alkane name is propane.

```
       CH3
        |
CH3 — C — CH3
        |
       CH3
```

Number the chain from either end.

```
       CH3
1     2|     3
CH3 — C — CH3
        |
       CH3
```

The substituents are two methyl groups on carbon number two. The substituents are named 2,2-dimethyl. The entire name is 2,2-dimethylpropane.

f. The longest continuous chain of carbon atoms has eight carbons, so the parent alkane name is octane.

```
                       CH3
1   2  3    4           |
CH3 CHCH2 CH — C — CH3
     |      |      |
    CH3   5 CH2   CH3
            |
          6 CH2 — CH2CH3
                  7    8
```

Begin numbering the chain from the left side, which is closest to the first branch.

```
                          CH3
 1   2  3    4             |
 CH3 CHCH2 CH ——  C—CH3
        |          | 5     |
      CH3      CH2   CH3
                   |      7   8
                6 CH2—CH2CH3
```

There are two groups, a methyl on carbon two, and a *tert*-butyl on carbon four. Assemble the group names in alphabetical order, with butyl before methyl. The group names are 4-*tert*-butyl-2-methyl. The entire molecule is named 4-*tert*-butyl-2-methyloctane.

8. One way to tell whether or not two molecules are structural isomers is to name them. If two molecules have the same name, they are not structural isomers. They are the same molecule. If two molecules have the same number of carbons and hydrogens, and they have different names, they are structural isomers. Molecule **a** has five carbons in the longest chain, and molecule **b** also has five carbons in the longest chain.

```
                                        CH3
                                         |
CH3CH—CH—CH3              CH3—CH    CH3
       |      |                           |        |
     CH3  CH2CH3                    CH2—CH
                                                   |
                                                 CH3

        a.                                  b.
```

Number both molecules from the left. Each molecule has two substituents, two methyl groups. The substituents on molecule **a** are 2,3-dimethyl, and on molecule **b** they are 2,4-dimethyl. The entire name for molecule **a** is 2,3-dimethylpentane, and for molecule **b** it is 2,4-dimethylpentane. The names are different. Each molecule has five carbons and twelve hydrogens, so they are structural isomers.

9. Writing structural formulas from names is the reverse of writing names from formulas. First, find the parent alkane name and write a carbon skeleton with the same number of carbons. Number the chain from either direction. Determine how many substituents there are, and where they are attached, then put them on the chain. Fill in hydrogens so that each carbon is surrounded by four other atoms.

a. 2-methyl-5-*sec*-butyloctane has the parent alkane name octane.

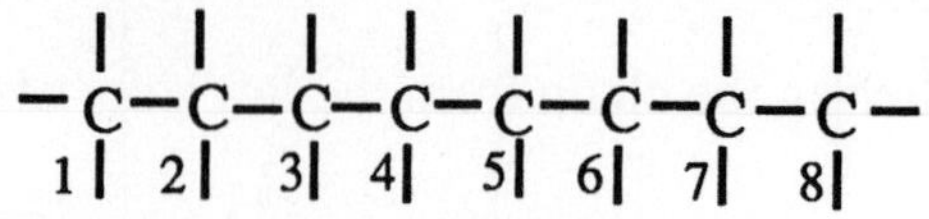

```
  |   |   |   |   |   |   |   |
 -C - C - C - C - C - C - C - C -
 1|  2|  3|  4|  5|  6|  7|  8|
```

2-methyl indicates that there is a methyl group on carbon two.

```
 1|  2|  3|  4|  5|  6|  7|  8|
 -C - C - C - C - C - C - C - C -
  |   |   |   |   |   |   |   |
     CH3
```

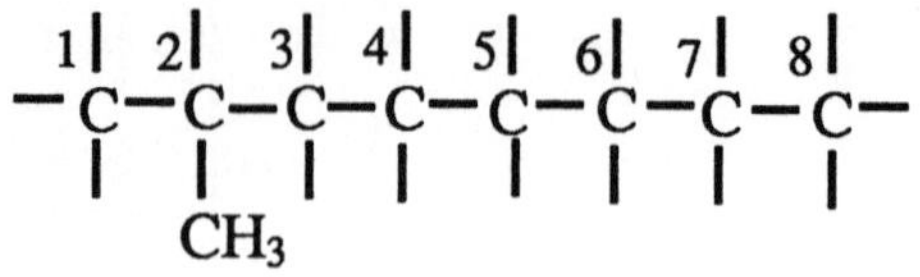

5-*sec*-butyl indicates that there is a *sec*-butyl group on carbon four.

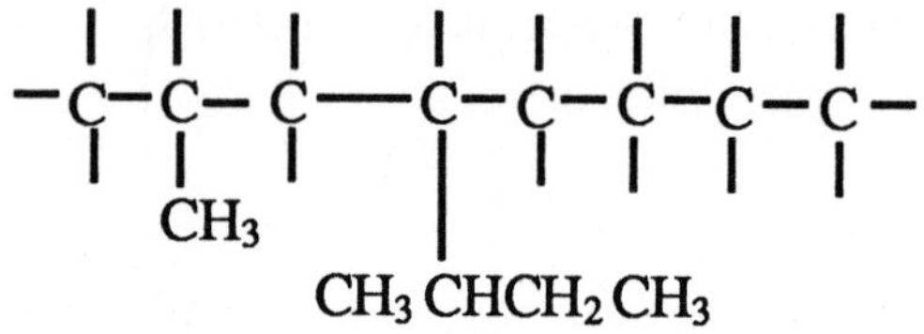

```
  |   |   |     |   |   |   |   |
 -C - C - C --- C - C - C - C - C -
  |   |   |     |   |   |   |   |
     CH3        |
          CH3 CHCH2 CH3
```

Now fill in the hydrogens so that each carbon is surrounded by four atoms.

```
CH3 CH CH2  CH CH2 CH2 CH2 CH3
     |       |
    CH3  CH3CH CH2CH3
```

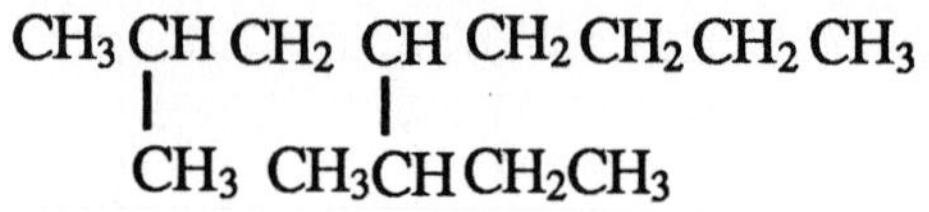

b. 2-methylpropane has the parent alkane name propane.

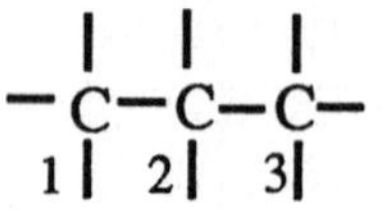

```
  |   |   |
 -C - C - C-
 1|  2|  3|
```

2-methyl indicates a methyl group on carbon two.

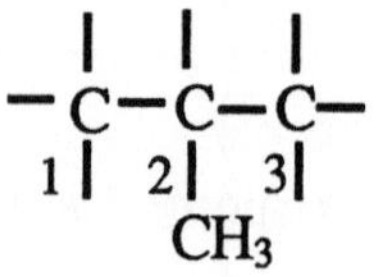

```
  |   |   |
 -C - C - C-
 1|  2|  3|
     CH3
```

Now fill in the hydrogens so that each carbon is surrounded by four atoms.

```
CH3 CHCH3
     |
    CH3
```

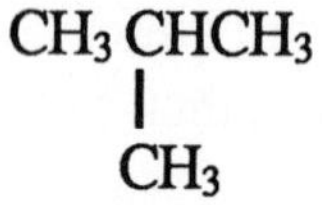

c. 3-ethylpentane has the parent alkane name pentane.

```
 1| 2| 3| 4| 5|
-C-C-C-C-C-
 |  |  |  |  |
```

3-ethyl indicates an ethyl group on carbon three.

```
 1| 2| 3| 4| 5|
-C-C-C-C-C-
 |  |  |  |  |
       CH2
       |
       CH3
```

Now fill in the hydrogens so that each carbon is surrounded by four atoms.

```
CH3 CH2 CHCH2 CH3
        |
        CH2
        |
        CH3
```

d. 2,2,4-trimethylhexane has the parent alkane name hexane.

```
 1| 2| 3| 4| 5| 6|
-C-C-C-C-C-C-
 |  |  |  |  |  |
```

2,2,4-trimethyl indicates that there are three methyl groups, two on carbon two and one on carbon four.

```
    CH3
 1| 2| 3| 4| 5| 6|
-C-C-C-C-C-C-
 |  |  |  |  |  |
    CH3   CH3
```

Now, fill in the hydrogen atoms so that each carbon is surrounded by four atoms.

```
    CH3
    |
CH3C- CH2 CHCH2CH3
    |      |
    CH3    CH3
```

10. a. 2-ethylbutane has the structure

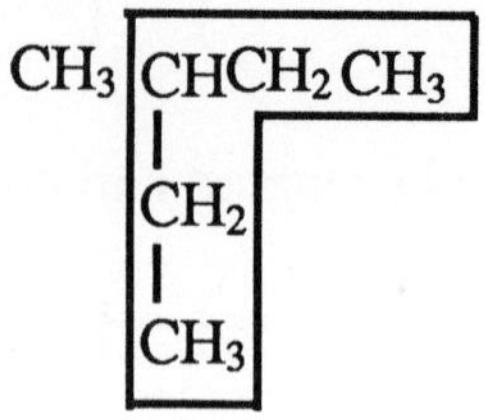

$$\begin{array}{l} CH_3\,\boxed{CHCH_2CH_3} \\ \quad\;\;| \\ \quad\;\;CH_2 \\ \quad\;\;| \\ \quad\;\;CH_3 \end{array}$$

The longest carbon chain has not four, but five carbons, so the correct alkane name is pentane. There is a methyl group on carbon three, so this is 3-methylpentane.

b. 1-methylpentane has the structure

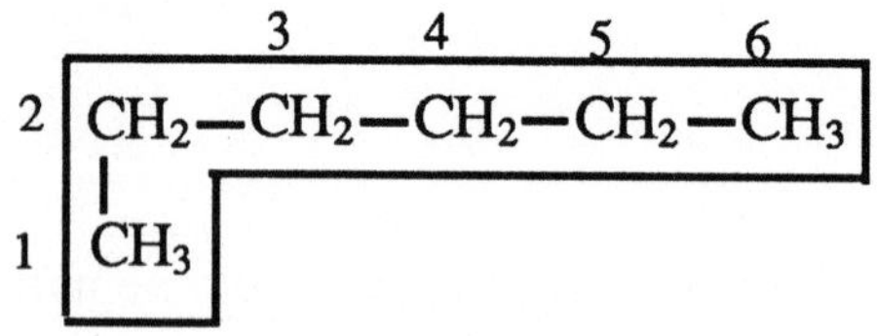

$$\begin{array}{ll} & \quad\;\; 3 \qquad\;\; 4 \qquad\;\; 5 \qquad\;\; 6 \\ 2 & CH_2{-}CH_2{-}CH_2{-}CH_2{-}CH_3 \\ & | \\ 1 & CH_3 \end{array}$$

The longest chain has six carbons, not five, so the correct name is hexane.

c. 4,4-dimethylhexane has the structure

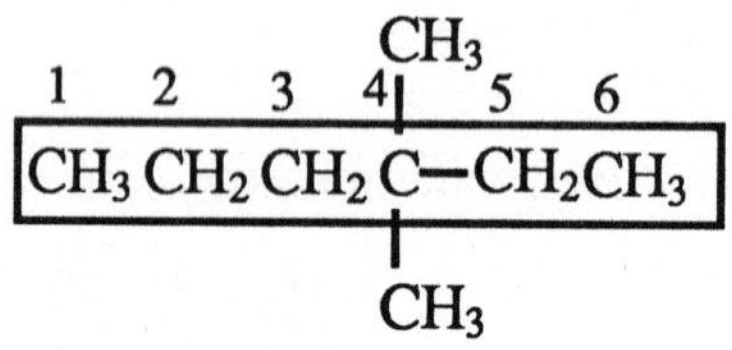

$$\begin{array}{c} \qquad\qquad\qquad CH_3 \\ 1 \;\; 2 \;\; 3 \;\; 4| \;\; 5 \;\; 6 \\ CH_3\,CH_2\,CH_2\,C{-}CH_2CH_3 \\ \qquad\qquad\quad | \\ \qquad\qquad\qquad CH_3 \end{array}$$

The longest chain has six carbons, so this is a hexane. The two methyl groups are on carbon four. Numbering the chain from the other end would put the methyl groups on carbon three. We want to number from the end which gives the substituents the lowest numbers, so 3,3-dimethylhexane would be the correct name.

d. 2-ethyl-2-methylheptane has the structure

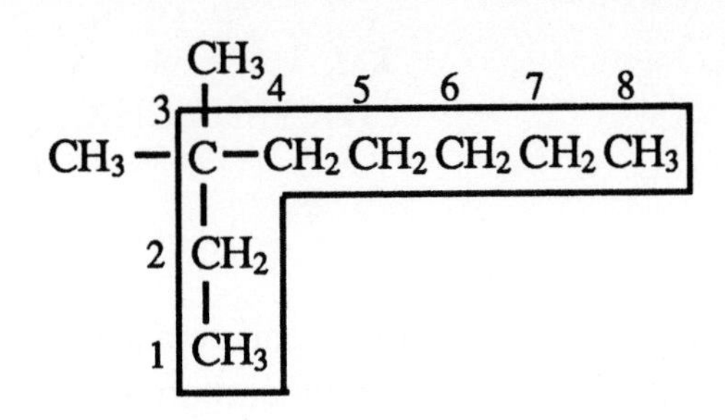

The longest chain has eight carbons, not seven, so the correct alkane name is octane. There are two methyl groups on carbon three, so this is 3,3-dimethyloctane.

11. Petroleum is a mixture of hydrocarbons containing molecules with various numbers of carbons in the chain. Usually, as the number of carbons in an alkane chain increases, the boiling point increases. Petroleum can be separated into different fractions by boiling.

12. First, draw the structure of pentane.

$$\overset{1}{CH_3}\overset{2}{CH_2}\overset{3}{CH_2}\overset{4}{CH_2}\overset{5}{CH_3}$$

Then, begin at one end and remove a hydrogen atom and replace it with a chlorine atom.

$$\underset{\underset{Cl}{|}}{\overset{1}{CH_2}}\overset{2}{CH_2}\overset{3}{CH_2}\overset{4}{CH_2}\overset{5}{CH_3}$$

There is only one structural isomer which can be produced from pentane by removing a hydrogen from carbon one. The order of bonding does not change by replacing either of the other two hydrogen atoms on carbon number one.

$$\underset{\underset{Cl}{|}}{\overset{1}{CH_2}}\overset{2}{CH_2}\overset{3}{CH_2}\overset{4}{CH_2}\overset{5}{CH_3} \text{ same as } Cl\overset{1}{CH_2}\overset{2}{CH_2}\overset{3}{CH_2}\overset{4}{CH_2}\overset{5}{CH_3} \text{ same as } \overset{\overset{Cl}{|}}{\overset{1}{CH_2}}\overset{2}{CH_2}\overset{3}{CH_2}\overset{4}{CH_2}\overset{5}{CH_3}$$

Now, remove a hydrogen from carbon two and replace it with a chlorine atom. There is one structural isomer with a chlorine atom on carbon two.

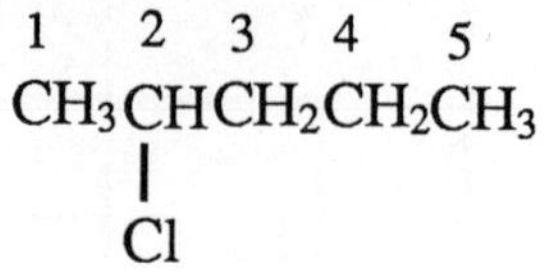

$$\begin{array}{l} \quad 1 \quad 2 \quad 3 \quad 4 \quad 5 \\ CH_3CHCH_2CH_2CH_3 \\ \quad\quad | \\ \quad\quad Cl \end{array}$$

Remove a hydrogen from carbon three and replace it with a chlorine atom. There is one structural isomer with a chlorine atom on carbon three.

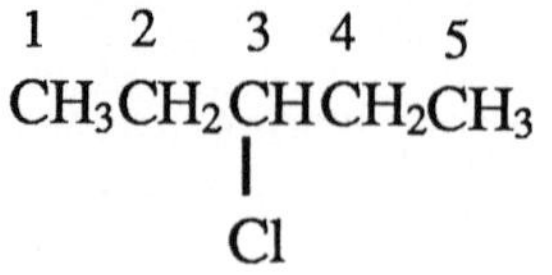

$$\begin{array}{l} \quad 1 \quad 2 \quad 3 \quad 4 \quad 5 \\ CH_3CH_2CHCH_2CH_3 \\ \quad\quad\quad\quad | \\ \quad\quad\quad\quad Cl \end{array}$$

Substituting a chlorine atom on carbon atoms four or five does not produce a new structural isomer, so there are three structural isomers.

13.

a. 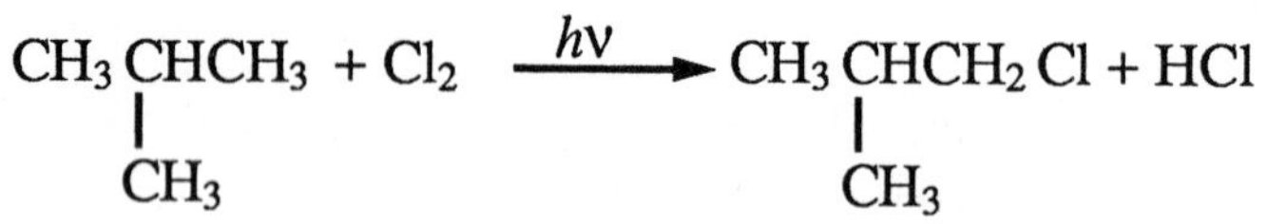 $$\begin{array}{l} CH_3CHCH_3 \\ \quad\; | \\ \quad\; CH_3 \end{array} + Cl_2 \xrightarrow{h\nu} \begin{array}{l} CH_3CHCH_2Cl \\ \quad\; | \\ \quad\; CH_3 \end{array} + HCl$$

In this reaction, a chlorine atom substitutes for a hydrogen atom.

b. $$CH_4 + 2O_2 \longrightarrow CO_2 + 2H_2O$$

This is an example of a combustion reaction.

c. $$CH_3CH_2CH_3 \xrightarrow[500^\circ C]{CrO_3} CH_3CH{=}CH_2 + H_2$$

This is an example of a dehydrogenation reaction.

14. The rules for naming alkenes and alkynes are given in section 20.7 of your textbook.

a. The longest chain which contains the double bond has five carbons.

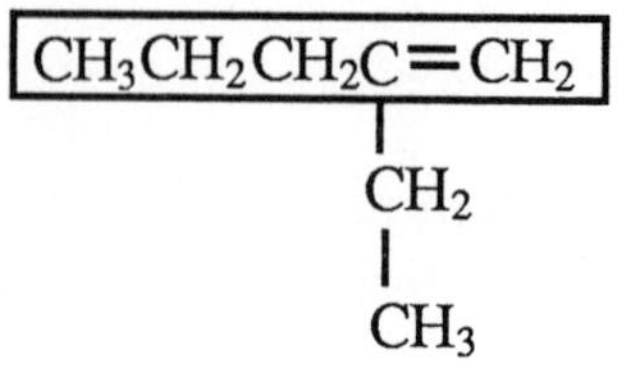

$$\begin{array}{l} \boxed{CH_3CH_2CH_2C{=}CH_2} \\ \quad\quad\quad\quad\quad\; | \\ \quad\quad\quad\quad\quad\; CH_2 \\ \quad\quad\quad\quad\quad\; | \\ \quad\quad\quad\quad\quad\; CH_3 \end{array}$$

Replace the -ane ending of pentane with -ene to produce pentene. Number the carbon chain from the end closest to the double bond.

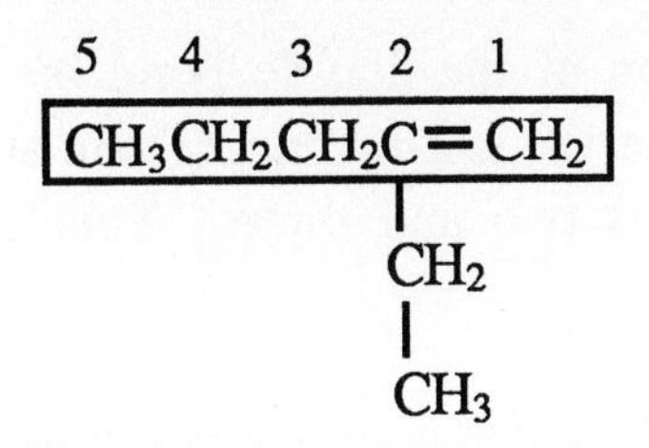

Give the location of the double bond by putting a number in front of the alkene name. This molecule is 1-pentene. There is an ethyl group on carbon two, so this molecule would be named 2-ethyl-1-pentene.

b. The longest chain which contains the triple bond has seven carbons.

```
CH3 CH—C≡C—CH2 CH—CH3
     |            |
    CH3          CH3
```

Replace the -ane ending of heptane with -yne to produce heptyne. Number the carbon chain from the end closest to the triple bond.

```
 1     2    3   4    5    6   7
CH3—CH—C ≡ C— CH2 CH CH3
     |              |
    CH3            CH3
```

Give the location of the triple bond by putting a three in front of heptyne. There are two methyl groups attached to the chain on carbons two and six. This molecule would be named 2,6-dimethyl-3-heptyne.

c. The longest chain which contains the double bond has four carbons.

```
CH3 CH=C—CH3
     |  |
   CH3 CH3
```

Replace the -ane ending of butane with -ene to produce butene. Number the carbon chain from either end to give the location of the double bond the smallest number.

```
 1   2    3 4
CH3 C = C CH3
    |   |
  CH3  CH3
```

Give the location of the double bond by putting a number in front of the alkene name. This molecule is 2-butene. There are two methyl groups attached to the main chain on carbons two and three. This molecule would be named 2,3-dimethyl-2-butene.

d. The longest chain which contains the triple bond has six carbons.

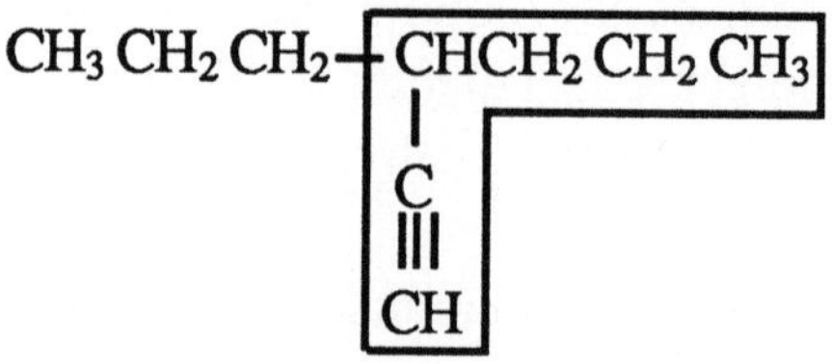

$$CH_3CH_2CH_2-\underset{\underset{\underset{\displaystyle CH}{|||}}{\displaystyle C}}{\underset{|}{C}}HCH_2CH_2CH_3$$

Replace the -ane ending of hexane with -yne to produce hexyne. Number the chain from the end closest to the triple bond.

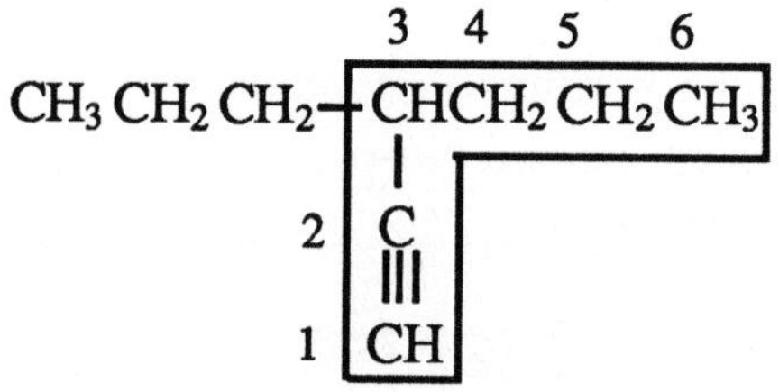

$$CH_3CH_2CH_2-\overset{3}{C}H\overset{4}{C}H_2\overset{5}{C}H_2\overset{6}{C}H_3 \quad (C\text{-}3 \text{ bears } -\overset{2}{C}\equiv\overset{1}{C}H)$$

Locate the triple bond with a number. This molecule would be 1-hexyne. There is a propyl group on carbon three, so the name is 3-propyl-1-hexyne.

15.

a. 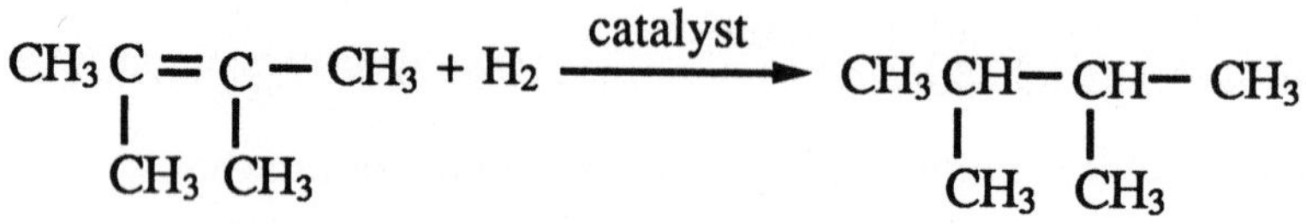 $$CH_3\underset{\underset{\displaystyle CH_3}{|}}{C}=\underset{\underset{\displaystyle CH_3}{|}}{C}-CH_3 + H_2 \xrightarrow{\text{catalyst}} CH_3\underset{\underset{\displaystyle CH_3}{|}}{C}H-\underset{\underset{\displaystyle CH_3}{|}}{C}H-CH_3$$

b. $$CH_3CH_2CH_2CH=CH_2 + Cl_2 \longrightarrow CH_3CH_2CH_2CHClCH_2Cl$$

c. 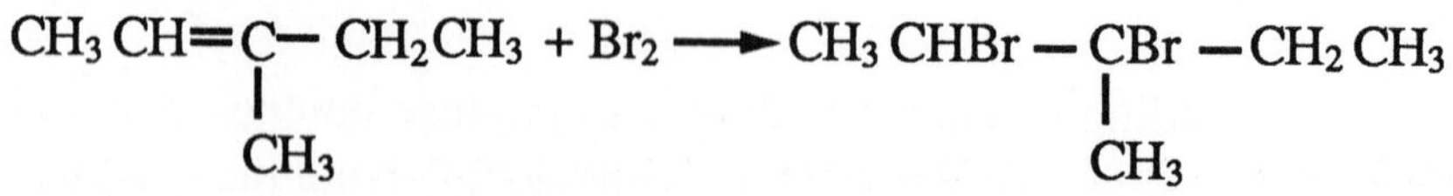 $$CH_3CH=\underset{\underset{\displaystyle CH_3}{|}}{C}-CH_2CH_3 + Br_2 \longrightarrow CH_3CHBr-\underset{\underset{\displaystyle CH_3}{|}}{C}Br-CH_2CH_3$$

16. a. This molecule is usually called toluene.

b. This molecule is named as a pentane with two substituents. There is a chlorine atom on carbon three and a phenyl group on carbon two. The name of this molecule is 3-chloro-2-phenylpentane.

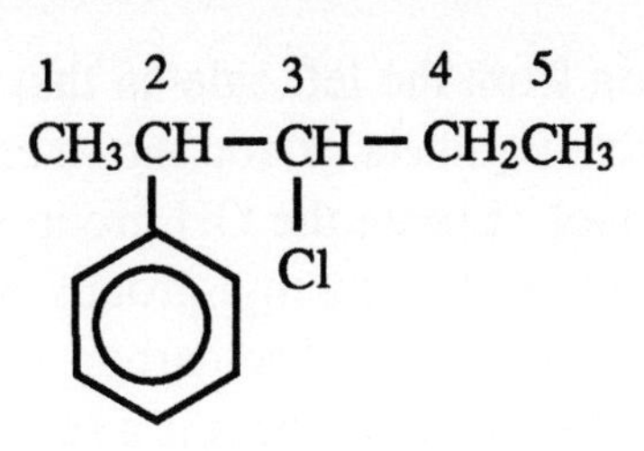

c. This molecule is chlorobenzene.
d. This molecule is nitrobenzene.

17.

a. 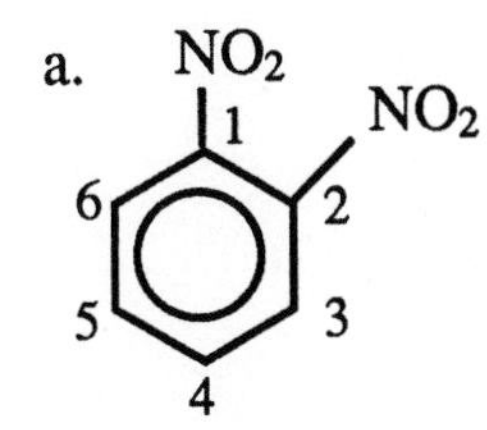

is 1,2-dinitrobenzene or *o*-dinitrobenzene

b. OH

Cl

is 4-chlorophenol or *p*-chlorophenol

c. 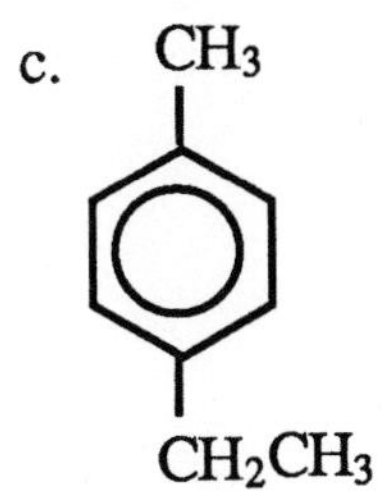

is 4-ethyltoluene or *p*-ethyltoluene

d. 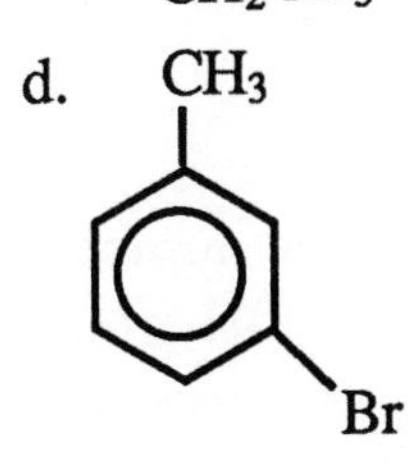

is 3-bromotoluene or *m*-bromotoluene

18. a. The longest chain which contains the OH has five carbons.

$$\begin{array}{c} \quad\ CH_3 \\ 1 \quad\ 2| \qquad 3\ 4\ 5 \\ CH_3-\underset{|}{C}-\underset{|}{C}HCH_2CH_3 \\ \quad\ CH_3\ \ OH \end{array}$$

Number the chain from the left side so that the OH group and the substituents have the lowest possible numbers. Drop the -e ending from pentane, and replace with -ol to produce pentanol. Locate the OH group with a number. This molecule would be 3-pentanol. There are two methyl groups on carbon two, so the entire name would be 2,2-dimethyl-3-pentanol. The carbon to which the OH is attached is bonded to two hydrocarbon fragments, so this is a secondary alcohol.

b. The longest chain which contains the OH has four carbons.

```
          1 CH3
            |
          2 |
   CH3 —  C — OH
            |
          3 CH2
            |
          4 CH3
```

Number the chain from the top so that the OH group has the lowest possible number. Drop the -e ending from butane and add -ol to produce butanol. Locate the OH group with a number. This molecule would be 2-butanol. There is a methyl group on carbon two, so the entire name is 2-methyl-2-butanol. The carbon to which the OH is attached is bonded to three hydrocarbon fragments, so this is a tertiary alcohol.

c. The longest chain which contains the OH group has six carbons.

```
              CH3
  6   5   4 |     3      2    1
  CH3 CH2 C  — CH — CH2 CH2 OH
            |     |
            Br    CH3
```

Begin numbering from the right to give the OH the lowest possible number. Drop the -e ending of hexane and add -ol to produce hexanol. Locate the OH group with a number. This molecule would be 1-hexanol. There are two methyl groups on carbons three and four, and a bromo group on carbon four. The entire name would be 4-bromo-3,4-dimethyl-1-hexanol. The carbon to which the OH is attached is bonded to one hydrocarbon fragment, so this is a primary alcohol.

d. The longest chain which contains the OH has eight carbons.

$$
\begin{array}{l}
\qquad\quad CH_3-CH-CH_3 \\
\quad\ \ 2\ \ 3\ \ 4\qquad 5\quad 6\quad 7\quad 8 \\
CH_3\,CHCH_2\,CH-CH_2\,CH_2\,CH_2\,CH_3 \\
\qquad\ | \\
\quad 1\ CH_2OH
\end{array}
$$

Begin numbering from the carbon which has the OH. Drop the -e ending of octane and add -ol to produce octanol. Locate the OH group with a number. This molecule would be 1-octanol. There is a methyl group on carbon two and an isopropyl group on carbon four. The entire name would be 2-methyl-4-isopropyl-1-octanol. The carbon to which the OH is attached is bonded to one hydrocarbon fragment, so this is a primary alcohol.

e. The longest chain has one carbon. Assign this carbon number one.

$$CH_3OH$$

Drop the -e ending of methane and add -ol to produce methanol. Because there is only one carbon, we do not need a number to locate the OH group. It can only be found on carbon one. There are no substituents so methanol is the entire name. This molecule is also known by the common names methyl alcohol and wood alcohol. The carbon to which the OH is attached is bonded to one hydrocarbon fragment, so this is a primary alcohol.

f. The longest chain which contains the OH has nine carbons.

$$
\begin{array}{l}
\qquad\qquad\quad 7\quad 8\quad 9 \\
CH_3\,CH_2\,CHCH_2\,CH_3 \\
\qquad\qquad\ |\quad 5\quad 4\quad 3\quad 2\ \ 1 \\
\qquad\quad 6\ CH_2\,CH_2\,CHCH_2\,CHCH_3 \\
\qquad\qquad\qquad\qquad |\qquad\quad | \\
\qquad\qquad\qquad\quad CH_2\quad\ OH \\
\qquad\qquad\qquad\qquad | \\
\qquad\qquad\qquad\quad CH_3
\end{array}
$$

Begin numbering from the right to give the OH the lowest possible numbr. Drop the -e ending of nonane and add -ol to produce nonanol. Locate the OH with a number. This molecule would be 2-nonanol. There are two ethyl groups on carbons four and seven. This molecule would be named 4,7-diethyl-2-nonanol. The carbon to which the OH is attached is bonded to two hydrocarbon fragments, so this is a secondary alcohol.

19.

a. methanol — commonly known as wood alcohol
b. phenol — used in the production of plastics
c. ethanol — found in alcoholic beverages
d. ethylene glycol — used in antifreeze

20.

a. (benzene ring)–CH_2OH $\xrightarrow{\text{oxidation}}$ (benzene ring)–C(=O)H

b. $CH_3CH(CH_3)CH_2OH$ $\xrightarrow{\text{oxidation}}$ $CH_3CH(CH_3)C(=O)OH$

c. $CH_3CH_2CH(OH)CH_2CH_3$ $\xrightarrow{\text{oxidation}}$ $CH_3CH_2C(=O)CH_2CH_3$

21. a. The longest chain which contains the carbonyl has four carbons. Begin numbering from the end which has the aldehyde group.

$$\overset{4}{CH_3}\,\overset{3}{C}(CH_3)_2-\overset{2}{CH}(CH_3)-\overset{1}{C}(=O)-H$$

Drop the -e ending of butane and add -al to produce butanal. Since the carbonyl in aldehydes is always found on an end, do not use a number to locate the carbonyl. There are three methyl groups, two on carbon three and one on carbon two. This molecule would be named 2,3,3-trimethylbutanal.

b. The longest chain which contains the carbonyl has six carbons. Begin numbering from the end which is closest to the ketone group.

$$\underset{1}{CH_3}\,\underset{2}{C}(=O)-\underset{3}{CH_2}\,\underset{4}{CH}(CH_2CH_3)\underset{5}{CH_2}\underset{6}{CH_3}$$

Drop the -e of hexane and add -one to produce hexanone. Locate the ketone carbonyl with a number, 2-hexanone. There is an ethyl group on carbon four, so the entire name would be 4-ethyl-2-hexanone.

c. The longest chain which contains the carbonyl has seven carbons. Begin numbering from the end which has the aldehyde group.

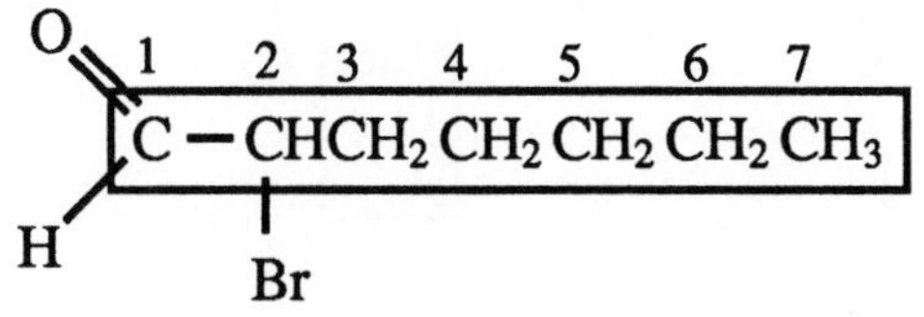

```
 O   1   2  3   4   5   6  7
  \\C — CHCH2 CH2 CH2 CH2 CH3
  /       |
 H        Br
```

Drop the -e ending of heptane and add -al to produce heptanal. There is a bromo group on carbon two, so the entire name is 2-bromoheptanal.

d. The longest chain which contains the carbonyl has six carbons. Begin numbering from the right to give the ketone the lowest number.

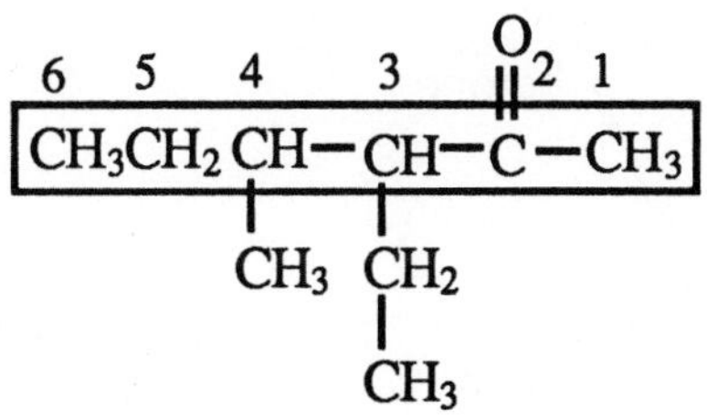

```
                       O
 6  5   4    3        ||2  1
 CH3CH2 CH — CH — C — CH3
         |     |
        CH3   CH2
               |
              CH3
```

Drop the -e ending of hexane and add -al to produce hexanal. Locate the ketone with a number. This molecule would be 3-hexanone. There is an ethyl group on carbon three, and a methyl group on carbon four. The entire name would be 3-ethyl-4-methylhexanal.

e. The longest chain which contains the carbonyl has four carbons. Begin numbering from the right to give the ketone carbonyl the lowest number.

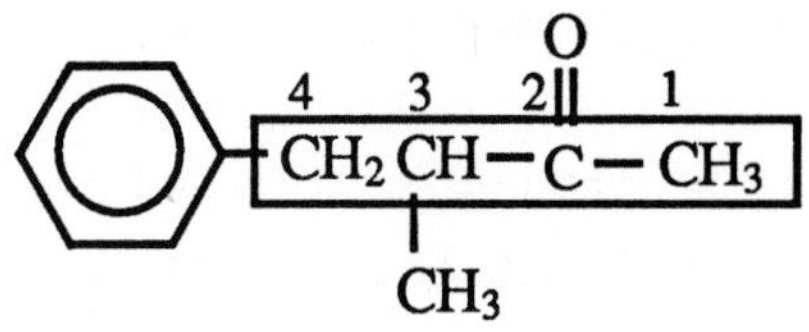

```
                      O
          4    3    2||   1
 (C6H5) — CH2 CH — C — CH3
               |
              CH3
```

Drop the -e of butane and add -one to produce butanone. Locate the ketone with a number. This molecule would be 2-butanone. There is a methyl group on carbon three and a phenyl group on carbon four. The name would be 3-methyl-4-phenyl-2-butanone.

f. The longest chain which contains the carbonyl has five carbons. Begin numbering from the end with the aldehyde group.

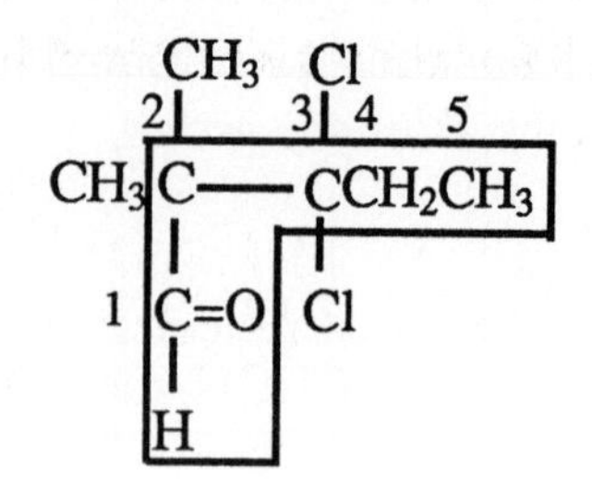

Drop the -e of pentane and add -al to produce pentanal. The aldehyde group does not need a number because it is always on the end. There are two methyl groups on carbon two, and two chloro groups on carbon three. The name would be 3,3-dichloro-2,2-dimethylpentanal.

22.

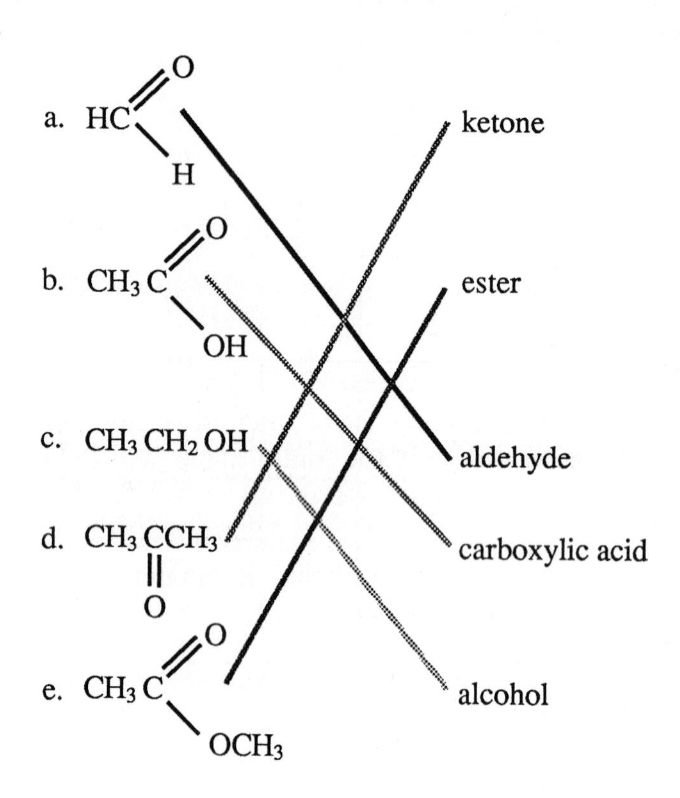

23.

a. $CH_3CH_2CH(CH_3)CH_2OH \xrightarrow{KMnO_4(aq)} CH_3CH_2CH(CH_3)C(=O){-}OH$

b. $CH_3CH_2CH_2CH_2CH_2OH \xrightarrow{KMnO_4(aq)} CH_3CH_2CH_2CH_2C(=O){-}OH$

24. a.

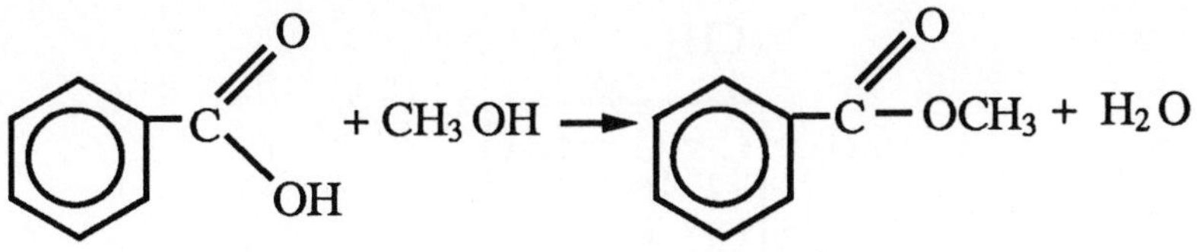

$$C_6H_5-C(=O)-OH + CH_3OH \rightarrow C_6H_5-C(=O)-OCH_3 + H_2O$$

b. The part of the ester which came from the alcohol is called methyl. The part of the ester which came from the carboxylic acid is called benzoate. The name is methylbenzoate.

25. a. The longest carbon chain which contains the carboxylic acid functional group has seven carbons. Number the chain from the end which has the carboxylic acid group.

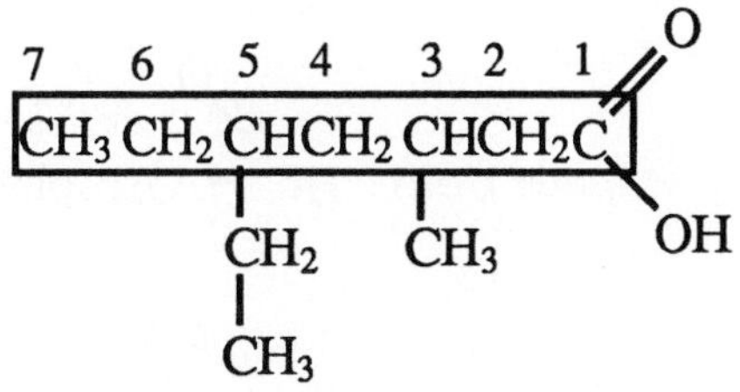

$$\overset{7}{CH_3}\overset{6}{CH_2}\overset{5}{CH}(CH_2CH_3)\overset{4}{CH_2}\overset{3}{CH}(CH_3)\overset{2}{CH_2}\overset{1}{C}(=O)OH$$

Drop the -e of heptane and add -oic acid to give heptanoic acid. There is a methyl group on carbon three and an ethyl group on carbon five. The name is 5-ethyl-3-methylheptanoic acid.

b. The longest chain of carbons which contains the carboxylic acid functional group has four carbons. Begin numbering from the right to give the carboxylic acid functional group the lowest number.

$$\underset{4}{CH_2}(Br)-\underset{3}{CH_2}\underset{2}{CH_2}\underset{1}{C}(=O)-OH$$

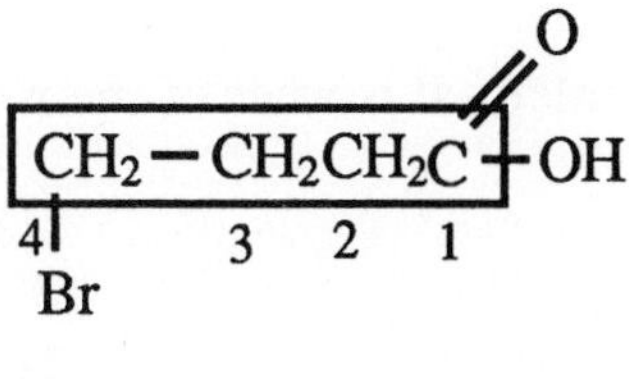

Drop the -e ending and add -oic acid to give butanoic acid. There is a bromo group on carbon four, so the name would be 4-bromobutanoic acid.

c. The longest chain of carbons which contains the carboxylic acid functional group has four carbons. Begin numbering from the right to give the carboxylic acid functional group the lowest number.

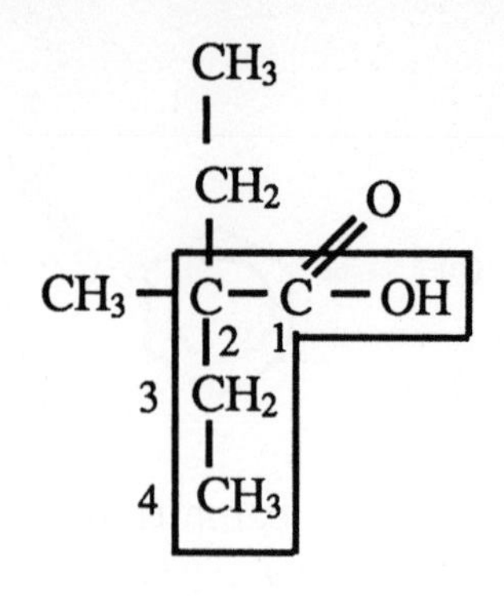

Drop the -e ending and add -oic acid to give butanoic acid. There is a methyl group and an ethyl group on carbon two. The entire name would be 2-ethyl-2-methylbutanoic acid.

26.

a. $HC(=O)-OCH_2CH_3$ is made from $HC(=O)-OH$ plus CH_3CH_2OH

b. $CH_3C(=O)-OCH_3$ is made from $CH_3C(=O)-OH$ plus CH_3OH

27. The polymer would have the structure below.

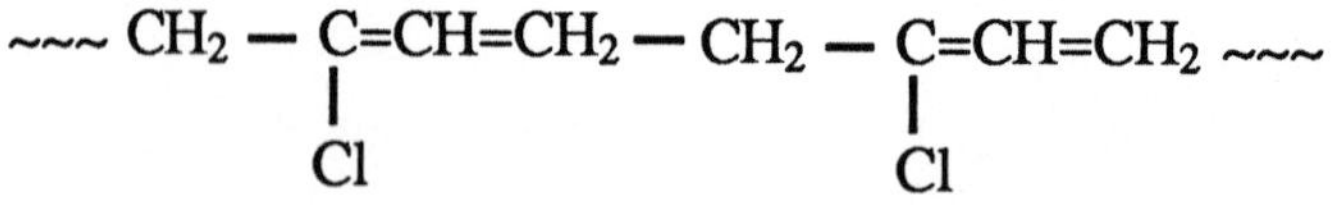

```
~~~ CH2 — C=CH=CH2 — CH2 — C=CH=CH2 ~~~
          |                 |
          Cl                Cl
```

PRACTICE EXAM

1. Which molecule is a structural isomer of the molecule below?

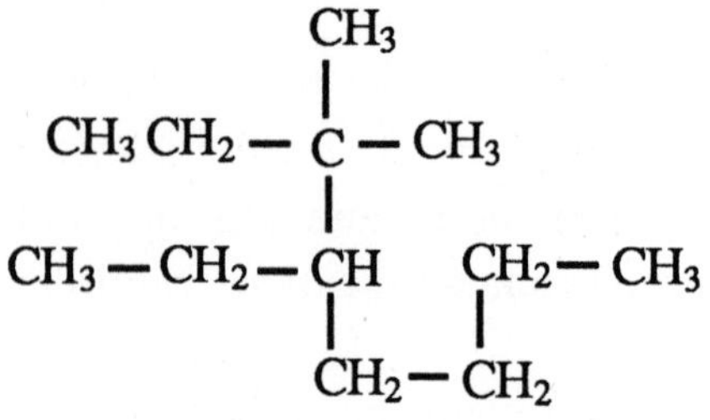

```
                 CH3
                  |
         CH3 CH2 — C — CH3
                  |
  CH3 — CH2 — CH     CH2 — CH3
               |      |
              CH2 — CH2
```

a.

```
              CH3
               |
        CH3   CH2            CH3
         |     |              |
CH3 CH2 C  —  CH — CH2 CH2 CH2
         |
        CH3
```

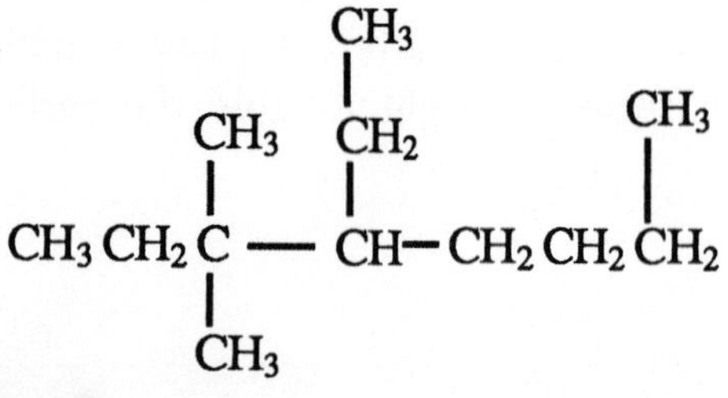

b.

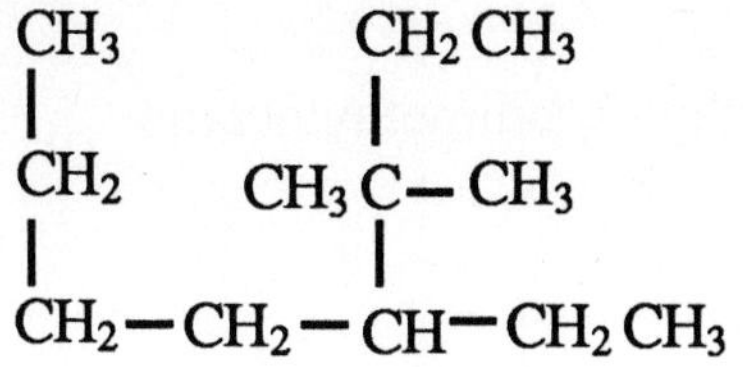

```
CH3           CH2 CH3
|             |
CH2      CH3 C— CH3
|             |
CH2—CH2—CH—CH2 CH3
```

c.

```
                CH3
                |
          CH3   CH2
          |     |
CH3 CH2 CH      CH—CH2 CH3
          |     |
CH3— CH—CH2
```

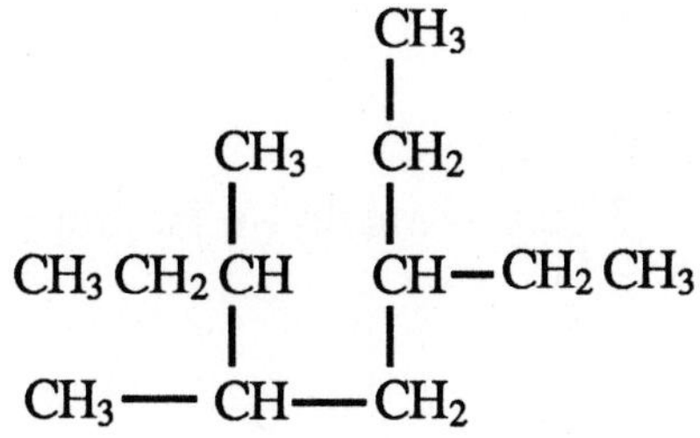

d.

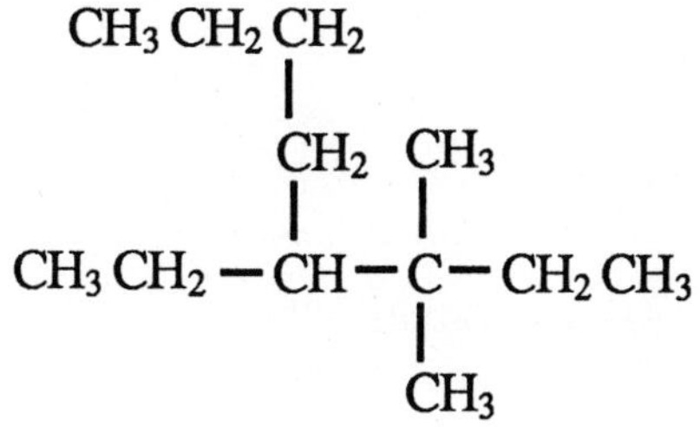

```
     CH3 CH2 CH2
             |
             CH2  CH3
             |    |
CH3 CH2 —CH—C—CH2 CH3
                  |
                  CH3
```

e.

```
        CH2—CH3
        |
        CH——CH2 CH2
        |           |
CH3—C— CH3     CH2
        |           |
        CH2         CH3
        |
        CH3
```

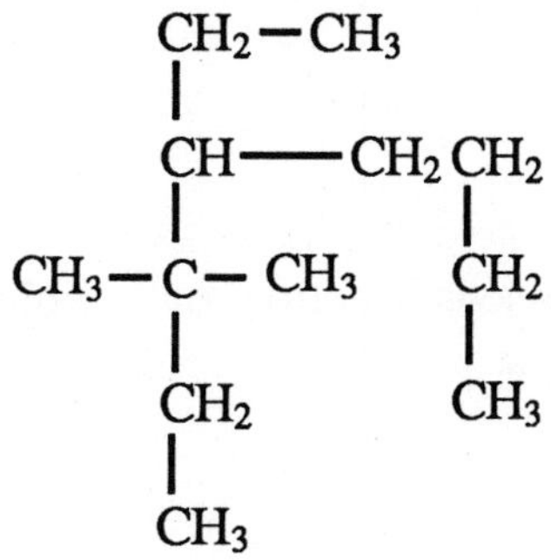

2. Which of the alkane names below is <u>not</u> correct?

a.

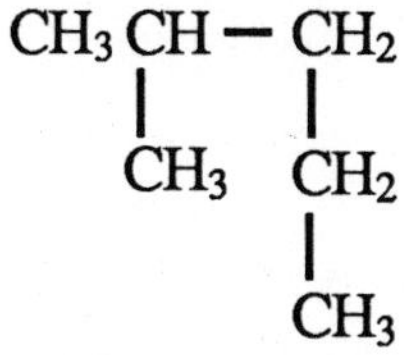

```
CH3 CH— CH2
     |      |
     CH3  CH2
            |
            CH3
```

1-ethyl-2-methylpropane

b.

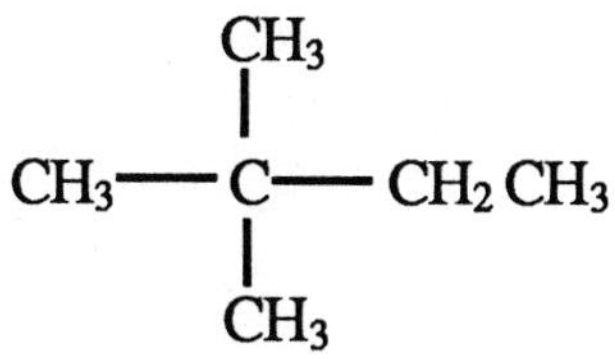

```
        CH3
        |
CH3——C——CH2 CH3
        |
        CH3
```

2,2-dimethylbutane

c.

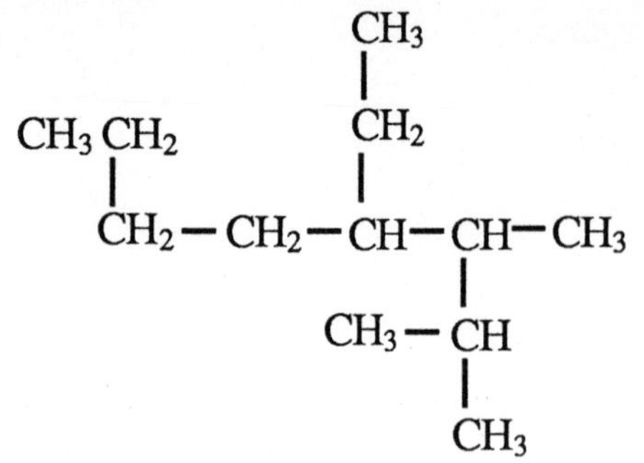

```
                  CH3
                   |
CH3 CH2           CH2
 |                 |
CH2 - CH2 - CH - CH - CH3
                  |
            CH3 - CH
                  |
                  CH3
```

4-ethyl-2,3-dimethyloctane

d.

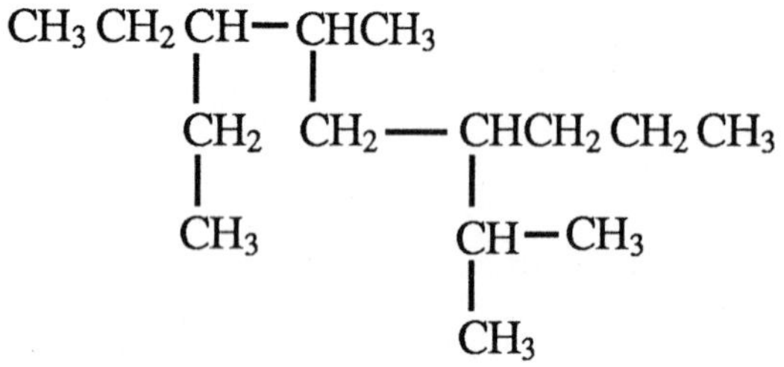

```
CH3 CH2 CH - CHCH3
         |    |
        CH2  CH2 -- CHCH2 CH2 CH3
         |           |
        CH3          CH - CH3
                     |
                     CH3
```

3-ethyl-4-methyl-6-isopropylnonane

e.

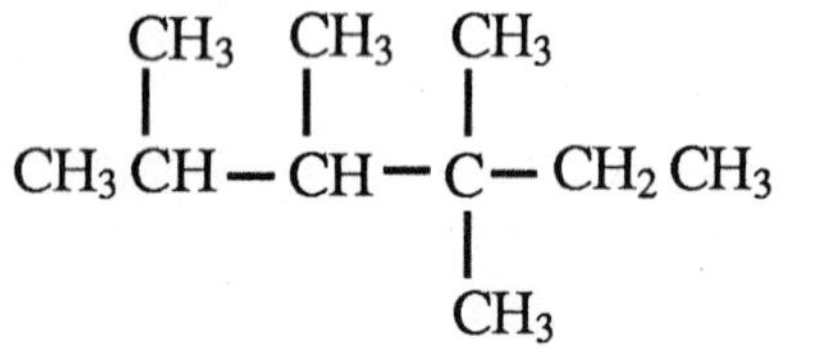

```
     CH3   CH3   CH3
      |     |     |
CH3 CH - CH  -  C - CH2 CH3
                  |
                 CH3
```

2,3,4,4-tetramethylhexane

3. Which statement about petroleum is not true?

   a. By the process of pyrolytic cracking, the kerosene fraction of petroleum can be converted to smaller molecules typical of gasoline.
   b. The C_{10}-C_{18} fraction of petroleum is the gasoline fraction.
   c. The petroleum fractions can be separated based on differences in boiling points.
   d. A popular gasoline additive in the recent past, tetraethyl lead, prevented engine knock.
   e. The most important use of hydrocarbons from the first oil wells was for lighting.

4. Which functional group/name pair is not correct?

   a. RCOOH/carboxylic acid
   b. RCH_2OH/alcohol
   c. RCHO/aldehyde
   d. ROR/ether
   e. RCOR/ester

5. Which of the alkene or alkyne names is <u>not</u> correct?

a. 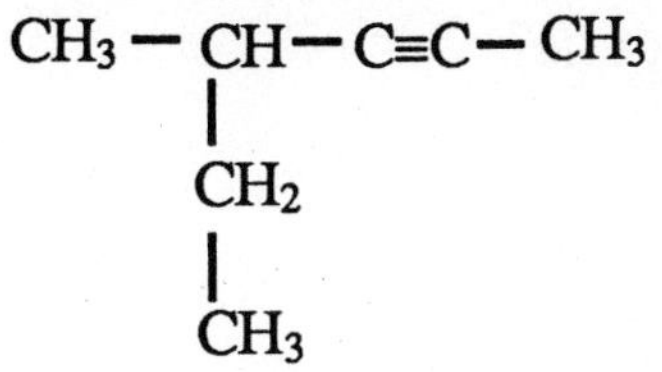
$$\begin{array}{l} CH_3-\underset{\displaystyle\underset{\displaystyle CH_3}{\overset{|}{\underset{|}{CH_2}}}}{CH}-C{\equiv}C-CH_3 \end{array}$$
4-ethyl-2-pentyne

b. 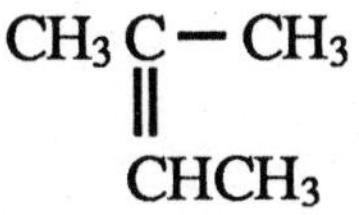
$$CH_3\underset{\displaystyle CHCH_3}{\underset{\|}{C}}-CH_3$$
2-methyl-2-butene

c. 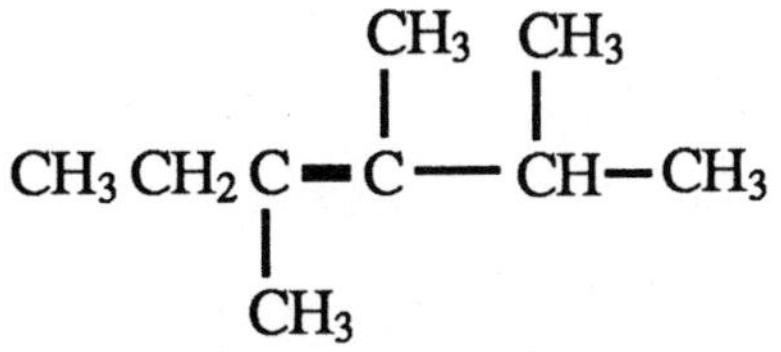
$$CH_3CH_2\underset{\displaystyle CH_3}{\underset{|}{C}}=\overset{\displaystyle CH_3}{\overset{|}{C}}-\overset{\displaystyle CH_3}{\overset{|}{CH}}-CH_3$$
2,3,4-trimethyl-3-hexene

d. 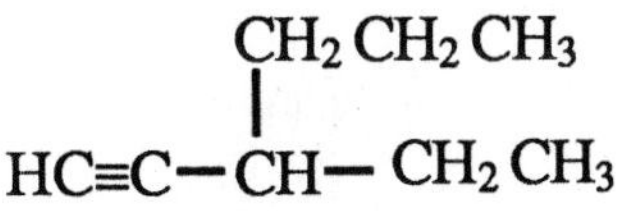
$$HC{\equiv}C-\overset{\displaystyle CH_2CH_2CH_3}{\overset{|}{CH}}-CH_2CH_3$$
3-ethyl-1-hexyne

e. 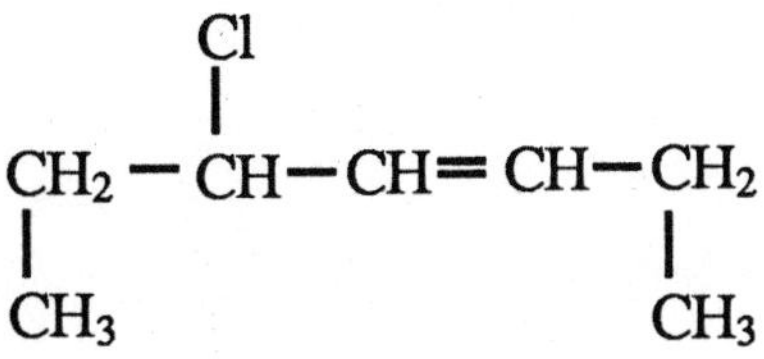
$$\underset{\displaystyle CH_3}{\underset{|}{CH_2}}-\overset{\displaystyle Cl}{\overset{|}{CH}}-CH=CH-\underset{\displaystyle CH_3}{\underset{|}{CH_2}}$$
5-chloro-3-heptene

6. Which reaction is an example of an addition reaction?

a. 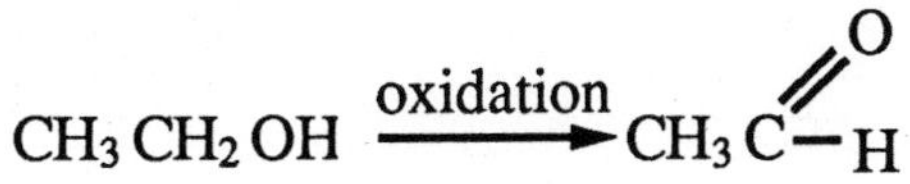 $CH_3CH_2OH \xrightarrow{\text{oxidation}} CH_3\overset{\displaystyle O}{\overset{\|}{C}}-H$

b. $CH_4 + 2O_2 \longrightarrow CO_2 + 2H_2O$

c. $CH_3CH_3 \xrightarrow[500^\circ]{Cr_2O_3} CH_2=CH_2 + H_2$

d. $CH_3CH_2CH_3 + Br_2 \xrightarrow{h\nu} CH_3CH_2CH_2Br + HBr$

e. $CH_3CH_2CH=CH_2 + H_2 \xrightarrow{\text{catalyst}} CH_3CH_2CH_2CH_3$

7. Which statement about the structure of benzene is <u>not</u> true?

   a. Benzene has two possible Lewis structures.
   b. Benzene is a planar molecule.
   c. Each carbon in benzene is bonded to two hydrogen atoms.
   d. The carbon-carbon double bonds in benzene can be represented by a circle inside the ring.
   e. All the bond angles around the atom in benzene are 120°.

8. Which statement about alcohols is true?

   a. Methyl alcohol is prepared by fermentation of sugars.
   b. Ethyl alcohol is prepared by the reaction of ethylene with water.
   c. 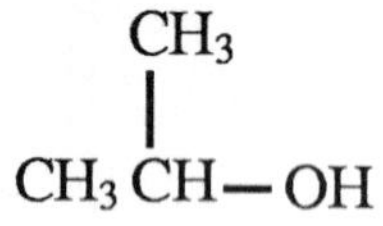 $CH_3\overset{\displaystyle CH_3}{\overset{|}{C}}H-OH$ is a primary alcohol.
   d. Ethylene glycol, which is used in antifreeze, has the structure $\begin{matrix}CH-OH\\ \|\\ CH-OH\end{matrix}$.
   e. 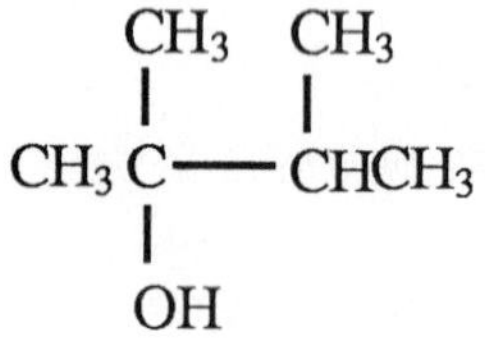 $CH_3\overset{\displaystyle CH_3}{\underset{\displaystyle OH}{C}}-\overset{\displaystyle CH_3}{\overset{|}{C}}HCH_3$ is named 2,3-dimethyl-3-butanol.

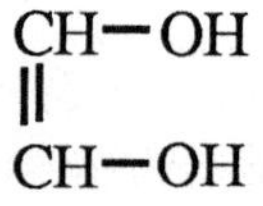

9. Which of the aldehydes or ketones below is named incorrectly?

   a. 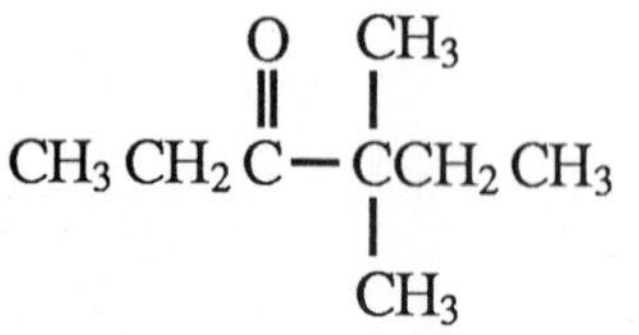 $CH_3CH_2\overset{\displaystyle O}{\overset{\|}{C}}-\overset{\displaystyle CH_3}{\underset{\displaystyle CH_3}{C}}CH_2CH_3$ 4,4-dimethyl-3-hexanone
   b. 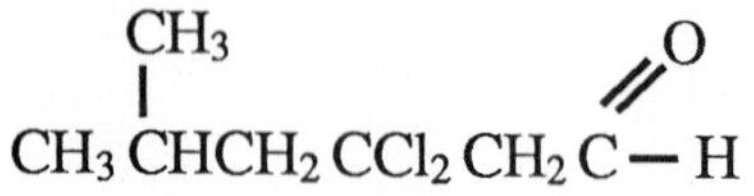 $CH_3\overset{\displaystyle CH_3}{\overset{|}{C}}HCH_2CCl_2CH_2\overset{\displaystyle O}{\overset{/\!/}{C}}-H$ 3,3-dichloro-5-ethyl pentanal
   c. 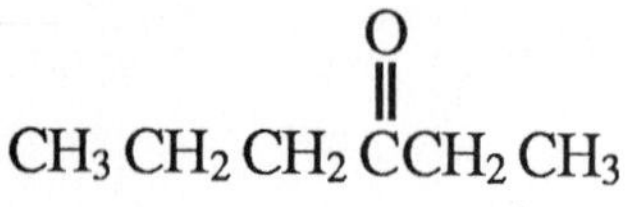 $CH_3CH_2CH_2\overset{\displaystyle O}{\overset{\|}{C}}CH_2CH_3$ ethyl propyl ketone

d. 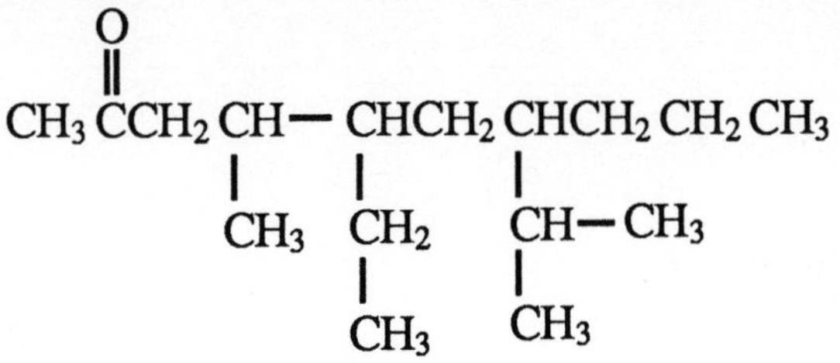 $CH_3\overset{O}{\overset{\|}{C}}CH_2\underset{CH_3}{\underset{|}{CH}}-\underset{\underset{CH_3}{\underset{|}{CH_2}}}{\underset{|}{CH}}CH_2\underset{\underset{CH_3}{\underset{|}{CH-CH_3}}}{\underset{|}{CH}}CH_2CH_2CH_3$ 5-ethyl-7-isopropyl-4-methyl-2-decanone

e. 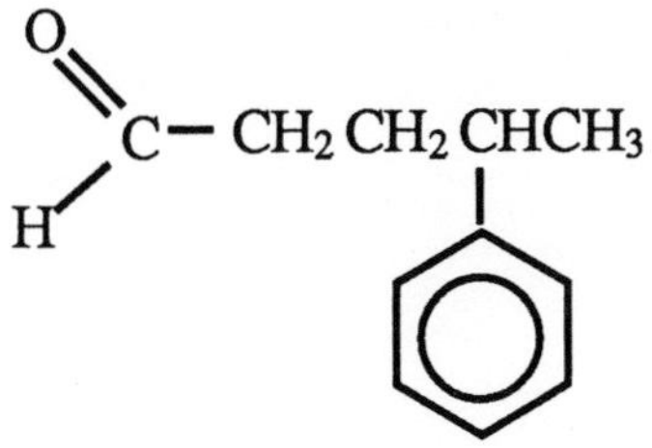 $H-\overset{O}{\overset{\|}{C}}-CH_2CH_2\underset{C_6H_5}{\underset{|}{CH}}CH_3$ 4-phenylpentanal

10. Which statement about polymers is not correct?

   a. In addition polymerization there are no products of the reaction other than the polymer itself.
   b. When propylene reacts to form polypropylene, the double bond is converted to a single bond.
   c. The characteristic feature of condensation polymers is that they are made from only one kind of monomer.
   d. Nylon is an example of a copolymer.
   e. A dimer is synthesized from two monomer units.

PRACTICE EXAM ANSWERS

1. c (20.3)
2. a (20.4)
3. b (20.5)
4. e (20.10)
5. a (20.7)
6. e (20.6, 20.7 and 20.13)
7. c (20.8)
8. b (20.11, 20.12)
9. b (20.14)
10. c (20.16)

CHAPTER 21: BIOCHEMISTRY

INTRODUCTION

Biochemistry, the chemistry of living systems, is a subject which concerns everyone. How do our bodies extract chemical energy from sugar and other substances? How can we find cures for diseases? The answers to these questions lie in the biochemistry of the human body. To begin to understand how the complex biochemical systems work, we need to know about the kinds of molecules which are important in living organisms. They are often large organic molecules, and they contain atoms and funtional groups with whose characteristics we are already familiar.

AIMS FOR THIS CHAPTER

1. Know which elements are important in living tissues. (Introduction)
2. Know the functions of proteins. (Section 21.1)
3. Understand how amino acids are linked together to form the primary structure of proteins. (Section 21.2)
4. Know the types of secondary structure in proteins, and the difference between primary and secondary structure. (Section 21.3)
5. Know the two general types of tertiary protein structure, and how tertiary and secondary protein structures differ. (Section 21.4)
6. Know what kinds of functions proteins serve in living systems, and how protein function can be disrupted. (Section 21.5)
7. Understand how the lock and key model of enzyme function works. (Section 21.6)
8. Be able to define monosaccharide, disaccharide, and polysaccharide, and know the properties of the common carbohydrates. (Section 21.7)
9. Know the structures of nucleotides, the monomers which make up DNA and RNA, and know how the monomers join together to produce the polymers. (Section 21.8)
10. Know how DNA and RNA function to produce individual proteins. (Section 21.8)
11. Know the characteristics of each of the four major lipid classes. (Section 21.9)

QUICK DEFINITIONS

Biochemistry — The study of the chemistry of living organisms. (Introduction)

Essential elements — Elements which are necessary to sustain life. (Introduction)

Trace elements — Essential elements found in small quantities in biologically important molecules. (Introduction)

Proteins — A class of natural polymers found in all living systems. (Section 21.1)

Fibrous proteins — Proteins which provide the basis for structure in an organism. (Section 21.1)

Globular proteins — Roughly spherical shaped proteins which do chemical work in an organism. (Section 21.1)

α-amino acids — Twenty or so different protein monomers which all contain an amino functional group, a carboxyl group, and an R group side chain which is different for each amino acid. (Section 21.2)

Side chains — The R groups on amino acids are called side chains. Side chains can be either polar, like water, or nonpolar, like hydrocarbons. (Section 21.2)

Dipeptide — Two amino acids joined together. (Section 21.2)

Peptide linkage — The kind of bond formed between two amino acids. The carboxyl group of one amino acid reacts with the amino group of the second amino acid. A water molecule is removed in the process. (Section 21.2)

$$-\overset{\overset{\displaystyle O}{\|}}{C}-\overset{\overset{\displaystyle H}{|}}{N}-$$

Polypeptide — Several amino acids joined together. (Section 21.2)

Primary structure — The structure of a protein described by the order in which the amino acids occur. (Section 21.2)

Secondary structure — The structure of a protein described by how the amino acids in the chain interact with each other to produce an α-helix or a pleated sheet. (Section 21.3)

α-helix — A protein secondary structure which is spiral shaped. (Section 21.3)

Pleated sheet — A protein secondary structure which is sheet-like with creases, or folds. (Section 21.3)

Tertiary structure — The three-dimensional shape of a protein. (Section 21.4)

Denaturation — The breaking down or unraveling of the tertiary structure of a protein. (Section 21.5)

Enzymes — Biological catalysts. Enzymes catalyze many reactions which produce energy and new tissue in organisms. (Section 21.5)

Lock-and-key model — A model which generally explains how many enzymes work. Suggests that the reactant molecule (the key) fits into a place on the enzyme which is of the same shape as the reactant (the lock). When the reactant and the enzyme come together, a reaction occurs to produce product. (Section 21.6)

Substrate — The reacting molecule in an enzymatic reaction. (Section 21.6)

Carbohydrates — A class of compounds constructed of simple sugars. (Section 21.7)

Monosaccharides — Simple sugars which are aldehydes or ketones, and which contain several hydroxyl (-OH) substituents. (Section 21.7)

Simple sugars — Also called monosaccharides. The units from which the large carbohydrate polymers are made. (Section 21.7)

Disaccharide — Two monosaccharides joined together. (Section 21.7)

Sucrose — Table sugar, a disaccharide made from glucose and fructose. (Section 21.7)

Glycoside linkage — The kind of bond which forms between two sugars. The result is a carbon-oxygen-carbon bond between the two sugars, and the elimination of a water molecule. (Section 21.7)

Polysaccharides — Carbohydrates made from many monosaccharides joined together by glycoside linkages. (Section 21.7)

Starch — A carbohydrate made from long polymers of glucose stored in plant tissues as a food reserve. Starch is a good food source for animals. The glucose-glucose bonds are of a different kind than those found in cellulose. (Section 21.7)

Cellulose — A carbohydrate made from long polymers of glucose. Cellulose is found as a structural component of plants and cannot be digested by humans or by most animals. The glucose-glucose bonds in cellulose are of a different kind than those in starch. (Section 21.7)

Glycogen — A carbohydrate made from long polymers of glucose. Glycogen is found in muscles and is used as a food reserve in animals. (Section 21.7)

Deoxyribonucleic acid — A large polymer which is made of a phosphate, the sugar deoxyribose, and organic bases. DNA carries the genetic information which is passed from cell to cell and generation to generation. (Section 21.8)

Ribonucleic acid — A polymer which is made of a phosphate, the sugar ribose (as opposed to deoxyribose in DNA) and organic bases. RNA is responsible for translating the information carried by the DNA into proteins. (Section 21.8)

Nucleotide — A building block unit in DNA and RNA which consists of a five-carbon sugar (ribose or deoxyribose), a nitrogen-containing organic base, and a phosphate group. (Section 21.8)

Protein synthesis — The production of proteins from individual amino acids. (Section 21.8)

Gene — A piece of DNA which carries the information for the synthesis of a specific protein. (Section 21.8)

Messenger RNA (mRNA) — A piece of RNA which is built from a specific length of DNA (gene). mRNA carries the genetic information to the site of protein synthesis. (Section 21.8)

Transfer RNA (tRNA) — A small piece of RNA which brings a specific amino acid to the site of protein synthesis. (Section 21.8)

Lipids — A group of substances characterized by their insolubility in water and their solubility in organic solvents. (Section 21.9)

Fats — Esters made from the alcohol glycerol and long chain carboxylic acids called fatty acids. (Section 21.9)

Fatty acids — Long chain carboxylic acids. (Section 21.9)

Triglycerides — Fats which are made from glycerol and three fatty acids. (Section 21.9)

Saponification — Decomposition of triglycerides by NaOH into glycerol and the salts of fatty acids (soaps). (Section 21.9)

Micelles Spherical aggregates of soap molecules in water. (Section 21.9)

Surfactant A substance which can cause greasy dirt (nonpolar materials) to be lifted from a surface to an aqueous solution. (Section 21.9)

Phospholipids Substances which have two fatty acids and one phosphate group bound to glycerol. This produces a polar phosphate group "head", and a nonpolar fatty acid "tail". (Section 21.9)

Waxes Esters made from long chain carboxylic acids and long chain alcohols. (Section 21.9)

Steroids A class of lipids with a characteristic ring structure. Different kinds of steroids have different substituents on the basic ring structure. (Section 21.9)

Cholesterol A lipid which is based on the steroid ring structure. Cholesterol is essential for humans, but can cause hardening of the arteries. (Section 21.9)

Adrenocorticoid hormones Hormones synthesized in the adrenal glands that are involved with regulatory functions. (Section 21.9)

Sex hormones Hormones involved with sexual function and characteristics. (Section 21.9)

Bile acids Substances produced from cholesterol in the liver that aid in the digestion of fats. (Section 21.9)

CONTENT REVIEW

21.1 PROTEINS

What Do Proteins Do?

Proteins perform much of the work in living organisms. Among their many functions are the transport of nutrients from one place to another, catalysis of thousands of reactions, and the regulation of body functions.

21.2 PRIMARY STRUCTURE OF PROTEINS

What Is the Primary Structure of Proteins?

Proteins are polymers made from many α-amino acids joined together. Amino acids are called α-amino acids because they all have an amino functional group on carbon two, which is also called the α-carbon. There are around twenty different amino acids commonly found in proteins. Each different amino acid has a carboxyl group and an amino group, and each one has a unique R group, or **side chain**. The R group can be as simple as a hydrogen or very complex.

$$R-\underset{NH_2}{\overset{H}{C}}-C(=O)-OH$$

Proteins are made when amino acids react with each other to form peptide linkages. The carboxyl group of one amino acid reacts with the amino group of another amino acid. A water molecule is removed during the process.

$$R-\underset{NH_2}{CH}-C(=O)-\boxed{OH \quad H}-N(H)-\underset{R}{CH}-C(=O)-OH \longrightarrow R-\underset{NH_2}{CH}-C(=O)-NH-\underset{R'}{CH}-C(=O)-OH + H_2O$$

The bond formed between two amino acids is called a **peptide linkage**. There is almost an endless number of proteins which could be formed from combinations of twenty different amino acids. Each different sequence of amino acids produces a different protein, so the order in which the amino acids occur is important. The order in which amino acids occur is called the **primary structure** of a protein.

21.3 SECONDARY STRUCTURE OF PROTEINS

What Is the Secondary Structure of Proteins?

The **secondary structure** of a protein is the shape the protein chain assumes. Two common secondary structures are the α-helix and the pleated sheet. In an **α-helix**, the amino acids spiral as in a staircase. In a **pleated sheet**, several amino acid chains form a sheet with creases (pleats).

21.4 TERTIARY STRUCTURE OF PROTEINS

What is the Tertiary Structure of Proteins?

The **tertiary structure** of a protein is the three-dimensional shape of the protein molecule. Some proteins are globular in shape, and others are elongated. A protein can have several areas of α-helix (secondary structure) which are separated from each other by bends. The protein folds at each bend, giving the molecule its tertiary structure.

21.5 FUNCTIONS OF PROTEINS

How Do Proteins Maintain Their Function?

Proteins continue to function as long as their tertiary structures remain intact. Any source of energy can cause **denaturation**, or the unraveling of the three-dimensional shape of a protein, and lead to the loss of function.

21.6 ENZYMES

Enzymes are proteins that catalyze specific biological reactions.

How Do Enzymes Catalyze Reactions?

Exactly how all enzymes work to speed a reaction along is not completely understood, but the lock-and-key model explains how some enzymes work. In the **lock-and-key model**, the reactant, also called the **substrate**, has a specific shape and plays the part of the key in the model. The substrate fits into the enzyme, which plays the role of the lock. A product is produced when the substrate and enzyme fit together. Only a specific substrate, one which fits the lock, will react. All other molecules do not produce product.

21.7 CARBOHYDRATES

What Are the Major Types of Carbohydrates?

Carbohydrates are a large class of biomolecules composed mainly of carbon, hydrogen and oxygen. They are a major source of food and serve as structural components in plants. Some carbohydrates are **simple sugars**, or **monosaccharides**, while others are long polymers made from many monosaccharides joined together. The monosaccharides all have features in common. All contain either an aldehyde or a ketone functional group, and all have one or more hydroxyl (-OH) groups. The number of carbons in a monosaccharide varies, but the most common ones have either five or six carbons. A typical monosaccharide is glucose, which has six carbons, an aldehyde functional group, and five hydroxyl groups.

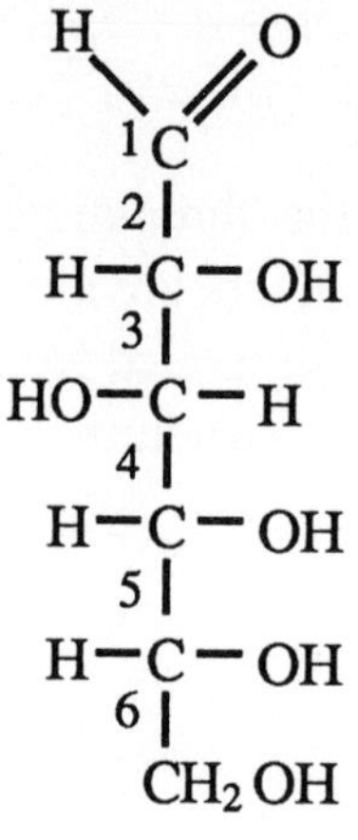

```
  H     O
   \\  //
    1C
    2|
  H—C—OH
    3|
 HO—C—H
    4|
  H—C—OH
    5|
  H—C—OH
    6|
    CH2OH
```

glucose

In aqueous solution, most monosaccharides exist as a ring structure, not as a chain. The ring form of glucose can be drawn as

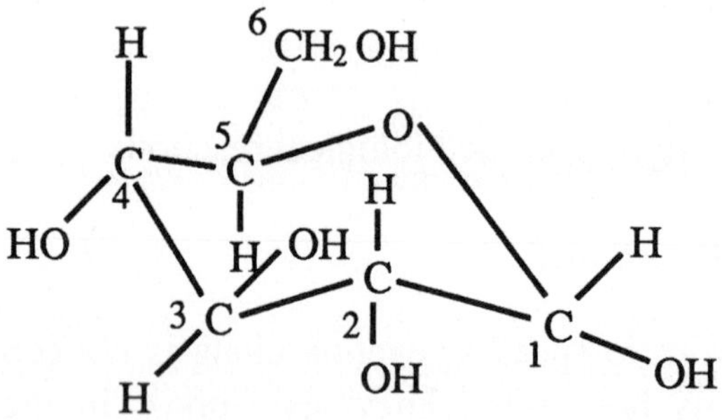

Two monosaccharides can join together with a C-O-C bond to form a **disaccharide**. A water molecule is eliminated in the process. The most common disaccharide is **sucrose**, or table sugar. Sucrose is made from the two monosaccharides fructose and glucose.

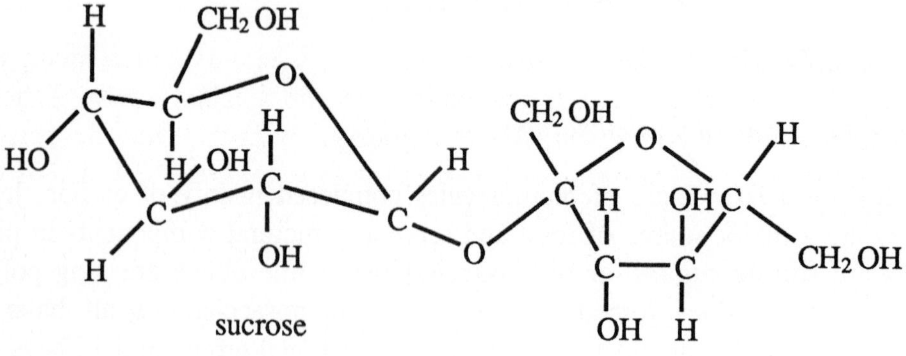

sucrose

Many monosaccharides can join together to form polymers called **polysaccharides**. **Starch** is a polysaccharide made from many glucose molecules, and is used as a food storage product in plants and as a food source for animals. **Cellulose** is another polysaccharide made of glucose monomers. In plants, cellulose forms part of the structure which holds plants upright. While starch is digestible by animals, cellulose is not. Both polymers are made of glucose, but the kind of bonding between the glucose units is different. A very small difference in bonding means that starch can be used as a food source by humans, while cellulose cannot.

21.8 NUCLEIC ACIDS

What Are the Types and Structures of Nucleic Acids?

Nucleic acids store the information necessary for life to continue from generation to generation. The nucleic acid which stores genetic information and transmits information from one generation to the next is **deoxyribonucleic acid**, or **DNA**. DNA is a large polymer containing many monomers. Another nucleic acid which helps translate the genetic information into proteins is **ribonucleic acid**, or **RNA**. RNA is also a polymer, but smaller than DNA. Both DNA and RNA are composed of smaller building blocks called nucleotides. A **nucleotide** is made from three parts, a five-carbon monosaccharide, a nitrogen-containing organic base, and a phosphate group. DNA and RNA do not have the same five-carbon monosaccharide. DNA contains the monosaccharide deoxyribose, while RNA contains ribose. The monosaccharides give the nucleic acids part of their names.

DNA consists of not one, but two polymer chains linked together. The two chains line up with each other so that the organic bases on each chain form a pair which can hydrogen bond with each other. Specific bases pair with other bases. If an adenine molecule is found on one chain, then the base on the other chain is always a thymine, and if the base on one chain is a guanine, then the other base is always a cytosine. The chains are twisted around to form a double helix, a spiral structure.

How Does DNA Function To Produce Proteins?

DNA is a long polymer, and the information to make proteins is stored along its chain. A specific piece of DNA which carries information for a protein is called a **gene**. When a protein is needed, a piece of RNA called **messenger RNA (mRNA)** is made using the gene as a pattern. The DNA pattern dictates which bases will be present in the mRNA. The mRNA travels to the site of protein synthesis. Another type of RNA, **transfer RNA (tRNA)**, brings individual amino acids to the same site. In this way, the information contained in the DNA determines how proteins are made.

21.9 LIPIDS

What Are Lipids?

Lipids do not contain a specific functional group, but are characterized by their insolubility in water, and their solubility in nonpolar organic solvents. There are several classes of lipids, including the fats or triglycerides, phospholipids, waxes, and steroids.

What Is the Structure Of the Triglycerides?

The most common fats are esters. Recall that esters form when an alcohol reacts with a carboxylic acid. The alcohol in fats is glycerol, a three-carbon alcohol with three hydroxyl groups. The carboxylic acids which are present in fats can vary, but most of them are long-chain carboxylic acids called **fatty acids**. Fats which contain glycerol and fatty acids are called **triglycerides**. Triglycerides will react with NaOH in a process called **saponification** to produce glycerol, and the sodium carboxylic acid salts, called soaps. This is the process used to make soap.

What Is the Structure Of the Phospholipids?

Phospholipids are made from glycerol and two fatty acids, but the third hydroxyl group has reacted with a phosphate group.

What Is the Structure Of the Waxes?

Waxes are esters made from monohydroxy alcohols and carboxylic acids. Both the alcohol and the carboxylic acids have long chains.

What Is the Structure of the Steroids?

There are many different steroids which function in the human body. All of them have a backbone of the ring structure below.

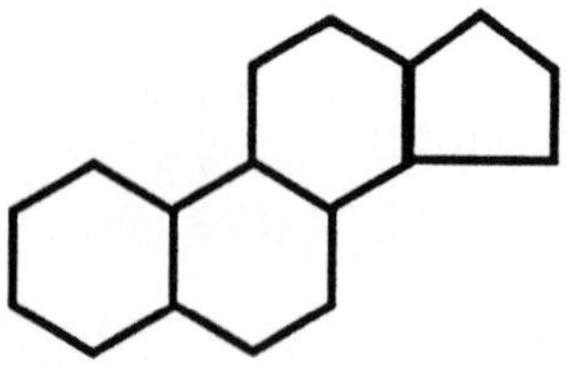

The steroid cholesterol is found in virtually all organisms and is the starting material for many other steroids.

LEARNING REVIEW

1. What functions do carbon, phosphorus and magnesium perform in the human body?

2. What are two major types of proteins and how do their functions differ?

3. What common features do the α-amino acids in proteins share and what feature varies?

4. Which of the amino acids below have hydrophilic and which have hydrophobic side chains?

   a. 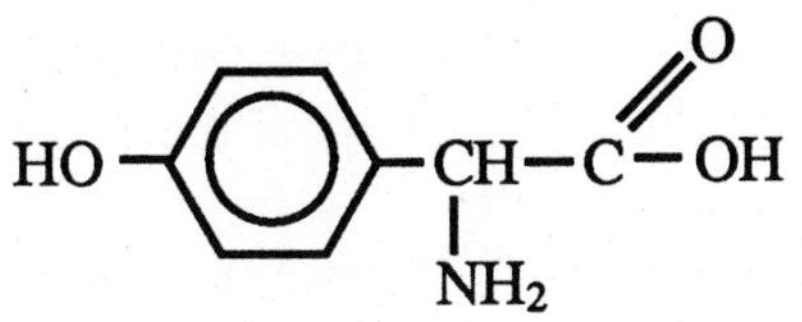 $HO-C_6H_4-CH(NH_2)-C(=O)-OH$

   b. $HO-C(=O)-CH_2\,CH(NH_2)-C(=O)-OH$

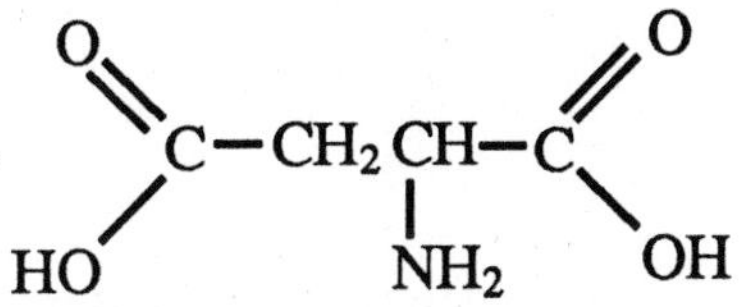

   c. $CH_3\,CH(CH_3)CH_2\,CH(NH_2)C(=O)OH$

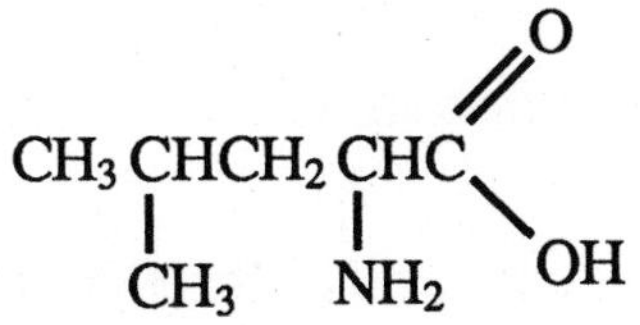

5. What is the structure of the dipeptide made from the two amino acids below?

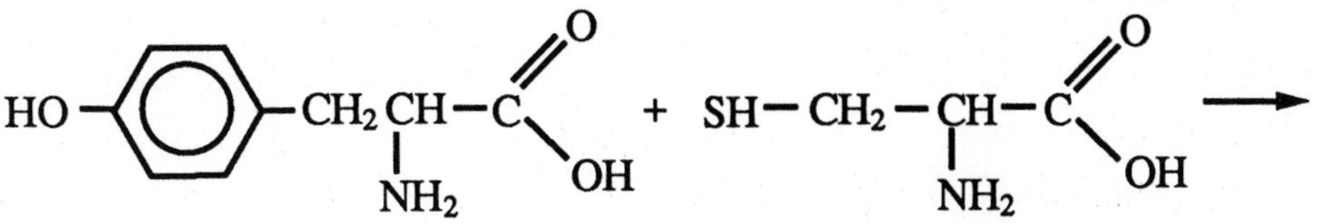

$$HO-C_6H_4-CH_2\,CH(NH_2)-C(=O)OH \; + \; SH-CH_2-CH(NH_2)-C(=O)OH \longrightarrow$$

6. Show the sequences of all tripeptides which can be made from the amino acids phenylalanine (phe), glycine (gly), proline (pro), and aspartic acid (asp).

7. Explain the difference between the primary, secondary and tertiary structures of proteins.

8. Two types of secondary structures are the α-helix and the pleated sheet. What are the major characteristics of each?

9. Which amino acid plays a special role in maintaining the tertiary structure of proteins?

10. Use a lock and key analogy to explain how an enzyme can catalyze a reaction.

11. What two functional groups are characteristic of monosaccharides?

12. Draw the structure of a tetrose which has an aldehyde functional group.

13. The disaccharide sucrose, or table sugar, is made from which two monosaccharides?

14. What difference between starch and cellulose causes starch to be a food source for humans, while cellulose is not?

15. a. What are the parts of a nucleotide?
    b. How do the nucleotides of RNA differ from nucleotides of DNA?

16. Show the structure of a nucleotide.

17. Show why the bases cytosine and guanine and the bases adenine and thymine pair with each other in DNA.

18. How is it thought that DNA reproduces itself?

19. How are proteins synthesized from the DNA code?

20. a. Draw the structure of the fat made from a molecule of glycerol and three molecules of oleic acid.
    b. Would this fat likely be a solid or a liquid at room temperature?

21. Explain how soap cleans away greasy dirt.

22. Molecules with which two functional groups combine to form wax molecules?

23. The steroids are a diverse group of molecules. What structural feature do they all have in common?

ANSWERS TO LEARNING REVIEW

1. Carbon is the backbone of all organic molecules in the body. Phosphorus is present in cell membranes and plays an important role in energy transfer in cells. Magnesium is required for proper functioning of some enzymes.

2. The two major types are fibrous and globular. Fibrous proteins provide structure and shape, while globular proteins perform chemical work.

3. All of the α-amino acids found in proteins have a carboxyl group, an amino group on carbon number two (the alpha carbon), and a side chain also on carbon two. The nature of the side chain varies.

4. a. The side chain on tyrosine is a benzene ring which has an OH substituted for one of the hydrogen atoms. The OH bond is polar, just as the OH bond in water is. The polarity of the OH bond makes the side chain of tyrosine hydrophilic.

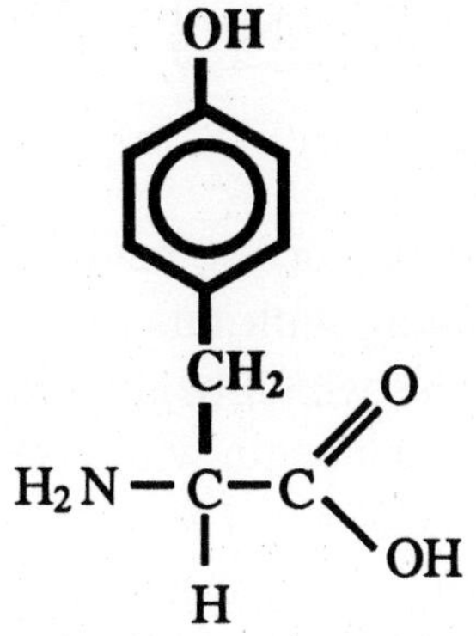

b. Aspartic acid has a side chain which has a CH_2 and a COOH. The OH bond is polar, and so is the CO bond. The polarity of the COOH makes the side chain of aspartic acid hydrophilic.

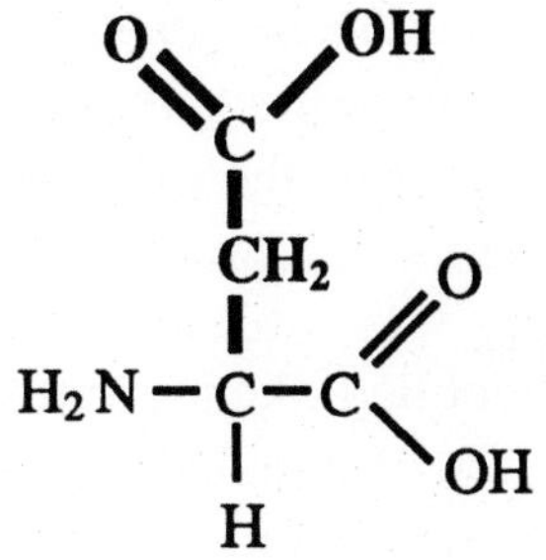

c. Leucine has a side chain composed entirely of carbon and hydrogen. Carbon-hydrogen bonds are not polar, so the side chain of leucine is hydrophobic.

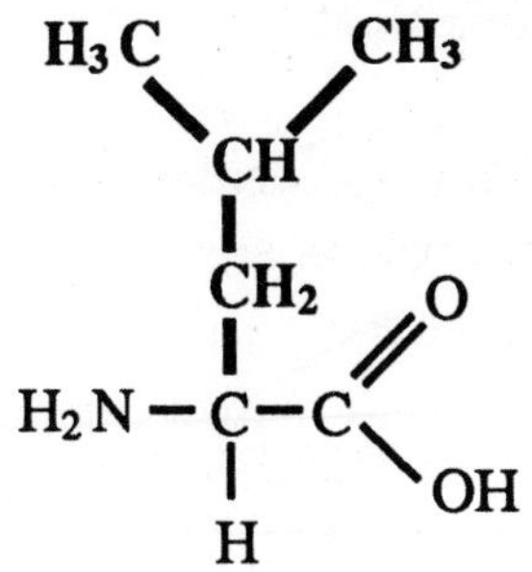

5.

$$\text{HO}-\langle\bigcirc\rangle-\text{CH}_2\text{CH}(\text{NH}_2)-\overset{\text{O}}{\overset{\|}{\text{C}}}-\boxed{\text{OH} + \text{H}}-\text{NH}-\text{CH}(\text{CH}_2\text{SH})-\overset{\text{O}}{\overset{\|}{\text{C}}}-\text{OH} \longrightarrow \text{OH}-\langle\bigcirc\rangle-\text{CH}_2\text{CH}(\text{NH}_2)-\overset{\text{O}}{\overset{\|}{\text{C}}}-\text{NH}-\text{CH}(\text{CH}_2\text{SH})-\overset{\text{O}}{\overset{\|}{\text{C}}}-\text{OH}$$

(Reactants as printed: $\text{HO}-C_6H_4-CH_2CH(NH_2)-C(=O)OH + SH-CH_2-CH(NH_2)-C(=O)OH$)

6. To help you determine all the sequences, begin with one of the amino acids, and write all the possible sequences beginning with this one amino acid. Then choose another amino acid and write all the sequences which begin with the second amino acid. Continue until you have written sequences which begin with each of the amino acids given.

phe-gly-pro	gly-phe-pro	pro-gly-phe	asp-phe-pro
phe-gly-asp	gly-phe-asp	pro-gly-asp	asp-phe-gly
phe-pro-gly	gly-pro-phe	pro-phe-asp	asp-pro-phe
phe-pro-asp	gly-pro-asp	pro-phe-gly	asp-pro-gly
phe-asp-pro	gly-asp-pro	pro-asp-phe	asp-gly-phe
phe-asp-gly	gly-asp-phe	pro-asp-gly	asp-gly-pro

7. The primary structure of proteins is the amino acid sequence. The secondary structure of proteins is the arrangement of the protein chain, or the arrangement of the amino acids to form an α-helix or a pleated sheet. How the α-helices or pleated sheets are bent in relation to one another is the tertiary structure of a protein.

8. In an α-helix the amino acids are coiled like a spiral staircase. This makes a long region of protein. Areas of protein with lots of α-helix are springy and elastic. In a pleated sheet, protein chains are bent back one or more times to form a sheet of protein chains. Pleated sheets are strong and resistant to stretching.

9. The amino acid cysteine helps proteins to maintain their unique tertiary structures. The SH side chains of two cysteine molecules can react to form a disulfide linkage which holds the protein chain in a fixed tertiary structure.

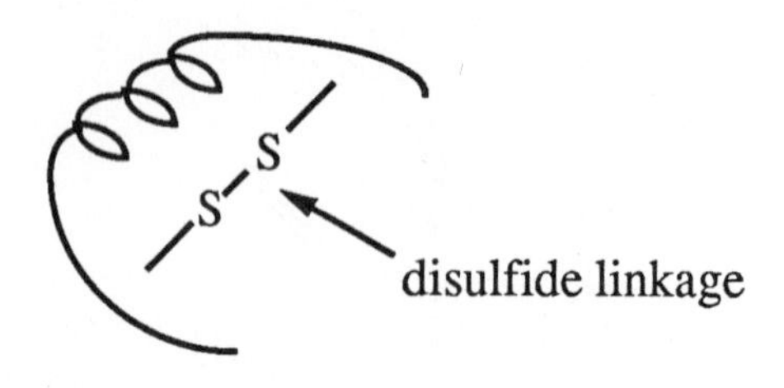

10. For some enzymes it is believed that the enzyme has a shape which fits the shape of the substrate, just like a key fits a lock. When the enzyme and the substrate join together, a reaction occurs. The enzyme then releases the product which has formed.

11. Monosaccharides all have hydroxyl and carbonyl functional groups. The carbonyl can either be an aldehyde or a ketone.

12. There are several tetroses with an aldehyde group. The structure of one is presented below.

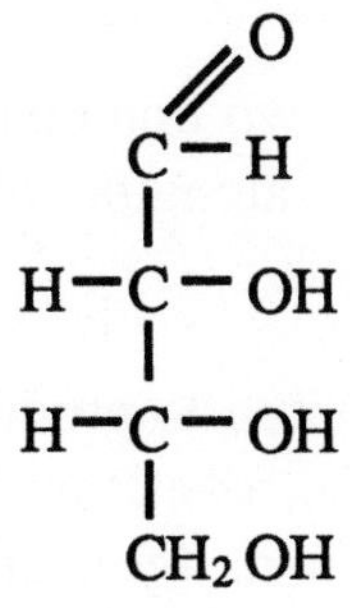

```
      O
     //
    C—H
    |
 H—C—OH
    |
 H—C—OH
    |
   CH2OH
```

13. Sucrose is made from the monosaccharides glucose and fructose.

14. Both starch and cellulose are made from glucose molecules. The way the glucose molecules are joined together in starch is different from the way they are joined in cellulose. Enzymes in our bodies can break the links between glucose molecules in starch, but not in cellulose.

15. a. All nucleotides are made from a phosphate group, a nitrogen-containing organic base, and a five-carbon sugar.

    b. In RNA, the sugar is ribose, while in DNA the sugar is deoxyribose. Some of the organic bases also differ between DNA and RNA. DNA and RNA have cytosine, adenine and guanine. Thymine is found only in DNA, and uracil is found only in RNA.

16.

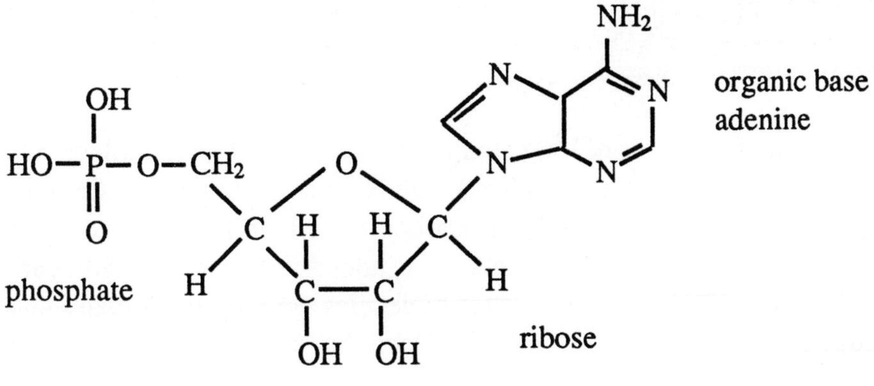

```
                                          NH2
                                          |
                                    N     C     N       organic base
     OH                            //  \ /  \\          adenine
     |                            ||    C    |
 HO—P—O—CH2        O              \\   /     /
     ||      \    /   \             N—C—N
     O        C  H   H  C  /
             / \ |   | / \
 phosphate  H   C———C     H
                |   |
                OH  OH        ribose
```

17. Cytosine and guanine pair with each other in DNA because the hydrogen bonding sites on the molecule are complementary. The three hydrogen bonds between the two molecules hold cytosine and guanine together. Adenine and thymine molecules on complementary DNA strands are also held together by hydrogen bonds. Two hydrogen bonds form between adenine and thymine molecules.

18. There is evidence to show that two complementary DNA chains unwind and that new strands are made by pairing bases along the old chains. The end result is two new chains, each complementary to the old one.

19. DNA stores the information required to produce proteins needed by organisms. A length of the DNA chain, called a gene, contains the information for one protein. The DNA transmits the information it contains by synthesizing a chain of RNA called messenger RNA (mRNA). Once the mRNA has been produced, it moves away from DNA to the site of protein synthesis. The mRNA and small bodies called ribosomes begin the synthesis of protein. Another kind of RNA, called transfer RNA (tRNA), brings up amino acids one by one to join them on the end of the growing protein chain. The mRNA contains the information which determines which amino acids join, and in what order.

20.

a.

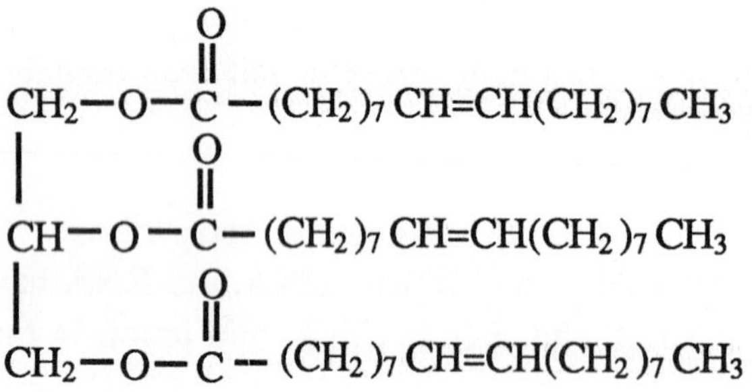

$$
\begin{array}{l}
CH_2-O-\overset{\overset{O}{\|}}{C}-(CH_2)_7CH=CH(CH_2)_7CH_3 \\
| \\
CH-O-\overset{\overset{O}{\|}}{C}-(CH_2)_7CH=CH(CH_2)_7CH_3 \\
| \\
CH_2-O-\overset{\overset{O}{\|}}{C}-(CH_2)_7CH=CH(CH_2)_7CH_3
\end{array}
$$

b. Fats made of unsaturated fatty acids are often liquids at room temperature. This fat would likely be a liquid at room temperature because oleic acid is an unsaturated fatty acid.

21. Soap molecules are the anions of long chain fatty acids. Each soap molecule has two parts. The carboxylate head has a negative charge and is attracted to polar water molecules. The remainder of the molecule is a long hydrocarbon tail which is not attracted to water. The hydrocarbon tail is hydrophobic. When soap is added to water, the hydrocarbon tail does not want to associate with the polar water molecules, so tails from many soap molecules associate together to form a soap micelle in which the polar heads face outward toward the water.

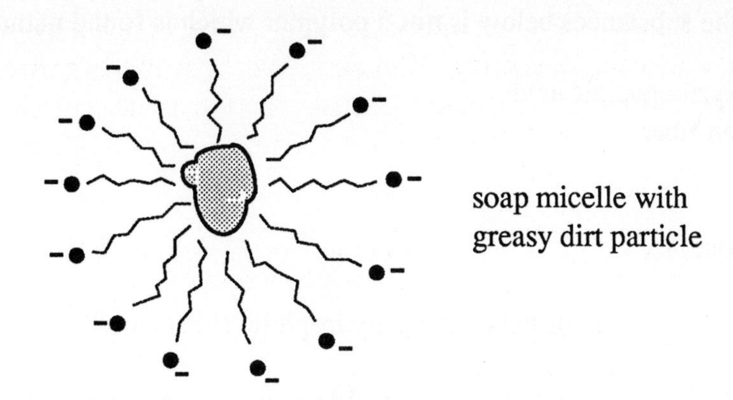
soap micelle with
greasy dirt particle

Much of the dirt we wish to remove is greasy dirt. This kind of dirt is not washed away by water, because it is hydrophobic. When soap micelles come in contact with greasy dirt, the dirt is lifted from the surface and enters the inside of the micelle. The hydrophobic dirt would rather be in contact with the hydrophobic hydrocarbon tails than with water. The soap micelles which contain the greasy dirt are washed away with water.

22. Waxes are esters with long carbon chains. Esters are made from a carboxylic acid and an alcohol.

23. All steroids have a basic ring structure called the steroid nucleus. Each steroid has different substituents on the rings.

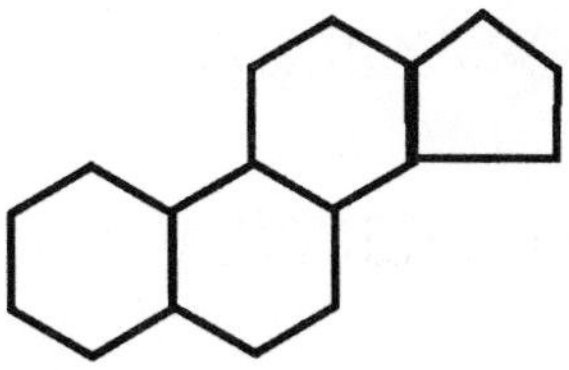

PRACTICE EXAM

1. Which of the elements below is **not** known to be an essential element?

   a. vanadium
   b. nickel
   c. fluorine
   d. magnesium
   e. chromium

2. Which of the substances below is **not** a polymer which is found naturally?

a. deoxyribonucleic acid
b. cotton fiber
c. hair
d. silk
e. sucrose

3. Which of the amino acids below has a hydrophilic side chain?

a. 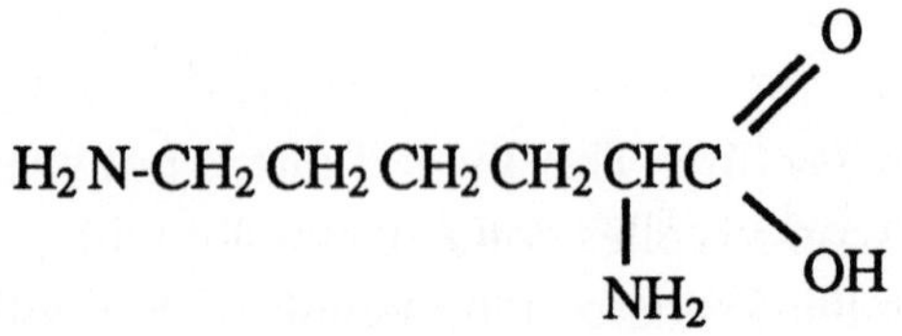 $H_2N\text{-}CH_2CH_2CH_2CH_2CH(NH_2)C(=O)OH$

b. $CH_3\text{—}S\text{—}CH_2CH_2CH(NH_2)C(=O)OH$

c. $CH_3CH_2CH(CH_3)\text{—}CH(NH_2)C(=O)OH$

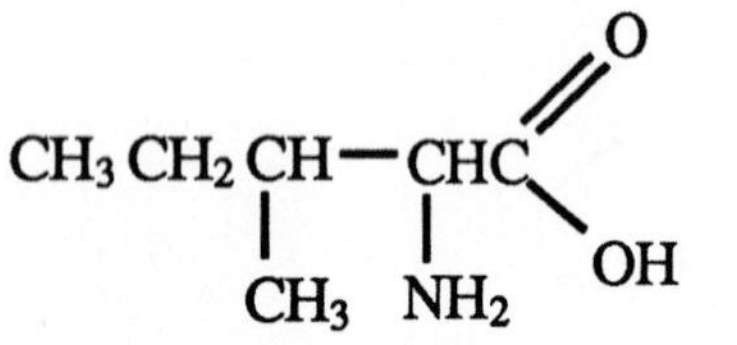

d. 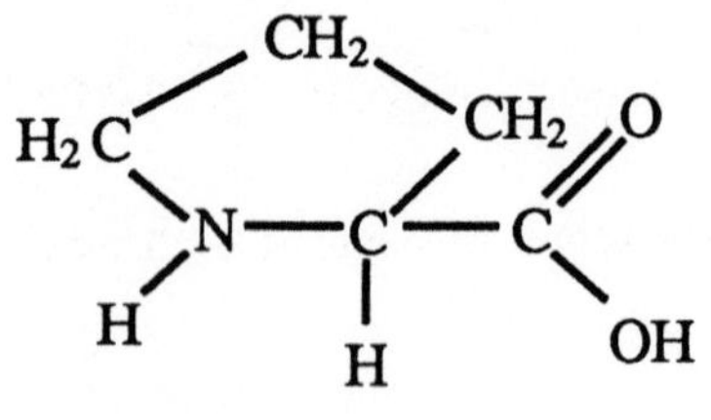 H_2C—CH_2—CH_2 ring with N(H)—C(H)—C(=O)OH (proline)

e. 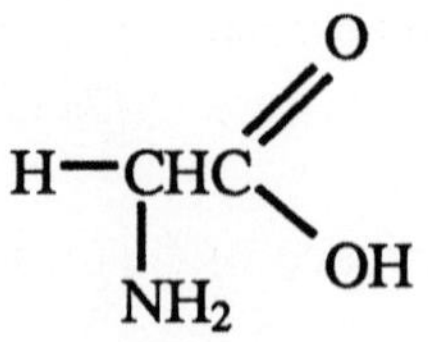 $H\text{—}CH(NH_2)C(=O)OH$

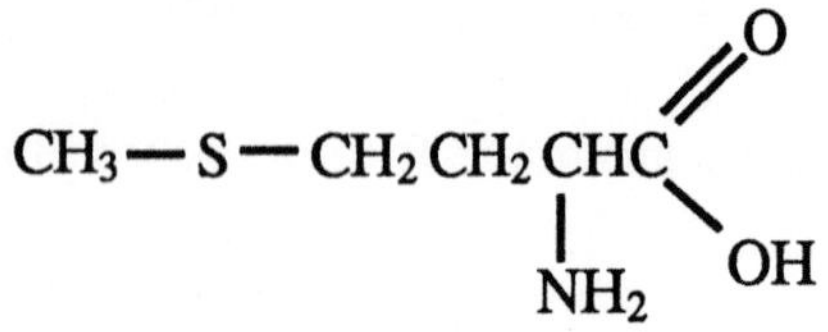

4. Which statement about amino acids and proteins is **not** true?

   a. The peptide link is formed from the carboxyl group of one amino acid and the amino group of another amino acid.
   b. A dipeptide contains from five to twenty amino acids.
   c. α-amino acids all have an amino functional group on carbon number two.
   d. Fibrous proteins occur in living systems where strength is required.
   e. Globular proteins in living systems usually perform work.

5. Which statement about proteins is **not** true?

   a. Proteins with different primary structures have different functions in the body.
   b. One kind of protein secondary structure is the pleated sheet.
   c. The overall shape of a protein is called the secondary structure.
   d. The amino acid cysteine helps to maintain the tertiary structure of a protein.
   e. The order in which amino acids occur is the primary structure of a protein.

6. Which of the following is **not** a function of proteins?

   a. Protect cells from foreign substances.
   b. Provide structural strength.
   c. Catalyze chemical reactions.
   d. Carry genetic information.
   e. Transport nutrients.

7. Which statement about carbohydrates is true?

   a. The monosaccharide arabinose is a hexose.
   b. Starch is a polysaccharide made from many fructose units.
   c. Sucrose, common table sugar, is a monosaccharide.
   d. Human digestive tracts possess the necessary enzyme to convert cellulose to glucose.
   e. Most monosaccharides form ring structures when present in aqueous solution.

8. Which statement about nucleic acids is <u>not</u> true?

   a. Thymine is found only in RNA.
   b. Nucleotides of DNA contain the five-carbon sugar deoxyribose.
   c. In a nucleotide, the phosphate group is attached directly to the monosaccharide.
   d. In DNA cytosine is always paired with guanine.
   e. When DNA replicates, one strand of the helix is a new piece of DNA and one strand is an old piece of DNA.

9. Which statement about protein synthesis is not true?

   a. An mRNA molecule is smaller than the DNA from which it was produced.
   b. A piece of DNA which contains the information to make one protein is called a gene.
   c. Proteins are synthesized from monosaccharides.
   d. There is a specific tRNA for each amino acid.
   e. In addition to nucleic acids, ribosomes are also necessary for protein synthesis.

10. Which statement about the composition of lipids is true?

   a. Phospholipids contain the alcohol ethylene glycol.
   b. Solid fats usually contain fatty acids such as linoleic or linolenic acid.
   c. Saponification produces an alcohol and salts of fatty acids.
   d. Waxes are carboxylic acids with large numbers of carbons.
   e. Many steroids contain rings with four carbons.

PRACTICE EXAM ANSWERS

1. b (Introduction)
2. e (21.1, 21.7, 21.8)
3. a (21.2)
4. b (21.1 and 21.2)
5. c (21.2, 21.3, 21.4)
6. d (21.5 and 21.8)
7. e (21.7)
8. a (21.8)
9. c (21.8)
10. c (21.9)